Advances in Remote Sensing for Forestry: Theory, Methods, Applications, and Validation

Advances in Remote Sensing for Forestry: Theory, Methods, Applications, and Validation

Guest Editors

Ying Yu
Xiguang Yang

MDPI

Basel • Beijing • Wuhan • Barcelona • Belgrade • Novi Sad • Cluj • Manchester

Guest Editors

Ying Yu
School of Forestry
Northeast Forestry University
Harbin
China

Xiguang Yang
School of Forestry
Northeast Forestry University
Harbin
China

Editorial Office
MDPI AG
Grosspeteranlage 5
4052 Basel, Switzerland

This is a reprint of the Special Issue, published open access by the journal *Forests* (ISSN 1999-4907), freely accessible at: https://www.mdpi.com/journal/forests/special_issues/4NBKJ4968T.

For citation purposes, cite each article independently as indicated on the article page online and as indicated below:

Lastname, A.A.; Lastname, B.B. Article Title. *Journal Name* **Year**, *Volume Number*, Page Range.

ISBN 978-3-7258-6700-4 (Hbk)
ISBN 978-3-7258-6701-1 (PDF)
https://doi.org/10.3390/books978-3-7258-6701-1

Contents

About the Editors

Ying Yu

Ying Yu is a full professor of Forestry at the School of Forestry, Northeast Forestry University, Harbin, China. She previously served as a visiting scholar at the Department of Geography, Faculty of Geography and Planning, University of Toronto. Her primary interests in research focus on the estimation of multi-scale structural and biochemical parameters of forests using multiple resource remote sensing data, as well as the estimation of carbon storage at local and large scales with the help of bio-geo-chemical models and advanced remote sensing technologies. She has presided over 3 projects funded by the National Natural Science Foundation of China. She served as the Principal Investigator (PI) for 2 sub-projects of the National Key R&D Program during the "13th Five-Year Plan" period and 2 projects grants from the Fundamental Research Funds for the Central Universities. In addition, she has participated in 5 national-level projects, including those supported by the NSFC and the National High-Tech R&D Program ("863" Program). Academically, she has published over 50 peer-reviewed papers, including more than 20 SCI-indexed papers and over 10 EI-indexed papers. She has also published 1 academic monograph and 2 academic textbooks. She was awarded the Second-Class Liang Xi Forestry Science and Technology Progress Award, the Third-Class Heilongjiang Provincial Science and Technology Progress Award, and the Second-Class Liangxi Excellent Youth Paper Award.

Xiguang Yang

Xiguang Yang is also a full professor of Forestry at the School of Forestry, Northeast Forestry University, Harbin, China. He is a PhD supervisor and an expert in forensic appraisal of environmental damage in Heilongjiang Province. He is classified as a Class E talent in Heilongjiang Province. His research interests include quantitative remote sensing, remote sensing parameterization of carbon cycle models, and application of remote sensing technologies in resource and environmental investigation and assessment. He has presided over more than 10 projects funded by the National Natural Science Foundation of China, sub-tasks of the National Key R&D Program, the China Postdoctoral Science Foundation, the Natural Science Foundation of Heilongjiang Province, and technical service projects. He has published over 40 peer-reviewed papers, including 20 SCI-indexed papers and 3 EI-indexed papers. He holds 2 granted invention patents and 5 utility model patents. Additionally, he has secured 5 software copyright registrations. He has published 3 academic monographs and 1 academic textbook. He has served as Editor-in-Chief of academic textbooks under the 14th Five-Year Plan of the National Forestry and Grassland Administration. He was awarded the First-Class and Second-Class Science and Technology Awards of the China Communications and Transportation Association, the Second-Class Liang Xi Forestry Science and Technology Progress Award, the Third-Class Heilongjiang Provincial Science and Technology Progress Award, and the Third-Class Liangxi Excellent Youth Paper Award.

forests

MDPI

Editorial

Advances in Remote Sensing for Forestry: Theory, Methods, Applications, and Validation

Xiguang Yang [1,2] and Ying Yu [1,2,*]

1 School of Forestry, Northeast Forestry University, Harbin 150040, China; yangxiguang@nefu.edu.cn
2 Key Laboratory of Sustainable Forest Ecosystem Management, Ministry of Education, Northeast Forestry University, Harbin 150040, China
* Correspondence: yuying@nefu.edu.cn; Tel.: +86-182-0461-8971

1. Introduction

Forests comprise the largest ecosystem in the world and provide significant value to natural ecosystems and human society [1]. They provide a habitat for a wide range of animal, plant, and microbial species and are invaluable in the maintenance of biodiversity [2]. Furthermore, they provide ecological and social services to human society, furthering our sustainable development. On the one hand, forests are a vital source of timber and various other products that are valuable to human society; on the other hand, forests deliver a range of essential ecological services, including soil and water conservation, windbreaks, sand stabilization, forest therapy, and health-enhancing recreational opportunities [3]. The forest ecosystem also plays a critical role in the fight against climate change. Its capacity to sequester atmospheric carbon dioxide via photosynthesis alleviates the effects of global warming [4], and forests are generally understood to be among the most economically viable and effective means of carbon sequestration. This is because the worldwide forest carbon sink is equal to almost half of the global emissions from burning fossil fuels [5]. However, recent research has indicated that the worldwide forest sink has been weakened by deforestation and advancing forest age [6]. Clarifying the precise nature of the forest carbon sink is thus of particular importance.

Remote sensing technology has become central to forestry research [7], having played a significant role in forestry investigation, forest cover, forest health, carbon storage, forest fire and pest monitoring, and forest dynamics over vast regions, even on global scale, and it is likely to become yet more essential in the future. This Special Issue, titled "Advances in Remote Sensing in Forestry: Theory, Methods, Applications and Validation," comprises 11 papers, focussing on emerging theories and methodologies in forest surveillance and monitoring, the application of remote sensing in forest ecosystem evaluation and biodiversity monitoring, and new technologies for monitoring carbon sinks in forest ecosystems. A couple of the studies included herein concern themselves with the theory underpinning remote sensing in forestry context [8,9]. Some of studies evaluate new sensors and equipment, such as Airborne or UAV LiDAR, which aid in forest resource investigation by reducing labor and time costs [10–12]. Additionally, emerging methods, such as deep learning, have been used to extract certain forest features from remote sensing data [13]. The most important application of remote sensing technology in forestry is the solving of real problems that need to be addressed in the forestry industry, such as the recognition of forest types, the detection of clear-cutting activities, and restoration assessment [14–16]. In meeting the challenges of global warming, forestry can also play a key role; scholars

 https://doi.org/10.3390/f17010073

have estimated the forest carbon stock using remote sensing in a large region and have evaluated the significant role played by forests as a global carbon sink [17,18]. The studies included herein showcase a number of theoretical innovations and the application of new methods to solve real problems in forestry. Our hope is that this Special Issue will serve as a crucial reference in promoting the application of remote sensing technology in the forestry domain.

2. Overview of Published Articles

Liu et al.'s article (Contribution 1) focuses on innovations in the application of remote sensing in forestry. The research address heterogeneous and discontinuous canopy structures and proposes an innovative inversion method combining radiative transfer models (RTMs), namely, PRO4SAIL, PRO4SAIL2 and PROGeoSail, with the random forest (RF) algorithm to estimate the canopy water content. A Gaussian copula model and prior knowledge are used to constrain the model parameters to alleviate the problem or poor positioning in remote sensing inversion. In the end, the authors compare the estimated results of the coupled PRO4SAIL, PRO4SAIL2, and PROGeoSail models. The results show that the coupled PROSPECT-5B and GeoSail models achieve the best performance, with an R2 of 0.68 and a RMSE of 0.15 $kg \cdot m^{-2}$. The forest canopy water content is estimated from MODIS products using the Google Earth Engine (GEE). The estimated results reveal the spatiotemporal heterogeneity of forest CWC across the contiguous US from west to east and show an increasing trend from January to August and a decreasing trend thereafter. This research demonstrates the superiority of 3D RTMs in forest canopy parameter inversion and provides an efficient, feasible technical scheme for forest CWC estimation in heterogeneous forest canopies, offering important references for forest drought stress monitoring and ecological protection against a background of global climate change.

Zhang et al.'s article (Contribution 2) is also focused on the estimation of forest features. The authors successfully described the three-dimensional (3D) distribution characteristics of Leaf Mass per Area (LMA) in the canopies of broad-leaved tree species, combining LiDAR and hyperspectral remote sensing data. In this study, the authors collected the leaf samples of eight dominant broad-leaved species by layer and direction and established four prediction models (PLS, LMM, SVM, and RF), integrating vegetation indices, DBH, and 3D leaf position parameters. The results show that the LMA indicates significant spatial heterogeneity: it will decrease with increasing canopy depth and increase with horizontal distance from the tree trunk. The authors also obtained some interesting results with respect to the LMA distribution in the canopy, revealing that it has no significant correlation with the orientation of leaf. Among the four models, random forest (RF) achieved the highest accuracy, with an R^2 ranging from 0.716 to 0.939 according to a 10-fold cross-validation method. The study presents an interesting study of high-precision 3D LMA prediction, and it could be extended to large-scale 3D LMA monitoring via LiDAR and optical remote sensing. At the same time, it offers a new perspective in clarifying canopy photosynthetic capacity and nutrient allocation mechanisms and could prove helpful in optimizing the vegetation biophysical parameter models used to estimate the carbon cycle study.

Su et al.'s article (Contribution 3) aims to solve the problems of incomplete lower canopy structure and sparse distribution in medium–low-density UAV point clouds caused by self-occlusion. The study proposes a tree DBH via the dual enhancement strategies of hierarchical random sampling and a spatially constrained loss function. The results show that this strategy avoids the loss of information in sparse areas and improves the structural bias of traditional methods. Moreover, it provides better results in trunk structure recovery and DBH measurement accuracy, as validated on FOR-instance and the Xiong'an dataset.

https://doi.org/10.3390/f17010073

This study provides a reference method for correcting the structural bias in medium–low-density point cloud processing.

You et al.'s article (Contribution 4) proposes a tree skeletonization method based on DBSCAN clustering to address the noise and occlusion defects of terrestrial laser scanning (TLS). The results show that the average positional deviation is 0.418 cm, and the average directivity deviation is 8.474 degrees. This method can be used to describe the constructed skeleton and tree's geometric structure and branch hierarchies, which is useful in accurately retrieving the tree parameter, the 3D model of the tree, and the forest inventory.

Li et al.'s article (Contribution 5) focuses on the application of new sensors, such as Airborne LiDAR, for estimating the forest inventory. They propose an error-in-variable simultaneous equation (SEq) for constructing compatible forest inventory estimation models. SEqs ensures consistency with forest mensuration principles by integrating stand volume, basal area, mean height, and diameter at breast height. The results show that SEqs' rRMSE increases by less than 2% across several forest species. Furthermore, this work showcases a forest inventory estimating method that can maintain both biophysical and mathematical compatibility. It will enhance the robustness and applicability of forest resource inventory and management models.

Ru et al.'s article (Contribution 6) proposes an automated forest monitoring system based on satellite imagery and deep learning to monitor the global deforestation. The authors integrate spatial pyramid pooling (SPP) modules into the U-Net architecture to segment forest areas without having to consider varying forest sizes, colors, and related factors. Compared with FCN, SegNet, and TernausNet, the model proposed in this study performs well in multi-scale feature extraction, with accuracies of 86.71%, 75.59%, and 82.88% on high-difficulty datasets. This work provides a reliable tool for detecting illegal logging and evaluating forest conservation efforts.

Xu et al.'s article (Contribution 7) considered an automatic extraction method for mapping the spatial distribution of Picea schrenkiana in the Tianshan Mountains. This work addresses the challenge of accounting for sample points, remote-sensing images, and classification features during high-resolution, large-scale mapping in complex terrain conditions by integrating Jeffries–Matusita (JM) distance and the random forest classifier in the Google Earth Engine (GEE). The results show that the approach achieves high accuracy, with an overall accuracy (OA) of 91.93% and a Kappa coefficient of 0.89. Furthermore, it provides a reliable method for coniferous forest mapping in complex terrains.

Pedraza et al.'s research (Contribution 8) generates annual cloud-free mosaic data from optical remote sensing data and employs a random forest algorithm for forest/non-forest classification to establish a forest dynamic monitoring and restoration assessment system for the period 1996–2021. The results indicate that the reduction in forest area was about 5587 ha over 25 years. This work provides a reference for quantifying forest dynamics and supporting forest conservation.

Choi et al.'s article (Contribution 9) also focus on the detection of forest dynamics. The authors propose a forest clear-cutting detection method via random forest (RF) classifier by integrating national spatial data and Sentinel-2 satellite imagery. Roughly 8164.0 ha of clear-cut areas, which accounted for 2.48% of the total forest area, were detected. These results fill a gap in the official statistical data and provide support for carbon sink management and sustainable forest management.

Mohamedi et al.'s article (Contribution 10) focuses on the issues of carbon peak and carbon neutrality in China and simulates China's forest net ecosystem productivity (NEP) from 2013 to 2023 using a 3-PGS model. The results show that the forests in question sequestered 1.71 ± 0.09 PgC over the decade, with an annual average of 0.156 ± 0.0071 PgC, and ev-

https://doi.org/10.3390/f17010073

ergreen broadleaf forests showed the highest NEP, as well as significant spatial–temporal heterogeneity across climatic regions in China. This research enriches the study of China's estimated forest carbon sink, provides evidence for the effectiveness of ecological restoration policies, and promotes the achievement of China's dual-carbon goal.

Liu et al.'s article (Contribution 11) concerns the spatial–temporal variations and driving mechanisms of vegetation NPP in mainland China from 2000 to 2022, employing the gravity center model, third-order partial correlation analysis, and geographical detector methods. The results show that the NPP decreases from southeast to northwest, and the gravity centers migrate southward in Northeast, Northwest, and North China. On the other hand, other regions show the opposite results, with gravity centers migrating northward. This research indicates that human activities dominated NPP growth in 2000–2010, while climate change became the primary driver in 2011–2022. It offers a reference for the impact of natural and anthropogenic factors in monitoring ecosystem restoration and ecological protection. It also contributes to the realization of China's carbon peak and carbon neutrality policies.

3. Conclusions

This Special Issue, titled "Advances in Remote Sensing in Forestry: Theory, Methods, Applications and Validation," collects 11 high-quality papers focussing on important scientific issues, systematically presenting the latest research on the application of remote sensing technology in forestry. Represented in this Special Issue are the core fields of forest canopy parameters inversion, forest structure reconstruction, the monitoring of forest dynamics, forest carbon sink estimation, and so on, alongside remote sensing equipment/platforms such as LiDAR and UAV hyper-spectral sensors. These studies not only enrich the field of remote sensing application in forestry but also offer solutions and references for the key issues in forestry management and ecological protection. This Special Issue also provides a number of scientific references for the furtherance of carbon peak and carbon neutrality policies. The studies published in this Special Issue demonstrate the latest applications of remote sensing technology in forestry, but it should be borne in mind that they cover only a fraction of the research being undertaken in this field. New technologies, such as artificial intelligence algorithms, big data mining technologies, and integrated space–air–ground monitoring technologies, have expanded the research horizon of the forestry domain. Our expectation is that this Special Issue and related work will motivate more scholars to engage with the ongoing research on remote sensing in forestry and promote the continuous innovation of new theories and methods in this domain.

Author Contributions: Conceptualization, X.Y. and Y.Y.; methodology, X.Y. and Y.Y.; writing—original draft preparation, X.Y. and Y.Y.; writing—review and editing, X.Y. and Y.Y. All authors have read and agreed to the published version of the manuscript.

Data Availability Statement: Not applicable.

Acknowledgments: As Guest Editors of the Special Issue "Advances in Remote Sensing in Forestry: Theory, Methods, Applications and Validation," we would like to express our deep appreciation to all the authors whose valuable work has been published herein. We would also like to thank the anonymous reviewers, whose contributions were invaluable to improving the quality of the papers published in this Special Issue.

Conflicts of Interest: The authors declares no conflicts of interest.

List of Contributions

https://doi.org/10.3390/f17010073

1. Liu, L.; Li, S.; Yang, W.; Wang, X.; Luo, X.; Ran, P.; Zhang, H. Forest Canopy Water Content Monitoring Using Radiative Transfer Models and Machine Learning. *Forests* **2023**, *14*, 1418.
2. Zhang, D.; Wang, Y.; Yang, X.; Yang, S.; Liu, Y.; Yu, Z.; Zhao, X. A Concise Approach to Characterizing the Distribution of Canopy Leaf Mass per Area in Broad-Leaf Species Based on Crown Three-Dimensional Position and Vegetation Index. *Forests* **2025**, *16*, 838.
3. Su, Y.; Chen, Z.; Xue, X. TreeDBH: Dual Enhancement Strategies for Tree Point Cloud Completion in Medium–Low Density UAV Data. *Forests* **2025**, *16*, 667.
4. You, L.; Sun, Y.; Liu, Y.; Chang, X.; Jiang, J.; Feng, Y.; Song, X. Tree Skeletonization with DBSCAN Clustering Using Terrestrial Laser Scanning Data. *Forests* **2023**, *14*, 1525.
5. Li, C.; Yu, Z.; Zhou, X.; Zhou, M.; Li, Z. Using the Error-in-Variable Simultaneous Equations Approach to Construct Compatible Estimation Models of Forest Inventory Attributes Based on Airborne LiDAR. *Forests* **2023**, *14*, 65.
6. Ru, F.X.; Zulkifley, M.A.; Abdani, S.R.; Spraggon, M. Forest Segmentation with Spatial Pyramid Pooling Modules: A Surveillance System Based on Satellite Images. *Forests* **2023**, *14*, 405.
7. Xu, F.; Xu, Z.; Xu, C.; Yu, T. Automatic Extraction of the Spatial Distribution of Picea schrenkiana in the Tianshan Mountains Based on Google Earth Engine and the Jeffries–Matusita Distance. *Forests* **2023**, *14*, 1373.
8. Pedraza, C.; Clerici, N.; Villa, M.; Romero, M.; Dueñas, A.S.; Rojas, D.B.; Quintero, P.; Martínez, M.; Kellndorfer, J. Monitoring Forest Dynamics and Conducting Restoration Assessment Using Multi-Source Earth Observation Data in Northern Andes, Colombia. *Forests* **2024**, *15*, 754.
9. Choi, S.-E.; Lee, S.; Park, J.; Lee, S.; Yim, J.; Kang, J. Detection and Analysis of Forest Clear-Cutting Activities Using Sentinel-2 and Random Forest Classification: A Case Study on Chungcheongnam-do, Republic of Korea. *Forests* **2024**, *15*, 450.
10. Mohamedi, F.J.; Yu, Y.; Yang, X.; Fan, W. Simulation of Carbon Sinks and Sources in China's Forests from 2013 to 2023. *Forests* **2025**, *16*, 1398.
11. Liu, Y.; Xu, M.; Guo, B.; Yang, G.; Li, J.; Yu, Y. Changes in the Vegetation NPP of Mainland China under the Combined Actions of Climatic-Socioeconomic Factors. *Forests* **2023**, *14*, 2341.

References

1. Pearce, D.W. The Economic Value of Forest Ecosystems. *Ecosyst. Health* **2001**, *7*, 284–296. [CrossRef]
2. Wang, L.; Wei, F.; Tagesson, T.; Fang, Z.; Svenning, J.-C. Transforming forest management through rewilding: Enhancing biodiversity, resilience, and biosphere sustainability under global change. *One Earth* **2025**, *8*, 101195. [CrossRef]
3. Singh, M.; Shahina, N.N.; Das, S.; Arshad, A.; Siril, S.; Barman, D.; Mog, U.; Panwar, P.; Shukla, G.; Chakravarty, S. Forest Resources of the World: Present Status and Future Prospects. In *Land Degradation Neutrality: Achieving SDG 15 by Forest Management*; Panwar, P., Shukla, G., Bhat, J.A., Chakravarty, S., Eds.; Springer Nature Singapore: Singapore, 2022; pp. 1–23.
4. Doelman, J.C.; Stehfest, E.; van Vuuren, D.P.; Tabeau, A.; Hof, A.F.; Braakhekke, M.C.; Gernaat, D.E.H.J.; van den Berg, M.; van Zeist, W.-J.; Daioglou, V.; et al. Afforestation for climate change mitigation: Potentials, risks and trade-offs. *Glob. Change Biol.* **2020**, *26*, 1576–1591. [CrossRef]
5. Pan, Y.; Birdsey, R.A.; Phillips, O.L.; Houghton, R.A.; Fang, J.; Kauppi, P.E.; Keith, H.; Kurz, W.A.; Ito, A.; Lewis, S.L.; et al. The enduring world forest carbon sink. *Nature* **2024**, *631*, 563–569. [CrossRef] [PubMed]
6. IPCC Working Group II. *Climate Change 2022: Impacts, Adaptation and Vulnerability*; Contribution of Working Group II to the Sixth Assessment Report of the Intergovernmental Panel on Climate Change; Cambridge University Press: Cambridge, UK, 2023.
7. Fassnacht, F.E.; White, J.C.; Wulder, M.A.; Næsset, E. Remote sensing in forestry: Current challenges, considerations and directions. *For. Int. J. For. Res.* **2023**, *97*, 11–37. [CrossRef]
8. Liu, L.; Li, S.; Yang, W.; Wang, X.; Luo, X.; Ran, P.; Zhang, H. Forest Canopy Water Content Monitoring Using Radiative Transfer Models and Machine Learning. *Forests* **2023**, *14*, 1418. [CrossRef]
9. Zhang, D.; Wang, Y.; Yang, X.; Yang, S.; Liu, Y.; Yu, Z.; Zhao, X. A Concise Approach to Characterizing the Distribution of Canopy Leaf Mass per Area in Broad-Leaf Species Based on Crown Three-Dimensional Position and Vegetation Index. *Forests* **2025**, *16*, 838.
10. Su, Y.; Chen, Z.; Xue, X. TreeDBH: Dual Enhancement Strategies for Tree Point Cloud Completion in Medium–Low Density UAV Data. *Forests* **2025**, *16*, 667. [CrossRef]
11. You, L.; Sun, Y.; Liu, Y.; Chang, X.; Jiang, J.; Feng, Y.; Song, X. Tree Skeletonization with DBSCAN Clustering Using Terrestrial Laser Scanning Data. *Forests* **2023**, *14*, 1525. [CrossRef]

 https://doi.org/10.3390/f17010073

12. Li, C.; Yu, Z.; Zhou, X.; Zhou, M.; Li, Z. Using the Error-in-Variable Simultaneous Equations Approach to Construct Compatible Estimation Models of Forest Inventory Attributes Based on Airborne LiDAR. *Forests* **2023**, *14*, 65.
13. Ru, F.X.; Zulkifley, M.A.; Abdani, S.R.; Spraggon, M. Forest Segmentation with Spatial Pyramid Pooling Modules: A Surveillance System Based on Satellite Images. *Forests* **2023**, *14*, 405. [CrossRef]
14. Xu, F.; Xu, Z.; Xu, C.; Yu, T. Automatic Extraction of the Spatial Distribution of Picea schrenkiana in the Tianshan Mountains Based on Google Earth Engine and the Jeffries–Matusita Distance. *Forests* **2023**, *14*, 1373. [CrossRef]
15. Pedraza, C.; Clerici, N.; Villa, M.; Romero, M.; Dueñas, A.S.; Rojas, D.B.; Quintero, P.; Martínez, M.; Kellndorfer, J. Monitoring Forest Dynamics and Conducting Restoration Assessment Using Multi-Source Earth Observation Data in Northern Andes, Colombia. *Forests* **2024**, *15*, 754. [CrossRef]
16. Choi, S.-E.; Lee, S.; Park, J.; Lee, S.; Yim, J.; Kang, J. Detection and Analysis of Forest Clear-Cutting Activities Using Sentinel-2 and Random Forest Classification: A Case Study on Chungcheongnam-do, Republic of Korea. *Forests* **2024**, *15*, 450. [CrossRef]
17. Mohamedi, F.J.; Yu, Y.; Yang, X.; Fan, W. Simulation of Carbon Sinks and Sources in China's Forests from 2013 to 2023. *Forests* **2025**, *16*, 1398. [CrossRef]
18. Liu, Y.; Xu, M.; Guo, B.; Yang, G.; Li, J.; Yu, Y. Changes in the Vegetation NPP of Mainland China under the Combined Actions of Climatic-Socioeconomic Factors. *Forests* **2023**, *14*, 2341. [CrossRef]

https://doi.org/10.3390/f17010073

forests

MDPI

Article

Simulation of Carbon Sinks and Sources in China's Forests from 2013 to 2023

Faris Jamal Mohamedi [1], Ying Yu [1,2,*], Xiguang Yang [1,2] and Wenyi Fan [1,2]

1 School of Forestry, Northeast Forestry University, Harbin 150040, China; mohamedifaris@nefu.edu.cn (F.J.M.); yangxiguang@nefu.edu.cn (X.Y.); fanwy@nefu.edu.cn (W.F.)
2 Key Laboratory of Sustainable Forest Ecosystem Management, Ministry of Education, School of Forestry, Northeast Forestry University, Harbin 150040, China
* Correspondence: yuying@nefu.edu.cn; Tel.: +86-182-0461-8971

Abstract: Chinese forest ecosystems are key carbon sinks that significantly contribute to lowering carbon emissions. Accurate Net Ecosystem Productivity (NEP) estimations are essential for evaluating their carbon sequestration capabilities and overall health. This study employed the Physiological Principles Predicting Growth-Satellites (3-PGS) and soil heterotrophic respiration models to simulate China's forest carbon sinks and sources distribution from 2013 to 2023. Then, climatic factors influencing NEP changes were examined through the application of a geographical detector model. The net carbon sequestered was 1.71 ± 0.09 PgC with an annual average of 0.156 ± 0.0071 PgC, signifying a substantial carbon sink in China's forest. The annual NEP was highest in evergreen broadleaf forests (352.12 gC m^{-2}) and lowest in deciduous needleleaf forests (148.31 gC m^{-2}). NEP in China's forests increased by a rate of 1.67 gC m^{-2} annually, with most regions exhibiting a 275.32 gC m^{-2} annual carbon sink. The geographical detector model analysis showed that solar radiation, precipitation, and vapor pressure deficit were the main drivers of NEP change, while temperature and frost days had a secondary influence. Furthermore, the interaction between solar radiation and temperature variables showed the greatest impact. This study can enhance the understanding of carbon sink and source distribution in China, serve as a reference for regional carbon cycle research, and provide key insights for policymakers in developing effective climate strategies.

Keywords: China's forest; NEP; 3-PGS; geographical detector model

1. Introduction

China has been the top carbon emitter globally since 2006, producing 10.67×10^9 tons of CO_2 in 2020, which makes up 30.65% of the world's total emissions [1]. Despite this, China's land, which covers about 6.5% of the world's landmass, plays an essential function in the global carbon cycle by capturing between 10 and 31% of the world's net CO_2 uptake [2]. Between 2010 and 2020, approximately 55.26% of China's land area maintained its status as a carbon sink, highlighting its important function in mitigating climate change [3]. In September 2020, China's government declared its goal to reach peak carbon emissions prior to 2030 and to attain carbon neutrality by 2060 [4]. According to the Intergovernmental Panel on Climate Change (IPCC), carbon neutrality is defined as achieving net zero CO_2 emissions when the amount of CO_2 entering the atmosphere is equal to the amount that is removed over a specified period [5]. In terrestrial landscapes, forests, being the largest carbon pool [6], accounted for over 90% of China's total carbon sinks between 1999 and

2014 [7], highlighting their prominent contribution to the carbon balance maintained by terrestrial ecosystems [8]. The concept of Net Ecosystem Production (NEP) is crucial for evaluating the dynamics of carbon sinks/sources, which indicates the net carbon fixation efficiency of atmospheric CO_2 absorbed by the ecological community [9]. NEP reflects the balance between carbon gain through photosynthesis (NPP, Net Primary Productivity) and carbon loss via soil heterotrophic respiration (R_h) [10]. NEP helps in explaining how ecosystems interact with the atmosphere by measuring the net carbon exchange [11].

Assessing the NEP change in the Chinese forest ecosystem is crucial in understanding China's 'carbon neutrality' goal and awareness of the global carbon flux in considering the existing changing climate patterns [12]. Recent advances in modeling approaches for carbon dynamics across ecosystems have enabled large-scale NEP estimation incorporating ecosystem process models, climate potential, and light use efficiency models (LUE) [13,14]. The LUE model simplifies plant physiological processes, focusing on how plants use light energy for photosynthesis; a commonly used model is the Carnegie–Ames–Stanford Approach (CASA) [15]. Ecosystem process models, like CENTURY and Community Land Model, provide productivity estimates using systematic modeling processes such as the decay of organic matter and cycling of nutrients, offering valuable insights into ecosystem physiology; however, their reliance on numerous input parameters, which are often challenging to measure and can compromise accuracy [16]. Several research initiatives have been made to assess and quantify NEP in Chinese forest ecosystems by utilizing a model approach, which is critical for understanding the forest carbon transformation. The use of models enhances the accuracy of NEP calculations, thereby refining the depiction of the carbon absorption effect and offering a thorough assessment of carbon sequestration in the forest ecosystem.

Chen et al. [17] assessed vegetation carbon dynamics in China by employing a CASA model and then identified key drivers influencing the changes using a geographical detector model between 1982 and 2020. Ma et al. [18] forecasted the sequestration capacity of forests in China from 2018 to 2060 using the CO2FIX model and inventory approach. He et al. [19] modeled terrestrial NEP shifts in China (1982 to 2010) and assessed their responses to climate fluctuation across various regions. Liu et al. [20] utilized inventory data and the CASA model along with a forest vegetation map to quantitatively estimate China's forests' NEP spatial distribution between 1984 and 2003. Piao et al. [21] examined carbon storage and emissions in the 1980s and 1990s by using an inventory-based approach along with remotely sensed vegetation indices, ecosystem simulations, and atmospheric CO_2 data inversion. Modeling forest carbon accumulation in China has advanced significantly due to extensive research efforts. Despite considerable progress in modeling forest carbon accumulation in China, uncertainties remain regarding the accuracy and regional variability of these estimates, which are impacted by factors like the forests' spatial-temporal heterogeneity due to China's large area and diversity in climate conditions, with alterations related to land use change [22] that greatly cause the variations of forest carbon sinks.

This study aims to bridge these gaps by utilizing a simplified, spatially based process 3-PGS model alongside a soil heterotrophic respiration model to simulate and provide a more accurate and comprehensive understanding of forest carbon sinks and sources distribution in China between 2013 and 2023. Additionally, the climatic influences on carbon NEP were assessed using a geographical detector model. This finding can inform the development of more effective carbon reduction policies and support China's ambitious carbon neutrality goals for 2060, thereby contributing to both regional and global efforts in mitigating climate change.

2. Materials and Methods

2.1. Area Description

China is ranked as the 3rd largest country worldwide with an enormous territory of 9.6 million km^2 and is in the eastern part of Asia [23]. The country features vast mountain ranges and plateaus, spanning multiple climatic zones with distinct geographical, natural, and human attributes, including a warm temperate semi-humid region (I), a mid temperate arid and semi-arid region (II, V), a northern subtropical humid region (III), a mid-temperate semi-humid region (IV), a plateau temperate semi-arid region (VI), and a marginal tropical humid region (VII). Forest ecosystems are distributed along a latitudinal gradient and include diverse types, including deciduous and evergreen coniferous forests, mixed forests, deciduous and evergreen broadleaf forests, rainforests, and monsoon forests. Together, forests cover 220 million hectares (22.96%), as reported in the 9th national forest inventory [24,25], as shown in Figure 1. Refer to Figure S1 for the climate information map.

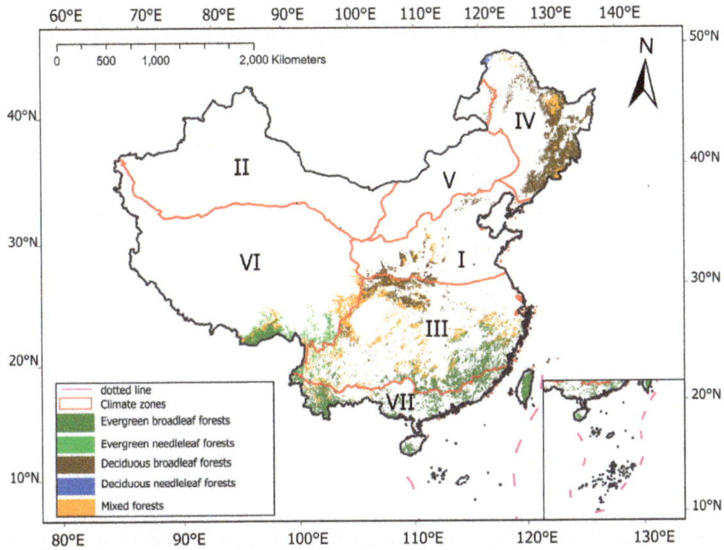

Figure 1. Forest Cover in China and Climate Zones Map.

2.2. Data Sources and Processing

Monthly data on precipitation, average air temperature, shortwave radiation, temperature extremes, dewpoint temperature, and frost frequency were retrieved from the ERA5-Land dataset of spatial resolution approximately 0.1° [26].

The MODIS normalized difference vegetation index (NDVI) product, providing monthly datasets in a grid format with a 0.05°/16-day spatiotemporal resolution [27], was used. To reduce atmospheric interferences on satellite imagery, the maximum value composite (MVC) technique was applied [28,29].

The MODIS Land Cover product (MCD12C1v061) from the International Geosphere-Biosphere Program (IGBP), with an annual temporal resolution and a spatial resolution of 0.05° [30], was used in this study. According to the IGBP land cover classification scheme, global land cover is categorized into 17 types, including 11 natural vegetation classes, 3 human-altered classes, and 3 non-vegetated classes. The reclassification tool in ArcGIS Pro software was employed to reclassify the original land cover data, resulting in five major forest categories found in China: evergreen needleleaf forest (ENF), deciduous broadleaf forest (DBF), deciduous needleleaf forest (DNF), evergreen broadleaf forest (EBF), and mixed forest (MF). The MODIS Land Cover product was chosen for its high overall accuracy, which reaches up to 66.42% at the pixel scale [31].

A soil fertility index (FR), soil texture classes, and the soil's water storage capacity were used to initialize the 3-PGS model (Tables S1 and S2) [32,33].

Due to the lack of NEP flux data for the study period (2013–2023), we evaluated the accuracy of the model's Net Primary Productivity (NPP) and soil heterotrophic respiration (R_h) estimates instead of directly validating NEP. MODIS NPP product MOD17A3HGFv061, with a 500 m pixel and annual resolution [34], was used to validate the model's NPP estimates. For R_h validation, we used 634 annual soil respiration data from Sun et al. [35] and 140 data from Song et al. [36] obtained through chamber methods (static and dynamic), gas chromatography (GC), and infrared gas analyzers. The Sun et al. data were stored as georeferenced points, which we overlaid onto our Rh raster to extract corresponding values. This dataset was then filtered by longitude, latitude, and elevation to remove duplicates, and points outside the Rh raster's pixel extent were excluded. The Song et al. data were provided as a list of geographic coordinates and forest type categories. By overlaying the land cover map of China with the Rh raster, we extracted Rh estimates based on the provided coordinates. After data filtering, a total of 200 estimates of annual forest soil CO_2 flux were obtained for Rh validation.

Using China as the area of interest (AOI), and applying filters for both time (2013 to 2023) and data quality, we developed codes to download the meteorological and MODIS data from the Google Earth Engine (GEE) platform. GEE is renowned for its efficiency and geospatial capabilities in processing large datasets [37–39]. Albers conical projection (equal-area), with 110° E as the central meridian and standard parallels at 25° N and 47° N, and data resampling to 0.05° spatial resolution through the bilinear interpolation method were applied in ArcGIS Map 10.4 software to ensure spatial analysis compatibility and consistency.

2.3. Research Methods

2.3.1. 3-PGS Model Description

Physiological Principles Predicting Growth using Satellites model, abbreviated as 3-PGS, is a spatial Light Use Efficiency (LUE)-based and simplified version of the 3-PG model [40,41]. This model estimates forest Net Primary Productivity (NPP) by integrating structural and functional processes using fundamental empirical principles. This allows for more scalable predictions of NPP without needing detailed field measurements. The main simplifications added to the model include using satellite data on vegetation indices at large spatial scales instead of stand-specific attributes. Additionally, the 3-PGS model eliminates the calculations of root turnover and detailed stand-growth models, making it computationally efficient [42]. Also, it uses a fixed fraction ratio (α) to approximate the relationship between NPP and Gross Primary Productivity (GPP). Monthly maximum potential Gross Primary Productivity (GPP) is linked to APAR (canopy-absorbed photosynthetically active radiation) in the model with maximum LUE, ε_{max}. It also incorporates climatic constraints (f_x) that reduce LUE based on three environmental modifiers: vapor pressure deficit (VPD), frost (f_d), and soil moisture stress (f_q). Importantly, f_x ranges from 0 (no production) to 1 (maximum production) [43] and uses the most limiting factor between VPD and f_q [41] to adjust the LUE. Therefore, the NPP was calculated according to the general LUE equation:

$$NPP = PAR \times \varepsilon_{max} \times f_{PAR} \times f_x \times \alpha \quad (1)$$

where NPP: Net Primary Productivity ($gC{\cdot}m^{-2}$ per month); ε_{max}: maximum LUE ($g\,Cm^{-2}MJ^{-1}$ APAR); f_x: climatic constraints; and α: fractional ratio of NPP to GPP. The portion of PAR captured by the canopy (f_{PAR}) is linearly related to NDVI [44,45] as follows:

$$f_{PAR} = m \times NDVI + C \tag{2}$$

Constants m and c were assigned values of 1.27 and −0.03, respectively [42] in this study. The value α varies across several ecosystems, ranging from 0.4 to 0.6 [46]. Then, in this study, α was adjusted for DNF (0.48), EBF (0.47), ENF (0.5853), DBF (0.4688), and MF (0.5488) [47].

2.3.2. Estimation of Soil Heterotrophic Respiration (R_h)

Soil R_h plays a crucial role in the terrestrial carbon cycle by representing the carbon released during the microbial breakdown of litter, detritus, and soil organic matter (SOM) [48]. R_h can be estimated using either direct methods (such as ground sampling and spatial interpolation) or indirect approaches. However, due to the vast territory and diverse ecosystems in China, variations in the key factors controlling R_h across various climate regions and the direct estimation approach may have a large margin of error [49]. The indirect estimation of the R_h approach entails the development of an empirical statistical model for soil respiration. This method initially estimates soil respiration (R_s) and subsequently employs the R_s–R_h model to derive R_h [50]. The indirect approach has been extensively utilized in large-scale applications owing to its effective parameterization techniques [51].

Then, this study calculated R_s from the density of carbon from surface soil, temperature, and precipitation data as suggested by Chen et al. [52].

$$R_S = 1.55e^{0.031T} \times \left(\frac{P}{P+0.68}\right)\left(\frac{SOC}{SOC+2.23}\right) \tag{3}$$

where R_S: annual soil respiration ($kgC{\cdot}m^{-2}{\cdot}yr^{-1}$); T; temperature (°C); P; rainfall (mm), both on an annual scale; and SOC: soil carbon density at a depth of 0 to 20 cm of the soil surface ($kgC{\cdot}m^{-2}{\cdot}yr^{-1}$). SOC values were assigned as follows: broadleaf forest (4.700), coniferous forest (3.770), and mixed forest (4.235) [53]. The R_h estimates in this study were estimated using the Zhang et al. [54] model.

$$R_h = 0.6163{R_S}^{0.7918} \tag{4}$$

2.3.3. NEP Estimation

NEP is derived by subtracting R_h from NPP. When NEP is positive, it reflects a carbon sink, while a negative value reflects sources [55].

$$NEP = NPP - R_h \tag{5}$$

where NEP: Net Ecosystem Productivity; NPP: Net Primary Productivity; and R_h is the estimated soil heterotrophic respiration.

2.3.4. The Geographical Detector Model

The GD model identifies and analyzes spatiotemporal differentiation, examining the interactions and explanatory potential of various driving factors [56,57]. It assesses the impact of explanatory elements on explained elements, assuming that significant relationships show spatial distribution similarities [58,59]. The GD model has four components: factor detector (FD), interactive detector (ID), ecological detector, and risk detector [56]. The FD and ID were major concerns in this study. The GD model in this study was used to explore the influence of climatic variables (predictors), including solar radiation, VPD, average temperature, precipitation, frost days, and the maximum and minimum temperature on NEP as the response variable.

Factor Detector

This component measures the influence of explanatory elements on explained elements through the influencing power value, q. A higher q-value (from 0 to 1) indicates a stronger spatial diversity of the response factor (NEP). A value of 1, signifying the NEP distribution pattern, is completely influenced by predictor factors, while 0 indicates no effect. According to the requirements of the GD model, which only processes discrete variables [56]; therefore, all predictor variables utilized for studying NEP spatial distribution were first stratified using the natural breakpoint approach. This strategy minimizes the average deviation within each class and maximizes the deviation between classes, as proposed by Jenks [60], thus avoiding human bias [61].

$$q = 1 - \frac{\sum_{i=1}^{L} N_i \sigma_i^2}{N\sigma^2} \quad (6)$$

where q: influencing power; N: total number of observations in study area; i = 1, 2, 3 . . ., L is the number of stratums of the climatic variables; N_i: number of observations present in stratum i; σ^2: variance of NEP; and σ_i^2: variance within stratum i.

Interaction Detector

The joint influence of factors, i.e., X_1 and X_2, on the explained variable is evaluated using this GD component, determining if their interaction strengthens or weakens its influence [57]. The interactions can be categorized based on the following judgement: if $q(x_1 \cap x_2) > q(x_1)$ or $q(x_2)$: enhance; if $q(x_1 \cap x_2) > q(x_1)$ and $q(x_2)$: enhance bivariate; if $q(x_1 \cap x_2) > q(x_1) + q(x_2)$: enhance nonlinear; if $q(x_1 \cap x_2) < q(x_1) + q(x_2)$: weaken; and if $q(x_1 \cap x_2) = q(x_1) + q(x_2)$: independent.

In this study, statistical computation and analysis were performed in RStudio 2024.12.0, while the GD model (www.geodetector.cn) and spatial distribution maps were implemented and generated in ArcGIS Pro 3.1.6, respectively.

3. Results

3.1. Accuracy Assessment of NPP and R_h

Statistical measures of R-squared (R^2) and root mean square error (RMSE) were calculated to assess the accuracy of the simulated NPP and estimated R_h. Using a sample size of 3720, a moderate positive correlation was found between the model NPP and MODIS NPP, with an R^2 value of 0.52 and RMSE of 126.17 gC·m^{-2}·yr^{-1}, as shown in Figure 2, suggesting that the model is reasonably accurate. At lower NPP values (<1000 gC·m^{-2}·yr^{-1}), many points fall above the 1:1 line, suggesting model overestimation, particularly in northeastern and central China. In contrast, at higher NPP values (>1000 gC·m^{-2}·yr^{-1}), fewer points fall below the 1:1 line, suggesting underestimation, while several points cluster around the 1:1 line, reflecting a good model fit. The gap between the blue and red lines widens as MODIS NPP increases, highlighting the model's difficulty in accurately representing high-productivity forests, especially in southern China. This gap arises due to limitations in model parameterization, missing soil factors, and uncertainties in remote sensing data.

Average forest NPP from the 3-PGS model was estimated to be 870 gC·m^{-2}·yr^{-1}. Across forest vegetation types, EBF had 1114.22 gC·m^{-2}·yr^{-1}, which exhibited the highest level of NPP, while ENF, MF, DBF, and DNF showed progressively lower values of 863.92,

837.92, 733.07, and 448.08 gC·m^{-2}·yr^{-1}, respectively. The subtropical and tropical environments of southern China are dominated by abundant rainfall and extended growing seasons, enhancing photosynthesis and boosting EBF and ENF productivity. Conversely, DNF, found in northern China and high-altitude areas, has the lowest NPP due to harsh temperatures and short growing seasons.

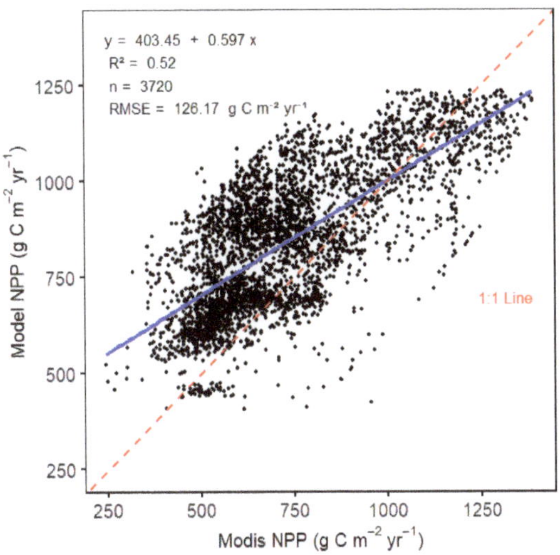

Figure 2. Validation of the NPP results.

The soil heterotrophic respiration (R_h) estimated results indicate a strong agreement, with observed R^2 and RMSE of 0.70 and 93.02 gC·m^{-2}·yr^{-1}, respectively, using a 200 sample size, with many sample points closely aligned along the 1:1 line (Figure 3), signifying minimal bias and high consistency (R_h) in estimations.

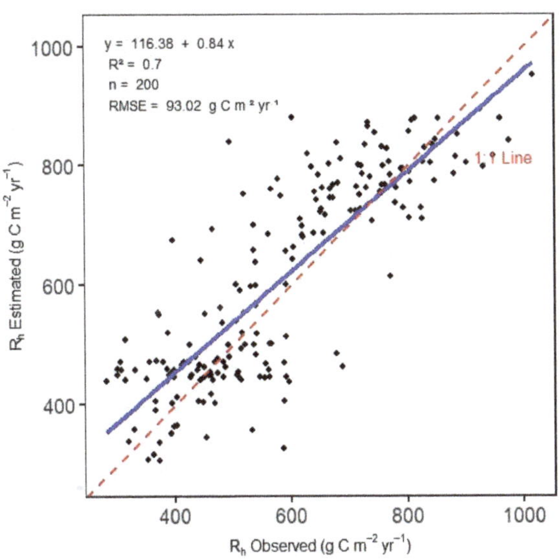

Figure 3. Validation of the soil heterotrophic respiration results.

The soil microbial ecosystem structure and function, with favorable precipitation and temperature conditions, are key factors driving variations of R_h. According to the results, R_h shows variation among different forest types. EBF exhibited the highest R_h at 805.62 gC·m^{-2}·yr^{-1}, MF had 590.07 gC·m^{-2}·yr^{-1}, and ENF at 526.60 gC·m^{-2}·yr^{-1}. Lower R_h estimates were shown by DBF and DNF, with 481.87 gC·m^{-2}·yr^{-1} and 296.87 gC·m^{-2}·yr^{-1}, respectively. The decrease in R_h across forest types appears to follow latitudinal gradients.

The annual soil respiration (R_S) estimated in this study was relatively high, averaging 997.36 ± 27.032 gC·m^{-2}·yr^{-1}, compared to previous studies: 851.88 ± 12.75 gC·m^{-2}·yr^{-1} [35], 917.73 gC·m^{-2}·yr^{-1} [36], and 975.50 gC·m^{-2}·yr^{-1} [62]. However, our estimate remains within the reported range of 0.1 and 1.6 kgC·m^{-2}·yr^{-1} [52], demonstrating the reliability of this study. The elevated R_S value observed here reflects recent ecosystem responses captured by the updated climate dataset spanning 2013 to 2023, highlighting potential increases in microbial activity and soil carbon flux under current climatic conditions.

3.2. NEP Spatiotemporal Variation in China's Forests

Carbon sinks and sources spatial distributions were estimated at a 5 km resolution from 2013 to 2023 (Figure S3) based on Equation (5). Over this period, China's forests operated as carbon sinks, fixing a net amount of 1.71 ± 0.09 PgC with an annual average of 0.156 ± 0.0071 PgC (155.86 TgC/year). It also indicated that the majority of forest regions function as sinks annually, corresponding to 275.32 gC·m^{-2}. Figure 4 illustrates the spatial distribution of NEP over a 5-year interval.

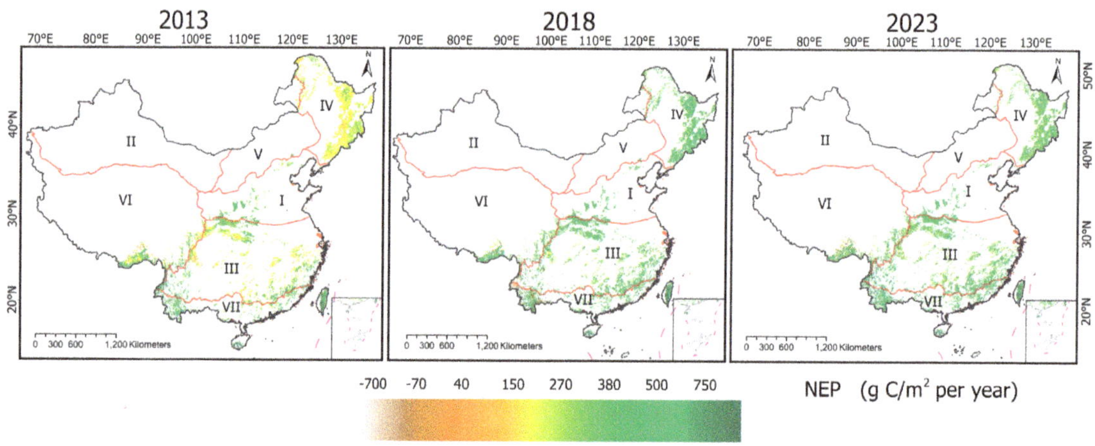

Figure 4. Spatial distribution of NEP in 2013, 2018, and 2023.

Despite variations in annual Net Ecosystem Production (NEP) among forest types, all exhibited positive NEP values over 100 gC·m^{-2}. EBF and ENF had the highest values, exceeding 300 gC·m^{-2}, reflecting greater carbon sequestration capacity. In contrast, DNF showed the lowest annual NEP (148.31 gC m^{-2}), followed by MF and DBF, indicating a comparatively smaller contribution to annual carbon storage (Figure 5). These differences are likely driven by variations in physiological traits, growth rates, and climatic adaptability. In addition, broadleaf forests contributed over 70% of total NEP (1.23 PgC), attributed to their high productivity, long growing season, and dense foliage, which enhances carbon uptake.

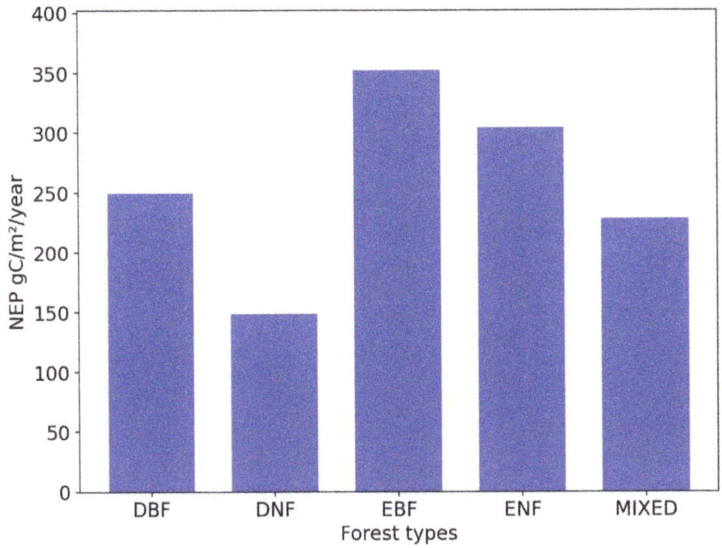

Figure 5. Annual NEP of each forest type.

The total amount of carbon released during the study was found to be 0.045 PgC, with an average of 4.07 TgC released each year. More than half of this release came from MF, followed by EBF at around 40%, while other forest types present a minor contribution. Although DNF and DBF showed minimal carbon source, they demonstrated greater stability and resilience in maintaining carbon balance compared to other forest ecosystems.

From 2013 to 2023, net carbon fixation across China's climatic regions exhibited notable spatial variability. The northern subtropical humid region (III) contributed the largest contribution at 0.531 PgC (31%), followed closely by the mid-temperate semi-humid region (IV) at 0.526 PgC (30.68%). The marginal tropical humid region (VII) contributed 0.313 PgC (18.29%), while the plateau temperate semi-arid region (VI) and warm temperate semi-humid region (I) contributed 0.187 PgC (10.91%) and 0.149 PgC (8.69%), respectively. Contributions from other regions (II and V) were comparatively minor at less than 2%. Although Regions III and IV exhibited similar levels of net carbon sequestration, their carbon source profiles differed significantly, with Region III showing substantially higher emissions. Among all regions, Region VI had the highest carbon source at 25.39 TgC (56.75%), followed by Region III with 13.50 TgC (30.17%) and Region VII with 5.78 TgC (12.93%). In contrast, Region IV contributed only 0.059 TgC (0.13%), and Regions I, II, and V had very low carbon source contributions. These regional dynamics are illustrated in Figure 6a.

On an annual average, all regions exhibited NEP values exceeding 200 g C/m^2. Region VII recorded the highest NEP (432.88 g C/m^2), followed by Region I (326.18 g C/m^2) and Region V (280.95 g C/m^2). Region III showed an NEP of 275.75 g C/m^2, while the remaining regions fell below 250 g C/m^2. These patterns suggest that Region VII is predominantly characterized by EBF, while DBF and DNF are concentrated in Region IV, contributing to a lower average NEP of 233.51 g C/m^2 (Figure 6b). While forests are more widespread in Regions III and IV, Region V has a higher average NEP due to concentrated areas with strong carbon sequestration. Its forests exhibit fewer but higher NEP values and lower carbon sources, boosting the average, whereas Regions III and IV have more dispersed, moderate NEP levels. Additionally, Region V's distribution of ENF (Figure 1) further elevates its NEP, explaining its superior sink performance despite smaller forest coverage.

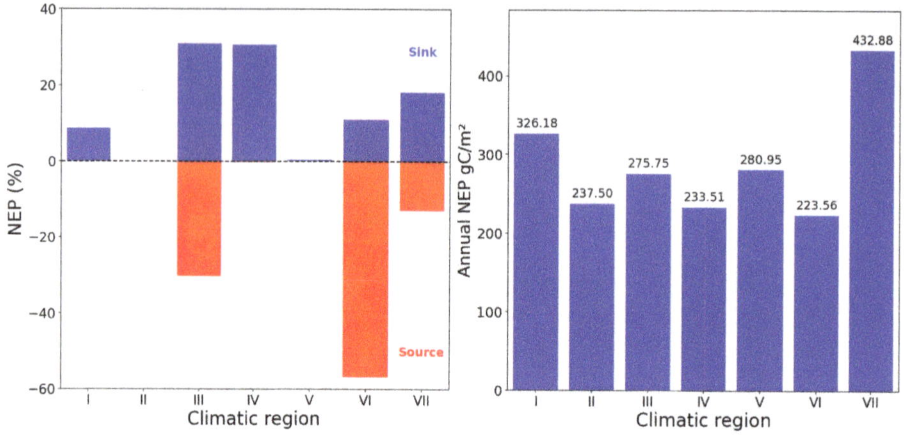

Figure 6. (**a**) NEP contribution by climatic regions; (**b**) annual NEP of each climatic region.

Figure 7 shows the average annual NEP variation for China's forests from 2013 to 2023. As NEP became more variable over time, the growth rate was estimated to be 1.67 $gC{\cdot}m^{-2}$ per year, with the lowest value observed in 2016 (255.09 $gC{\cdot}m^{-2}$) while the highest was recorded in 2019 (296.10 $gC{\cdot}m^{-2}$), resulting in a mean of 275.32 $gC{\cdot}m^{-2}$.

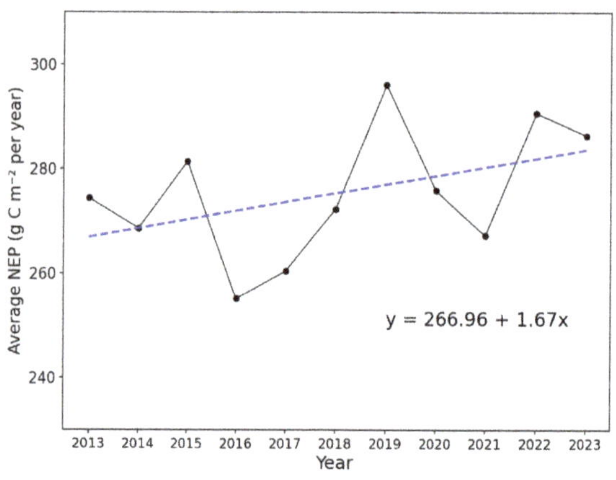

Figure 7. Average annual NEP of forest in China during 2013–2023.

3.3. Climatic Factors Contribution to NEP Variation

First case: The GD model assessed the influence of seven climatic variables on NEP in forest ecosystems over the entirety of China. In the GD model, all variables significantly impacted NEP change at a *p*-value of < 0.01, indicating a clear relationship between the climatic factors and NEP. Solar radiation, precipitation, and VPD were recognized as the key drivers of temporal NEP variations in China's forests from 2013 to 2023. Solar radiation showed the highest impact, having a 0.104 q-value, followed by precipitation at 0.031 and VPD at 0.028. Other climatic variables, such as frost days (0.013), T_{avg} (0.014), T_{min} (0.015), and T_{max} (0.018), had relatively weaker effects on NEP, as illustrated in Figure 8a.

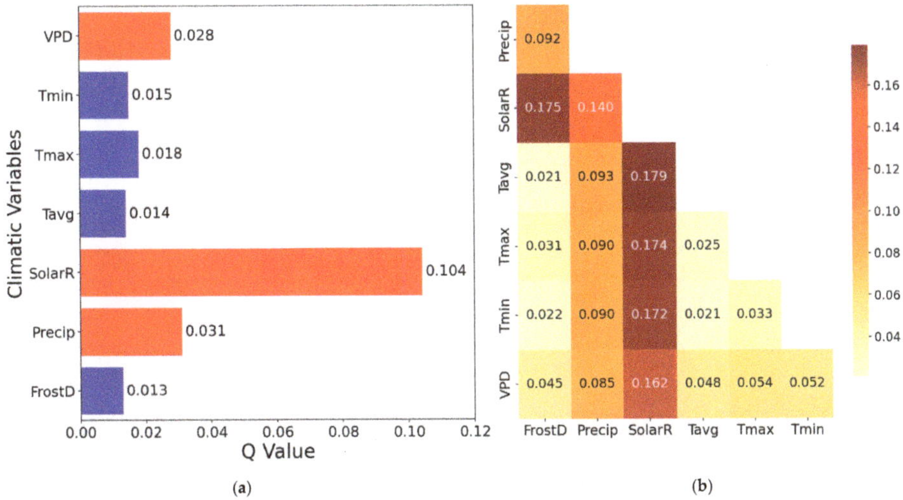

Figure 8. Results of geographical detector model. (**a**) Factor detector; (**b**) interaction detector.

The interaction detector estimated q-values for pairs of climatic variables as illustrated in Figure 8b. Factor interaction had higher q-statistic values than individual factors, indicating that NEP dynamics are driven mainly by combined influences. The interaction of the solar radiation and T_{avg} (0.179) had the strongest effect on vegetation productivity, followed by solar radiation and frost days (0.175), and solar radiation and T_{max} (0.174). Other interactions are as follows: solar radiation and T_{min} (0.172), solar radiation and VPD (0.162), and solar radiation and precipitation (0.14). Other climate factor interactions had a contribution of <0.1 q-value, indicating that their effects are very low in influencing NEP change. The interaction between solar radiation, temperature, and precipitation is vital for plant growth and carbon sequestration. The results also showed that all factor pairs, except four that showed bivariate enhancement (FrostD + T_{avg} (0.021), FrostD + T_{min} (0.022), T_{avg} + T_{max} (0.025), and T_{avg} + T_{min} (0.021), exhibited nonlinear enhanced interactions. This suggests that NEP dynamics are influenced by complex climatic interdependencies rather than isolated drivers.

Second case: The GD model was applied to each climatic region in China to assess the impact of climatic variables on NEP of forest ecosystems (Figure 9). Region I: VPD (0.171), precipitation (0.096), T_{max} (0.06), and solar radiation (0.052) exhibit the highest q-values (>0.05), indicating a stronger influence on NEP in this region. T_{avg} (0.033), frost days (0.026), and T_{min} (0.007) show a relatively smaller but still notable impact. This is consistent with the fact that as latitude increases, the effect of these three factors tends to strengthen, especially in higher latitudes where temperature variation and frost conditions become more prominent. Region II and V: To ensure the GD model runs effectively, sufficient observations are needed. Therefore, Region II was combined with Region V due to its very small forest cover. Then, frost days (0.713), T_{avg} (0.695), and VPD (0.68) showed a strong effect on NEP. Frost days had a higher q-value as most of the region experiences winter conditions. Meanwhile, other factors such as precipitation, T_{max}, T_{min}, and solar radiation play a more significant role in influencing the ecosystem's carbon dynamics, particularly during the summer and spring seasons. Region III: Solar radiation (0.199), precipitation (0.062), and T_{avg} (0.018) are the dominant factors influencing NEP, highlighting the significant roles of sunlight and moisture in driving forest ecosystem productivity. Minimal effects have been observed from T_{max} (0.017), T_{min} (0.014), and VPD (0.007), suggesting that temperature is less of a limiting factor in this region's ecosystem dynamics. Similarly, frost days (0.001) shows a negligible impact, indicating that the frost factor is not as influential in this climate zone. Region IV: Precipitation (0.073) remains a key factor supporting forest productivity, providing essential moisture during and after winter for biological processes. Frost days (0.036) and T_{min} (0.033) play a crucial role in limiting biological activity and NEP, especially since the northern part of this climate region experiences winter conditions. Other factors (Solar radiation, T_{max}, T_{avg}, and VPD) have a notable effect on NEP. Region VI: VPD (0.095), T_{max} (0.057), solar radiation (0.044), and precipitation (0.043) have the highest q-values, making them the most influential factors on NEP in this region. On the other hand, T_{avg} (0.02), T_{min} (0.006), and frost days (0.01) have weaker effects on NEP, with q-values less than 0.03, suggesting that temperature-related factors, although still important, are less critical in limiting forest productivity compared to moisture-related factors in this region. Region VII: The factors influencing NEP are solar radiation (0.129), T_{max} (0.062), T_{avg} (0.049), VPD (0.046), T_{min} (0.041), and precipitation (0.032). Among these, solar radiation stands out with the highest q-value, indicating its dominant role in influencing NEP in this region. T_{max}, T_{avg}, VPD, and T_{min} have moderate impacts, suggesting that temperature-related factors also influence the growing season and ecosystem dynamics. Precipitation is another significant factor, showing the importance of rainfall in supporting photosynthesis and forest growth. Lastly, frost days has the least influence, reflecting that frost is not a major limiting factor for NEP in this region, likely due to milder winter conditions.

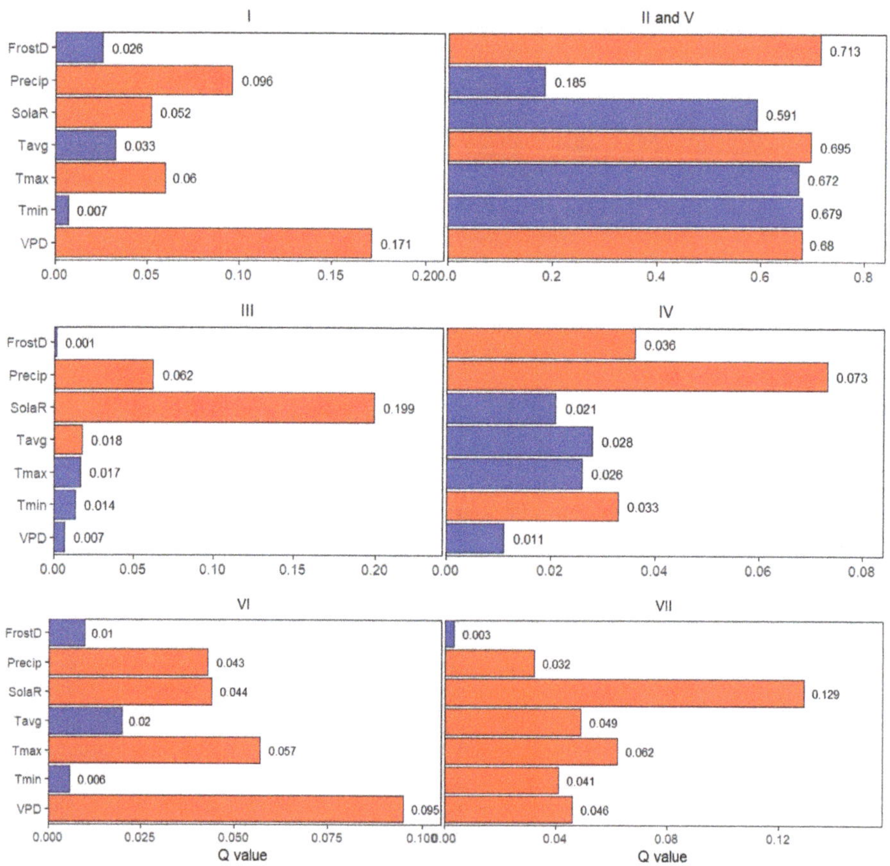

Figure 9. Factor detector results for climatic regions in China.

The interaction effects of climate variables on NEP in China's forest ecosystems for each climate region were analyzed, providing clear evidence that paired factors have a greater influence than isolated ones (Figure S5).

4. Discussion

This research simulates China's forests' Net Primary Productivity (NPP) from 2013 to 2023, offering valuable insights into the fundamental carbon balance. The study's average NPP estimate of 870 $gC·m^{-2}·yr^{-1}$ is higher compared to earlier estimates of around 840 $gC·m^{-2}·yr^{-1}$ [63,64]. Although the average differences in NPP across studies are not large, the slight difference of 30 $gC·m^{-2}·yr^{-1}$ between this study and previous research highlights the advancement brought about by the use of the 3-PGS model, which accounts for a broad range of physiological, environmental, and ecological mechanisms that may not have been incorporated in other models. This suggests that China's forest ecosystems have made a more significant contribution to carbon storage than previously understood, improving our ability to model and predict their role in the global carbon cycle. The difference in NPP estimates can largely be attributed to the use of more recent climate, moderate-resolution data from the years 2013 to 2023, enabling the model to incorporate the latest climatic conditions, forest management practices, and ecosystem dynamics. As latitude increases, climatic factors limit carbon production, lowering NPP. Warmer and more humid environments in the south promote higher forest productivity, while the colder, drier north reduces it, reflecting the strong influence of climate on forest carbon storage [65,66]. The 3-PGS model's process-based approach is a major strength, allowing it to model physiological processes such as photosynthesis, transpiration, and overall growth, which directly influence NPP and are sensitive to climatic variability. Additionally, Mao et al. [67] found that the long-term overall average NPP for various forest types, accurately represented by models, is above 500 $gC·m^{-2}·yr^{-1}$. When comparing with the current estimate of 870 $gC·m^{-2}·yr^{-1}$, which comfortably exceeds Mao et al. [67]'s estimation, this provides strong evidence that China's forests have recently made a significant contribution to carbon sequestration. The 3-PGS model also demonstrates significant improvements, as shown by the comparison of the annual average NPP and its range for each forest type with previous research [64,67–69] as summarized in Table 1.

Table 1. Average annual NPP comparison between this study, MODIS NPP, and other findings.

	Forest Type				
	EBF	**ENF**	**Mixed**	**DBF**	**DNF**
This study	1114.22 (256.74–1518.63)	863.92 (176.20–1304.74)	837.92 (6.82–1417.33)	733.07 (340.69–1368.23)	448.08 (287.06–491.58)
MOD17A3HGF.061 [34]	1087.83 (211.86–1913.64)	813.84 (48.19–1860.58)	752.49 (60.6–1846.42)	631.08 (117.56–1509.74)	508.23 (354.73–613.15)
Ji et al., 2020 [64]	1058 (942–1211)	934 (851–1053)	860 (790–960)	759 (699–810)	590 (568–618)
Zhu et al., 2007 [68]	986 (407–1913)	367 (179–806)	257–1098	643 (114–1669)	439 (179–824)
Liang et al., 2023 [69]	893.00	948.90	530.10	495.70	352.70
Mao et al., 2010 [67]	891	510	-	743	546

The wider range of the model's NPP values (i.e., 256.74–1518.63 $gC·m^{-2}·yr^{-1}$ for EBF, 176.20–1304.74 $gC·m^{-2}·yr^{-1}$ for ENF, and 6.82–1417.33 $gC·m^{-2}·yr^{-1}$ for MF) in this study compared to other estimate ranges further reflects the flexibility and adaptive nature of the 3-PGS model. The wide range observed here emphasizes the model's ability to represent the heterogeneity in forest conditions across China, which other models might not have captured. The variability in NPP values across different forest types reflects real-world differences in forest structure, species diversity, and environmental influences that are fundamental to understanding forest productivity and carbon sequestration. The wide range observed here emphasizes the model's ability to represent the heterogeneity in forest conditions across China, which other models might not have captured.

Understanding the carbon cycle in ecosystems, particularly in forests, requires knowledge of NEP estimation [70,71]. NEP provides a measure of an ecosystem's capacity to both sequester and emit carbon, thereby serving as a valuable metric for assessing its ecological integrity across diverse regions. Evaluating carbon sequestration and emission processes is key to understanding whether an ecological structure serves as a carbon sink or source, and ongoing assessments of these processes are essential for projecting global carbon flow [11,72,73]. By providing more precise NEP estimates for China's forests, this research helps refine the global carbon budget and supports the design of more targeted mitigation strategies for addressing global climate change.

Our annual average NEP estimate of 0.156 ± 0.0071 PgC (or 155.86 TgC) for China's forests from 2013 to 2023 is higher compared to earlier studies that employed process-based models and various multi-source spatial data in evaluating China's NEP status, including estimates by He et al. (0.118 PgC/yr) [19], Yang et al. (0.11 PgC/yr) [74], and Zhang et al. (0.134 PgC/yr) [54]. Our estimate also falls within the reported range of 0.10 to 0.24 PgC/yr [3,21,75] for model-based approaches, thereby confirming the high accuracy of this study. The Global Carbon Project, 2023 (https://globalcarbonatlas.org/emissions/carbon-emissions/, accessed on 26 August 2025) provides China's annual data on anthropogenic CO_2 emissions at 11.903 Gt CO_2. This closely corresponds with the 11.4 Gt CO_2 (or 3.1 Gt C) projection for 2022 presented by Friedlingstein et al. (2022) [76]. Our analysis of forest Net Ecosystem Production (NEP) reveals an annual carbon sink of 155.86 Tg C yr^{-1}, which offsets approximately 4.8% of the total emissions. This forest-specific estimate is within the 3–8% range indicated by ecosystem process models, although it is lower than that derived from inventory approaches (~10%) [21,77] and that proposed by atmospheric inversions, which reaches up to 45% [78].

The increased annual NEP value observed in this study, compared to earlier research, can primarily be attributed to the Chinese government's large-scale vegetation restoration initiatives implemented in the late 20th century [22]. Efforts such as afforestation, farmland restoration to forests, and ecological protection policies have collectively led to notable enhancements in carbon sequestration nationwide [79,80]. The Global Canopy Program's soil erosion control in the Loess Plateau and natural forest conservation projects are among the initiatives implemented to combat land degradation, desertification, and soil erosion [81]. These initiatives have substantially increased China's carbon sequestration capacity by enhancing forest cover, mitigating soil erosion, and rehabilitating degraded lands. As a result, the ecosystem's ability to sequester and store carbon has improved, directly supporting China's strategic goal of achieving carbon neutrality. The elevated NEP estimates in this study provide empirical evidence of the effectiveness of these ecological restoration efforts, demonstrating that China's terrestrial ecosystems are functioning as increasingly robust carbon sinks [82]. This progress underscores China's commitment to reducing net carbon emissions and highlights the vital role of forest and land management in advancing toward national carbon neutrality targets [83]. The study suggests that these efforts need to be intensified, particularly in areas that have been less affected by large-scale restoration.

To measure China's climate's influence on forest NEP during 2013–2023, the GD model was utilized due to its ability to effectively elucidate causal relationships between two variables, which are in two dimensions [56]. Variations in NEP across China's forests are primarily driven by solar radiation, precipitation, and VPD, with temperature and frost days playing a less significant role.

Due to cold temperatures, solar radiation has little influence in Region IV during the winter months, but plants can efficiently use radiation to boost productivity in spring and summer [84]. In Region VII and the southern part of Region VI, high humidity and cloud cover influence solar radiation by enhancing its penetration into the forest canopy, which improves light use efficiency (LUE) when direct sunlight is limited. The enhanced LUE under these conditions contributes to the high productivity of carbon sequestration observed in tropical and subtropical areas year-round [85–87]. The consistent rainfall in southern China (Regions III, VI, and VII) sustains tropical and subtropical vegetation, resulting in high NEP as plants have continuous access to water throughout the year. Conversely, the northern part of Region V experiences lower precipitation, particularly during winter, leading to seasonal droughts that restrict forest growth and overall productivity [88].

In arid regions, including I, II, and V, high VPD due to low humidity and fluctuating temperatures accelerates evapotranspiration, which in turn reduces soil moisture and plant producing capacity [89]. VPD has minimal impact on NEP in Regions VII, III, and southeastern VI, where frequent rainfall and high humidity support both consistent moisture retention and NEP [88]. Although warmer spring and summer temperatures typically enhance the growing season, the low temperatures in Region IV limit NEP by diminishing soil microbial respiration rates. In Regions V and I, higher temperatures lead to increased VPD, which subsequently reduces absorption efficiency [90]. While warmer temperatures in Regions VII, III, and southeastern VI generally enhance Net Ecosystem Productivity (NEP), extreme heat and drought conditions can diminish carbon storage capabilities [91]. The occurrence of frosts significantly impacts northern and high-altitude forests, specifically in Regions I, II, IV, and V, by shortening the growing season and thereby reducing NEP [92]. These findings suggest that climate change mitigation strategies should include regional assessments of temperature thresholds to better understand how different ecosystems may respond to future climate scenarios. The application of the GD model in this study, along with its corresponding results, aligns closely with findings reported in previous research [17,19,93], reinforcing the model's reliability in assessing climate–NEP relationships. A key advancement of this study lies in the application of the 3-PGS model on simulating the current NEP estimate, which supports the significant success of China's ongoing initiatives to increase forest cover, achieve carbon neutrality, and realize its goal of establishing a fully developed green, low-carbon, and circular economy by 2060 [94]. Our findings not only validate the effectiveness of large-scale ecological restoration programs but also offer actionable insights for optimizing policies to enhance carbon sequestration. Also, the use of the GD model at both the national scale and across distinct climatic zones enables a more comprehensive and spatially resolved understanding of how NEP in China's forest ecosystems responds to diverse climate conditions. Future research should investigate the long-term impacts of disturbances, including land-use changes and fire risk, on carbon storage, and refine models to integrate finer-scale remote sensing data along with additional climate and soil variables, such as the effects of nitrogen, in order to provide more precise NEP estimates.

Limitation

The model's reliability was confirmed by the Net Primary Productivity (NPP) validation, which proved reasonable performance with an R^2 of 0.52 and an RMSE of 126.17 gC m^{-2} for a sample size (n) of 3720. However, despite this moderate agreement between the model and MODIS data, the model's limitations must still be acknowledged. The 3-PGS model has limited capacity to incorporate the effects of disturbances, including fires, pest outbreaks, and land use changes. Since these factors can significantly alter forest carbon dynamics, their exclusion may reduce the model's predictive accuracy. To improve NEP estimation, it is important to account for disturbances like fires, land use changes, and pest outbreaks. Additionally, using high-resolution climate, soil, and vegetation data will help reduce errors and improve model accuracy.

5. Conclusions

Combining the 3-PGS, soil heterotrophic respiration, and GD models allowed for the simulation of forests' NEP distribution across China, while also addressing the influence of climate variables on NEP changes from 2013 to 2023. The main conclusions are as follows:

- This study provides a comprehensive assessment of China's forest carbon sequestration potential from 2013 to 2023, with a focus on Net Ecosystem Production (NEP). The findings indicate that China's forests accumulated 1.71 $\pm$ 0.09 PgC over the study period, with an annual average NEP of 0.156 $\pm$ 0.0071 PgC (155.86 TgC/year). These results highlight significant variability in NEP across different forest types, with southern China contributing over 45% of the total sequestration. This underscores the influence of regional climate and ecological conditions on forest productivity. The findings also demonstrate the growing role of China's forests as carbon sinks, emphasizing their capacity to mitigate carbon emissions. While the increase in NEP over time highlights the effectiveness of China's forest ecosystems in carbon sequestration, the study also reveals that forest-based sequestration alone cannot fully offset the country's total emissions. This highlights the necessity of a comprehensive, integrated approach to achieving China's carbon neutrality goals by 2060. Forest management must be combined with other ecosystem-based solutions, including afforestation, sustainable agriculture, and advancements in carbon capture technologies.
- The study further identifies solar radiation, precipitation, and vapor pressure deficit (VPD) as the primary climatic drivers of NEP variations in China's forests, with temperature and frost days playing a secondary role. The interaction between these climatic factors underscores the importance of region-specific strategies for climate-resilient forest management. Given the variability in forest productivity across China, targeted interventions should prioritize areas with low carbon sequestration potential, aiming to enhance carbon storage in these regions while also optimizing carbon sequestration in areas with high potential, particularly in tropical and subtropical zones.
- This study's findings not only contribute to China's carbon neutrality strategies but also enhance global understanding of forest carbon dynamics. The insights provided will be valuable for policymakers seeking to optimize forest management practices, inform regional climate strategies, and contribute to global carbon cycle models. As such, this research lays the foundation for future efforts aimed at maximizing the role of forests in mitigating climate change and achieving long-term environmental sustainability.

Supplementary Materials: The following supporting information can be downloaded at: https://www.mdpi.com/article/10.3390/f16091398/s1.

Author Contributions: Conceptualization, F.J.M. and Y.Y.; paper writing, F.J.M.; paper reviewing and editing, F.J.M., Y.Y. and X.Y.; supervision, Y.Y. and X.Y.; funding acquisition, W.F. All authors have read and agreed to the published version of the manuscript.

Funding: Special Science Fund for Carbon Neutrality of Northeast Forestry University (HFW220100054); The National Key Research and Development Program of China (2023YFD2201704).

Data Availability Statement: All the data used in this study are publicly and freely available.

Acknowledgments: I express my gratitude to Nicholas Coops for his assistance in utilizing the data to execute the model. The guidance provided was absolutely essential.

Conflicts of Interest: The authors declare no conflicts of interest.

References

1. Ritchie, H.; Rosado, P.; Roser, M. Greenhouse Gas Emissions. Our World Data 2020. Available online: https://ourworldindata.org/greenhouse-gas-emissions (accessed on 26 August 2025).
2. Piao, S.; He, Y.; Wang, X.; Chen, F. Estimation of China's Terrestrial Ecosystem Carbon Sink: Methods, Progress and Prospects. *Sci. China Earth Sci.* **2022**, *65*, 641–651. [CrossRef]
3. Tian, H.; Melillo, J.; Lu, C.; Kicklighter, D.; Liu, M.; Ren, W.; Xu, X.; Chen, G.; Zhang, C.; Pan, S. China's Terrestrial Carbon Balance: Contributions from Multiple Global Change Factors. *Glob. Biogeochem. Cycles* **2011**, *25*, GB1007. [CrossRef]
4. The Chinese Foreign Ministry (TCFM). *Statement by Xi Jinping, President of the People's Republic of China, at the General Debate of the 75th Session of the United Nations General Assembly*; The Chinese Foreign Ministry (TCFM): Beijing, China, 2021.
5. Masson-Delmotte, V.; Zhai, P.; Pörtner, H.-O.; Roberts, D.; Skea, J.; Shukla, P.R. *Global Warming of 1.5 C: IPCC Special Report on Impacts of Global Warming of 1.5 C Above Pre-Industrial Levels in Context of Strengthening Response to Climate Change, Sustainable Development, and Efforts to Eradicate Poverty*; Cambridge University Press: Cambridge, UK, 2022; ISBN 1009157949.
6. Xu, D.-F.; Wang, R.-H.; Li, Y.-X.; Hui, Z. Review on Carbon Cycle in Terrestrial Ecosystem and Its Influenced Factors. *Chin. J. Agrometeorol.* **2009**, *30*, 519.
7. Zhang, P.; He, J.; Hong, X.; Zhang, W.; Qin, C.; Pang, B.; Li, Y.; Liu, Y. Carbon Sources/Sinks Analysis of Land Use Changes in China Based on Data Envelopment Analysis. *J. Clean. Prod.* **2018**, *204*, 702–711. [CrossRef]
8. Lü, F.C.; Ma, J.Y.; Cao, Y. Carbon Fluxes Simulation of China′ s Typical Forest Ecosystem Based on FORCCHN Model. *Acta Ecol. Sin.* **2022**, *42*, 2810–2821.
9. Zhang, C.; Huang, N.; Wang, L.; Song, W.; Zhang, Y.; Niu, Z. Spatial and Temporal Pattern of Net Ecosystem Productivity in China and Its Response to Climate Change in the Past 40 Years. *Int. J. Environ. Res. Public Health* **2022**, *20*, 92. [CrossRef]
10. Woodwell, G.M.; Whittaker, R.H.; Reiners, W.A.; Likens, G.E.; Delwiche, C.C.; Botkin, D.B. The Biota and the World Carbon Budget: The Terrestrial Biomass Appears to Be a Net Source of Carbon Dioxide for the Atmosphere. *Science* **1978**, *199*, 141–146. [CrossRef]
11. Pathak, K.; Malhi, Y.; Sileshi, G.W.; Das, A.K.; Nath, A.J. Net Ecosystem Productivity and Carbon Dynamics of the Traditionally Managed Imperata Grasslands of North East India. *Sci. Total Environ.* **2018**, *635*, 1124–1131. [CrossRef]
12. Qi, S.; Zhang, H.; Zhang, M. Net Primary Productivity Estimation of Terrestrial Ecosystems in China with Regard to Saturation Effects and Its Spatiotemporal Evolutionary Impact Factors. *Remote Sens.* **2023**, *15*, 2871. [CrossRef]
13. Wang, J.; Wu, C.; Wang, X.; Zhang, X. A New Algorithm for the Estimation of Leaf Unfolding Date Using MODIS Data over China's Terrestrial Ecosystems. *ISPRS J. Photogramm. Remote Sens.* **2019**, *149*, 77–90. [CrossRef]
14. Field, C.B.; Behrenfeld, M.J.; Randerson, J.T.; Falkowski, P. Primary Production of the Biosphere: Integrating Terrestrial and Oceanic Components. *Science* **1998**, *281*, 237–240. [CrossRef] [PubMed]
15. Zheng, D.; Prince, S.; Wright, R. Terrestrial Net Primary Production Estimates for 0.5 Grid Cells from Field Observations—A Contribution to Global Biogeochemical Modeling. *Glob. Change Biol.* **2003**, *9*, 46–64. [CrossRef]
16. Chen, P.F. Monthly NPP Dataset Covering China's Terrestrial Ecosystems at North of 18° N (1985–2015). *J. Glob. Change Data Discov.* **2019**, *3*, 34–41.
17. Chen, Y.; Xu, Y.; Chen, T.; Zhang, F.; Zhu, S. Exploring the Spatiotemporal Dynamics and Driving Factors of Net Ecosystem Productivity in China from 1982 to 2020. *Remote Sens.* **2023**, *16*, 60. [CrossRef]
18. Ma, X.; Wu, L.; Zhu, Y.; Wu, J.; Qin, Y. Simulation of Vegetation Carbon Sink of Arbor Forest and Carbon Mitigation of Forestry Bioenergy in China. *Int. J. Environ. Res. Public Health* **2022**, *19*, 13507. [CrossRef] [PubMed]

19. He, H.; Wang, S.; Zhang, L.; Wang, J.; Ren, X.; Zhou, L.; Piao, S.; Yan, H.; Ju, W.; Gu, F. Altered Trends in Carbon Uptake in China's Terrestrial Ecosystems under the Enhanced Summer Monsoon and Warming Hiatus. *Natl. Sci. Rev.* **2019**, *6*, 505–514. [CrossRef] [PubMed]
20. Liu, S.; Zhou, T.; Wei, L.; Shu, Y. The Spatial Distribution of Forest Carbon Sinks and Sources in China. *Chin. Sci. Bull.* **2012**, *57*, 1699–1707. [CrossRef]
21. Piao, S.; Fang, J.; Ciais, P.; Peylin, P.; Huang, Y.; Sitch, S.; Wang, T. The Carbon Balance of Terrestrial Ecosystems in China. *Nature* **2009**, *458*, 1009–1013. [CrossRef]
22. Zhou, Y.; Li, X.; Liu, Y. Land Use Change and Driving Factors in Rural China during the Period 1995–2015. *Land. Use Policy* **2020**, *99*, 105048. [CrossRef]
23. Sun, Q.; Miao, C.; Duan, Q. Extreme Climate Events and Agricultural Climate Indices in China: CMIP5 Model Evaluation and Projections. *Int. J. Climatol.* **2016**, *36*, 43–61. [CrossRef]
24. Qin, H.; Jin, X.; Guo, K. An Overview of China's Vegetation and Plant Diversity. In *Conservation and Reintroduction of Rare and Endangered Plants in China*; Springer: Singapore, 2020; pp. 3–19.
25. National Forestry and Grassland Administration. *China Forest Resources Report 2014–2018*; China Forestry Publishing House: Beijing, China, 2019; pp. 357–469.
26. Muñoz-Sabater, J.; Dutra, E.; Agustí-Panareda, A.; Albergel, C.; Arduini, G.; Balsamo, G.; Boussetta, S.; Choulga, M.; Harrigan, S.; Hersbach, H. ERA5-Land: A State-of-the-Art Global Reanalysis Dataset for Land Applications. *Earth Syst. Sci. Data* **2021**, *13*, 4349–4383. [CrossRef]
27. Didan, K. MODIS/Terra Vegetation Indices Monthly L3 Global 0.05 Deg CMG V061. Available online: https://www.earthdata.nasa.gov/data/catalog/lpcloud-mod13c2-061 (accessed on 26 August 2025).
28. Holben, B.N. Characteristics of Maximum-Value Composite Images from Temporal AVHRR Data. *Int. J. Remote Sens.* **1986**, *7*, 1417–1434. [CrossRef]
29. Tian, J.; Zhu, X.; Chen, J.; Wang, C.; Shen, M.; Yang, W.; Tan, X.; Xu, S.; Li, Z. Improving the Accuracy of Spring Phenology Detection by Optimally Smoothing Satellite Vegetation Index Time Series Based on Local Cloud Frequency. *ISPRS J. Photogramm. Remote Sens.* **2021**, *180*, 29–44. [CrossRef]
30. Friedl, M.; Sulla-Menashe, D. MODIS/Terra+ Aqua Land Cover Type Yearly L3 Global 0.05 Deg CMG V061. Available online: https://www.earthdata.nasa.gov/data/catalog/lpcloud-mcd12c1-061 (accessed on 26 August 2025).
31. Zeng, T.; Zhang, Z.; Zhao, X.; Wang, X.; Zuo, L. Evaluation of the 2010 MODIS Collection 5.1 Land Cover Type Product over China. *Remote Sens.* **2015**, *7*, 1981–2006. [CrossRef]
32. Xie, Y.; Wang, H.; Lei, X. Application of the 3-PG Model to Predict Growth of *Larix Olgensis* Plantations in Northeastern China. *Ecol. Manag.* **2017**, *406*, 208–218. [CrossRef]
33. Feng, X.; Liu, G.; Chen, J.M.; Chen, M.; Liu, J.; Ju, W.M.; Sun, R.; Zhou, W. Net Primary Productivity of China's Terrestrial Ecosystems from a Process Model Driven by Remote Sensing. *J. Environ. Manag.* **2007**, *85*, 563–573. [CrossRef]
34. Running, S.; Zhao, M. MOD17A3HGF MODIS/Terra Net Primary Production Gap-Filled Yearly L4 Global 500 m SIN Grid V006. Available online: https://www.earthdata.nasa.gov/data/catalog/lpcloud-mod17a3hgf-006 (accessed on 26 August 2025).
35. Sun, H.; Xu, Z.; Jia, B. A Compiled Soil Respiration Dataset at Different Time Scales for Forest Ecosystems across China from 2000 to 2018. *Earth Syst. Sci. Data* **2022**, *14*, 2951–2961. [CrossRef]
36. Song, X.; Peng, C.; Zhao, Z.; Zhang, Z.; Guo, B.; Wang, W.; Jiang, H.; Zhu, Q. Quantification of Soil Respiration in Forest Ecosystems across China. *Atmos. Environ.* **2014**, *94*, 546–551. [CrossRef]
37. Kumar, L.; Mutanga, O. Google Earth Engine Applications since Inception: Usage, Trends, and Potential. *Remote Sens.* **2018**, *10*, 1509. [CrossRef]
38. Tamiminia, H.; Salehi, B.; Mahdianpari, M.; Quackenbush, L.; Adeli, S.; Brisco, B. Google Earth Engine for Geo-Big Data Applications: A Meta-Analysis and Systematic Review. *ISPRS J. Photogramm. Remote Sens.* **2020**, *164*, 152–170. [CrossRef]
39. Gorelick, N.; Hancher, M.; Dixon, M.; Ilyushchenko, S.; Thau, D.; Moore, R. Google Earth Engine: Planetary-Scale Geospatial Analysis for Everyone. *Remote Sens. Environ.* **2017**, *202*, 18–27. [CrossRef]
40. Coops, N.C.; Waring, R.H.; Landsberg, J.J. Assessing Forest Productivity in Australia and New Zealand Using a Physiologically-Based Model Driven with Averaged Monthly Weather Data and Satellite-Derived Estimates of Canopy Photosynthetic Capacity. *For. Ecol. Manag.* **1998**, *104*, 113–127. [CrossRef]
41. Landsberg, J.J.; Waring, R.H. A Generalised Model of Forest Productivity Using Simplified Concepts of Radiation-Use Efficiency, Carbon Balance and Partitioning. *Ecol. Manag.* **1997**, *95*, 209–228. [CrossRef]
42. Coops, N.C.; Waring, R.H. Estimating Forest Productivity in the Eastern Siskiyou Mountains of Southwestern Oregon Using a Satellite Driven Process Model, 3-PGS. *Can. J. Res.* **2001**, *31*, 143–154. [CrossRef]

43. McMurtrie, R.E.; Gholz, H.L.; Linder, S.; Gower, S.T. Climatic Factors Controlling the Productivity of Pine Stands: A Model-Based Analysis. *Ecol. Bull.* **1994**, *43*, 173–188.
44. Sellers, P.J. Canopy Reflectance, Photosynthesis, and Transpiration, II. The Role of Biophysics in the Linearity of Their Interdependence. *Remote Sens. Environ.* **1987**, *21*, 143–183. [CrossRef]
45. Kumar, M.; Monteith, J.L. Remote Sensing of Plant Growth. In *Plants and the Daylight Spectrum*; Smith, H., Ed.; Academic Press: London, UK, 1982.
46. Landsberg, J.J.; Waring, R.H.; Williams, M. The Assessment of NPP/GPP Ratio. *Tree Physiol.* **2020**, *40*, 695–699. [CrossRef] [PubMed]
47. Zhang, Y.; Yu, G.; Yang, J.; Wimberly, C.; Zhang, X.; Tao, J.; Jiang, Y.; Zhu, J. Climate-Driven Global Changes in Carbon Use Efficiency. *Glob. Ecol. Biogeogr.* **2014**, *23*, 144–155. [CrossRef]
48. Ryan, M.G.; Law, B.E. Interpreting, Measuring, and Modeling Soil Respiration. *Biogeochemistry* **2005**, *73*, 3–27. [CrossRef]
49. Xie, W.; Chen, S.-T.; Hu, Z.-H. Factors Influencing the Variability in Soil Heterotrophic Respiration from Terrestrial Ecosystem in China. *Huan Jing Ke Xue* **2014**, *35*, 334–340.
50. Shi, Z. Spatial-Temporal Simulation of Vegetation Carbon Sink and Its Influential Factors Based on CASA and GSMSR Model in Shaanxi Province. Ph.D. Thesis, Northwest A&F University, Xianyang, China, 2015.
51. Bond-Lamberty, B.; Wang, C.; Gower, S.T. A Global Relationship between the Heterotrophic and Autotrophic Components of Soil Respiration? *Glob. Change Biol.* **2004**, *10*, 1756–1766. [CrossRef]
52. Chen, S.; Huang, Y.; Zou, J.; Shi, Y.; Lu, Y.; Zhang, W.; Hu, Z. Interannual Variability in Soil Respiration from Terrestrial Ecosystems in China and Its Response to Climate Change. *Sci. China Earth Sci.* **2012**, *55*, 2091–2098. [CrossRef]
53. Xie, X.L.; Sun, B.; Zhou, H.Z.; Li, Z.P.; Li, A.B. Estimation and Spatial Distribution Analysis of Soil Organic Carbon Density and Storage in China. *Acta Pedol. Sin.* **2004**, *1*, 35–43.
54. Zhang, M.; Huang, X.J.; Chuai, X.; Xie, X.; Zhu, Z.; Wang, Y. Spatial Distribution and Changing Trends of Net Ecosystem Productivity in China. *Geogr. Geo-Inf. Sci.* **2020**, *36*, 69–74.
55. Fang, J.; Chen, A.; Peng, C.; Zhao, S.; Ci, L. Changes in Forest Biomass Carbon Storage in China between 1949 and 1998. *Science* **2001**, *292*, 2320–2322. [CrossRef]
56. Wang, J.F.; Xu, C.D. Geodetector: Principle and Prospective. *Acta Geogr. Sin.* **2017**, *72*, 116–134.
57. Wang, J.F.; Li, X.H.; Christakos, G.; Liao, Y.L.; Zhang, T.; Gu, X.; Zheng, X.Y. Geographical Detectors-Based Health Risk Assessment and Its Application in the Neural Tube Defects Study of the Heshun Region, China. *Int. J. Geogr. Inf. Sci.* **2010**, *24*, 107–127. [CrossRef]
58. Wang, J.-F.; Zhang, T.-L.; Fu, B.-J. A Measure of Spatial Stratified Heterogeneity. *Ecol. Indic.* **2016**, *67*, 250–256. [CrossRef]
59. Wang, J.F.; Hu, Y. Environmental Health Risk Detection with GeogDetector. *Environ. Model. Softw.* **2012**, *33*, 114–115. [CrossRef]
60. Jenks, G.F. The Data Model Concept in Statistical Mapping. *Int. Yearb. Cartogr.* **1967**, *7*, 186–190.
61. Liu, Y.S.; Yang, R. The Spatial Characteristics and Formation Mechanism of the County Urbanization in China. *Acta Geogr. Sin.* **2012**, *67*, 1011–1020.
62. Chen, G.S.; Yang, Y.S.; Lv, P.P.; Zhang, Y.P.; Qian, X.L. Regional Patterns of Soil Respiration in China's Forests. *Acta Ecol. Sin.* **2008**, *28*, 1748–1761.
63. Zhuang, Q.; Zhang, T.; Xiao, J.; Luo, T. Quantification of Net Primary Production of Chinese Forest Ecosystems with Spatial Statistical Approaches. *Mitig. Adapt. Strat. Glob. Change* **2009**, *14*, 85–99. [CrossRef]
64. Ji, Y.; Zhou, G.; Luo, T.; Dan, Y.; Zhou, L.; Lv, X. Variation of Net Primary Productivity and Its Drivers in China's Forests during 2000–2018. *Ecosystems* **2020**, *7*, 15. [CrossRef]
65. Gu, F.; Zhang, Y.; Huang, M.; Tao, B.; Guo, R.; Yan, C. Effects of Climate Warming on Net Primary Productivity in China during 1961–2010. *Ecol. Evol.* **2017**, *7*, 6736–6746. [CrossRef]
66. Yuan, Z.; Wang, Y.; Xu, J.; Wu, Z. Effects of Climatic Factors on the Net Primary Productivity in the Source Region of Yangtze River, China. *Sci. Rep.* **2021**, *11*, 1376. [CrossRef] [PubMed]
67. Mao, J.; Dan, L.; Wang, B.; Dai, Y. Simulation and Evaluation of Terrestrial Ecosystem NPP with M-SDGVM over Continental China. *Adv. Atmos. Sci.* **2010**, *27*, 427–442. [CrossRef]
68. Zhu, W.-Q.; Pan, Y.-Z.; Zhang, J.-S. Estimation of Net Primary Productivity of Chinese Terrestrial Vegetation Based on Remote Sensing. *Chin. J. Plant Ecol.* **2007**, *31*, 413.
69. Liang, L.; Wang, Q.; Qiu, S.; Geng, D.; Wang, S. NEP Estimation of Terrestrial Ecosystems in China Using an Improved CASA Model and Soil Respiration Model. *IEEE J. Sel. Top. Appl. Earth Obs. Remote Sens.* **2023**, *16*, 10203–10215. [CrossRef]
70. Aubinet, M.; Hurdebise, Q.; Chopin, H.; Debacq, A.; De Ligne, A.; Heinesch, B.; Manise, T.; Vincke, C. Inter-Annual Variability of Net Ecosystem Productivity for a Temperate Mixed Forest: A Predominance of Carry-over Effects? *Agric. Meteorol.* **2018**, *262*, 340–353. [CrossRef]

71. Zhao, J.; Ma, J.; Zhu, Y. Evaluating Impacts of Climate Change on Net Ecosystem Productivity (NEP) of Global Different Forest Types Based on an Individual Tree-Based Model FORCCHN and Remote Sensing. *Glob. Planet. Change* **2019**, *182*, 103010. [CrossRef]
72. Bloom, A.A.; Exbrayat, J.-F.; Van Der Velde, I.R.; Feng, L.; Williams, M. The Decadal State of the Terrestrial Carbon Cycle: Global Retrievals of Terrestrial Carbon Allocation, Pools, and Residence Times. *Proc. Natl. Acad. Sci. USA* **2016**, *113*, 1285–1290. [CrossRef] [PubMed]
73. Zhang, J.; Hao, X.; Hao, H.; Fan, X.; Li, Y. Climate Change Decreased Net Ecosystem Productivity in the Arid Region of Central Asia. *Remote Sens.* **2021**, *13*, 4449. [CrossRef]
74. Yang, Y.; Ma, Y.; Jiang, H.; Zhu, Q.; Liu, J.; Peng, C. Evaluating the Carbon Budget Pattern of Chinese Terrestrial Ecosystem from 1960 to 2006 Using Integrated Biosphere Simulator. *Acta Ecol. Sin.* **2016**, *36*, 3911–3922. [CrossRef]
75. Zeng, J.; Zhou, T.; Wang, Q.; Xu, Y.; Lin, Q.; Zhang, Y.; Wu, X.; Zhang, J.; Liu, X. Spatial Patterns of China's Carbon Sinks Estimated from the Fusion of Remote Sensing and Field-Observed Net Primary Productivity and Heterotrophic Respiration. *Ecol. Inf.* **2023**, *76*, 102152. [CrossRef]
76. Friedlingstein, P.; O'sullivan, M.; Jones, M.W.; Andrew, R.M.; Gregor, L.; Hauck, J.; Le Quéré, C.; Luijkx, I.T.; Olsen, A.; Peters, G.P. Global Carbon Budget 2022. *Earth Syst. Sci. Data* **2022**, *14*, 4811–4900. [CrossRef]
77. Liu, S.; Yang, Y.; Shen, H.; Hu, H.; Zhao, X.; Li, H.; Liu, T.; Fang, J. No Significant Changes in Topsoil Carbon in the Grasslands of Northern China between the 1980s and 2000s. *Sci. Total Environ.* **2018**, *624*, 1478–1487. [CrossRef]
78. Wang, J.; Feng, L.; Palmer, P.I.; Liu, Y.; Fang, S.; Bösch, H.; O'Dell, C.W.; Tang, X.; Yang, D.; Liu, L.; et al. Large Chinese Land Carbon Sink Estimated from Atmospheric Carbon Dioxide Data. *Nature* **2020**, *586*, 720–723. [CrossRef]
79. Shi, Y.; Zhang, B.; Liang, L.; Wang, S.; Zhang, H.; Sun, H.; Han, X. Unfolding the Effectiveness of Ecological Restoration Programs in Enhancing Vegetation Carbon Sinks across Different Climate Zones in China. *Resour. Conserv. Recycl.* **2025**, *212*, 107974. [CrossRef]
80. Yue, C.; Xu, M.; Ciais, P.; Tao, S.; Shen, H.; Chang, J.; Li, W.; Deng, L.; He, J.; Leng, Y.; et al. Contributions of Ecological Restoration Policies to China's Land Carbon Balance. *Nat. Commun.* **2024**, *15*, 9708. [CrossRef]
81. Li, J. A Simulation Approach to Optimizing the Vegetation Covers under the Water Constraint in the Yellow River Basin. *Policy Econ.* **2021**, *123*, 102377. [CrossRef]
82. Zhang, X.M.; Brandt, M.; Yue, Y.M.; Tong, X.W.; Wang, K.L.; Fensholt, R. The Carbon Sink Potential of Southern China after Two Decades of Afforestation. *Earths Future* **2022**, *10*, e2022EF002674. [CrossRef]
83. Zhang, Y.; Zhao, Y.; Chen, Q.; Zhu, Y.; Liu, B.; Zhang, X.; Yin, X. Capacity of Forests and Grasslands to Achieve Carbon Neutrality in China. *Forests* **2024**, *15*, 1060. [CrossRef]
84. Piao, S.; Fang, J.; He, J. Variations in Vegetation Net Primary Production in the Qinghai-Xizang Plateau, China, from 1982 to 1999. *Clim. Change* **2006**, *74*, 253–267. [CrossRef]
85. Rap, A.; Scott, C.E.; Reddington, C.L.; Mercado, L.; Ellis, R.J.; Garraway, S.; Evans, M.J.; Beerling, D.J.; MacKenzie, A.R.; Hewitt, C.N.; et al. Enhanced Global Primary Production by Biogenic Aerosol via Diffuse Radiation Fertilization. *Nat. Geosci.* **2018**, *11*, 640–644. [CrossRef]
86. Zhang, X.; Zhou, Y.; He, W.; Ju, W.; Liu, Y.; Bi, W.; Cheng, N.; Wei, X. Land Cover Change Instead of Solar Radiation Change Dominates the Forest GPP Increase during the Recent Phase of the Shelterbelt Program for Pearl River. *Ecol. Indic.* **2022**, *136*, 108664. [CrossRef]
87. Sun, Z.; Wang, X.; Yamamoto, H.; Tani, H.; Zhong, G.; Yin, S.; Guo, E. Spatial Pattern of GPP Variations in Terrestrial Ecosystems and Its Drivers: Climatic Factors, CO_2 Concentration and Land-Cover Change, 1982–2015. *Ecol. Inf.* **2018**, *46*, 156–165. [CrossRef]
88. Ye, X.; Liu, F.; Zhang, Z.; Xu, C.; Liu, J. Spatio-Temporal Variations of Vegetation Carbon Use Efficiency and Potential Driving Meteorological Factors in the Yangtze River Basin. *J. Mt. Sci.* **2020**, *17*, 1959–1973. [CrossRef]
89. Xie, S.; Mo, X.; Hu, S.; Liu, S. Contributions of Climate Change, Elevated Atmospheric CO_2 and Human Activities to ET and GPP Trends in the Three-North Region of China. *Agric. Meteorol.* **2020**, *295*, 108183. [CrossRef]
90. Keenan, T.F.; Gray, J.; Friedl, M.A.; Toomey, M.; Bohrer, G.; Hollinger, D.Y.; Munger, J.W.; O'Keefe, J.; Schmid, H.P.; Wing, I.S.; et al. Net Carbon Uptake Has Increased through Warming-Induced Changes in Temperate Forest Phenology. *Nat. Clim. Change* **2014**, *4*, 598–604. [CrossRef]
91. Piao, S.; Sitch, S.; Ciais, P.; Friedlingstein, P.; Peylin, P.; Wang, X.; Ahlström, A.; Anav, A.; Canadell, J.G.; Cong, N. Evaluation of Terrestrial Carbon Cycle Models for Their Response to Climate Variability and to CO_2 Trends. *Glob. Change Biol.* **2013**, *19*, 2117–2132. [CrossRef] [PubMed]
92. Wang, T.; Fan, G.; Zhang, H.; Shen, X. Temporal and Spatial Variability of Ground Frost Indices in Northeast China. *Atmosphere* **2024**, *15*, 817. [CrossRef]

93. Zhou, Y.; Wei, X.; Wang, Y.; He, W.; Dong, Z.; Zhang, X.; Liu, Y.; Nguyen, N.T.; Ju, W. Solar Radiation Variation Weakened the Boost of Gross Primary Production by Vegetation Restoration in China's Most Forestry Engineering Areas during 2001–2020. *Environ. Res. Lett.* **2024**, *19*, 014082. [CrossRef]
94. Wang, Y.; Guo, C.; Chen, X.; Jia, L.; Guo, X.; Chen, R.; Zhang, M.; Chen, Z.; Wang, H. Carbon Peak and Carbon Neutrality in China: Goals, Implementation Path and Prospects. *China Geol.* **2021**, *4*, 720–746. [CrossRef]

forests

MDPI

Article

A Concise Approach to Characterizing the Distribution of Canopy Leaf Mass per Area in Broad-Leaf Species Based on Crown Three-Dimensional Position and Vegetation Index

Depeng Zhang [1,2], Yueqi Wang [1], Xiguang Yang [1,2,*], Shengtao Yang [3], Yuanyuan Liu [3], Zijuan Yu [3] and Xingcai Zhao [3]

1 School of Forestry, Northeast Forestry University, Harbin 150040, China; z2022010073@nefu.edu.cn (D.Z.); 2022220345@nefu.edu.cn (Y.W.)

2 Key Laboratory of Sustainable Forest Ecosystem Management, Ministry of Education, Northeast Forestry University, Harbin 150040, China

3 Heilongjiang Province Natural Resources Rights and Interests Investigation and Monitoring Institute, Harbin 150036, China; yangshengtao0928@gmail.com (S.Y.); liuyuanyuan0219@gmail.com (Y.L.); yuzijuan0524@gmail.com (Z.Y.); tddx123@gmail.com (X.Z.)

* Correspondence: yangxiguang@nefu.edu.cn; Tel.: +86-150-0452-2757

Abstract: Leaf mass per area (LMA) represents the allocation of carbon resources per unit leaf area, which is closely related to the photosynthetic capacity of tree leaves. Clarifying the distribution features of LMA is very useful in understanding nutrient and energy transmission and photosynthetic capacity in the canopy. To this aim, the leaf samples of varied forest types were collected, and LMA and related spectral data were measured. The Partial Least Squares (PLS), Linear Mixed Models (LMM), Support Vector Machine Regression (SVM), and Random Forest (RF) methods were used to establish a new model of three-dimension LMA prediction by using vegetation index, DBH, and the vertical and horizontal position of leaves in this study. The results found that the LMA varies significantly with the change in the spatial position of the leaves and horizontal distance to the tree trunk. Statistically speaking, changes in LMA were not significantly related to the direction where the leaves were located. The best model of 3D LMA estimation was RF with a 10-fold R^2 value of 0.939. Compared to the RF model, the maximum and minimum of R^2 of 10-fold testing of other models increased by 23.75% and 55.87%. The results indicated that RF has a strong generalization ability and can predict the LMA distribution in 3D with a high accuracy. This study showed a reference for LMA 3D feature distribution and is helpful in clarifying the photosynthetic capacity of the canopy.

Keywords: leaf mass per area; three-dimensional position; vegetation index; random forests; vertical canopy layers

1. Introduction

Recent scientific observations reveal a troubling rise in extreme weather phenomena [1], drawing unprecedented attention to the pressing issue of carbon sinks [2]. In this context, the Leaf Economic Spectrum (LES) of vegetation provides a clear reflection of the forest carbon sink, which plays a large role in terrestrial ecosystems [3]. LMA (Leaf Mass per Area) is a crucial component of the leaf economic spectrum, representing the amount of dry matter resources invested by plants per unit leaf area and describing the characteristics of the resource allocation strategies of the leaf. The plant LMA is a utilization pattern of carbon jointly determined by genes and environment, which can reflect the ecological niche of plants [4]. LMA has been proposed to determine the construction cost of a single leaf in the growth model, while biomass allocation has also be determined by leaf area and LMA in previous research [5]. LMA is defined by the leaf dry mass and the leaf area, which are significantly related to the photosynthetic process. Therefore, LMA is an important parameter for estimating plant leaf photosynthetic capacity and constructing vegetation nutrient budget models [6–9]. The LMA of plants can reflect the maximum photosynthetic rate and adaptive capacity of plants to their environment [7,10,11]. And it has been used to represent the responses of plants to stress in some studies [12,13]. LMA varies greatly among different vegetation types, and evergreen plants often have a larger LMA due to their thicker leaves compared with deciduous plants [9]. However, there is still a huge space for improvement in the accuracy of tree species identification for individual trees over larger areas. Even within the same species, LMA can be influenced by environmental factors such as soil composition, water availability, and CO_2 concentration [11,14,15]. Therefore, estimating the LMA distribution among a canopy at a large scale is significant.

Traditional analyses tend to consider the canopy as a big-leaf-structure and disregard the differential physiological and environmental resource gradients within the canopy. Consequently, more recent studies have attempted to explain within-canopy heterogeneity in leaf traits [16–18]. Thus, developing a new method to give a 3D estimate of LMA in canopy would be very useful for variations in three-dimensional LMA changes within the canopy. LMA is adjusted by plants according to the light gradient in the canopy, affecting the canopy nitrogen distribution [19]. According to prior investigations, the spatial distribution of leaf LMA exhibits a robust correlation with the distribution of leaf N per area (Narea) [20,21]. An accurate estimation of three-dimensional leaf mass per area (LMA) will enhance our understanding of nitrogen allocation strategies in plants, thereby allowing for a more precise assessment of photosynthetic efficiency. Results of previous research on the vertical distribution of LMA within the plant canopy showed that there was a notable decline in LMA from the upper to lower canopy parts [22]. This trend has been attributed to gradient fluctuations in light and water potential within the woody tissue in the canopy, and the water potential changes have even more of an effect on LMA variation than changes in light [23]. Besides this, seasonal variation is also an important factor affecting leaf traits [24]. Meanwhile, other studies have focused on the temporal and spatial dynamics of LMA within plant canopies, and results indicate that spatial correlations are more important than seasonal influences in LMA estimation [25]. It is beneficial that accurate three-dimensional (3D) LMA estimation is essential for assessing vegetation canopies.

Remote sensing technology has the ability for estimation at a local or large scale. As a part of remote sensing, the UVA platform has also been widely used for parameter estimation or monitoring environmental features. UAV-derived canopy height models (CHMs) using RGB imagery to estimate solar radiation showed a high consistency with hemispherical photography with an R^2 of 0.7. And UAV photogrammetry has been proposed as an alternative for forest light environment assessment [26]. This information was then merged with regularly obtained LMA data from imaging spectroscopy, ultimately yielding a higher-precision estimation of canopy LMA [27]. In other studies, the vertical canopy was divided into nine equal parts, resulting in three layers: upper, middle, and bottom. Leaves are randomly collected from each layer. The result was basically consistent with those of previous studies [18]. Nonetheless, research delving into the 3D distribution of LMA within vegetation canopies has been limited, with most studies solely focused on vertical shifts. Previous research on specific leaf area (SLA), the reciprocal of LMA, has investigated leaf area's vertical and horizontal distribution [28]. A model of LMA for a specific layer at the top of the canopy has been established, which allows us to consider the model as based on the vegetation index related with LMA [29]. Hence, this paper offers an approach based on vertical and horizontal parameters. By incorporating leaf position within the canopy, description of the tree size (DBH and tree height) and vegetation index related to LMA, we generate 3D LMA data which accurately reflect the LMA distribution throughout the canopy.

This study aims to develop an LMA inversion model applicable to most broad-leaf tree species within the vegetation canopy in the region by deploying Hs (the height of the sample point), HD (horizontal distance from the sample point to the trunk), tree height and DBH (diameter at breast height), which represent the 3D position within the canopy and tree size as reference indicators respectively. This study used the normalized parameter RDINC (relative depth into the crown)—based on tree height, subbranch height, and sampling height—and mHD (mean relative horizontal distance)—based on the E-W diameter or N-S diameter. By combining the acquired vegetation index related with LMA data from remote sensing with the LMA distribution in this study, a more precise prediction of LMA values at various positions within the canopy could be generated. Such an approach might increase the accuracy of models that utilize LMA as a parameter for vegetation assessment.

2. Materials and Methods

2.1. Study Area

Maoershan Forest Farm (45°02′20″–45°18′16″ N, 127°18′00″–127°41′06″ E), located in the northwest of Shangzhi City, Heilongjiang Province (Figure 1), in the northeast of China, was founded in 1958. This region is mostly mountainous and hilly, with slopes ranging between 10 and 15°. The highest altitude can reach 799 m, the lowest can reach 246 m, and the mean altitude is around 426 m. The climate of this region belongs to the middle temperate continental monsoon climate zone. January is the coldest month, with an average temperature of −20.2 °C, while July records the highest average temperature of 21.6 °C. The mean annual temperature stands at 2.5 °C. The mean annual precipitation is around 400~650 mm. The annual sunshine duration is around 2300 h. The vegetation cover is mainly the natural secondary forest formed after the destruction of the Primitive Korean pine forest. The main tree species include the following: *Pinus koraiensis* Siebold et Zuccarini, *Pinus sylvestris* var. *mongolica* Litv., *Picea koraiensis* Nakai, *Fraxinus mandschurica* Rupr., *Juglans mandshurica* Maxim, *Quercus mongolica* Fisch. ex Ledeb., *Tilia amurensis* Rupr., *Ulmus davidiana* Planch. var. *japonica* (Rehd.) Nakai, and *Betula platyphylla* Sukaczev, along with more than 10 additional species.

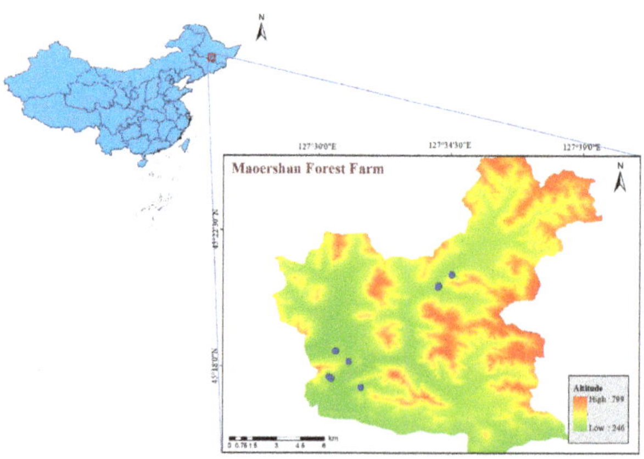

Figure 1. Location of study area and sample point distribution.

2.2. Data Acquisition

The experiment was carried out in the Maoershan Forest Farm from July to August 2021. During field sampling, eight rectangular plots (each 0.06 ha, 20 m × 30 m) were established, with each plot representing a single target species. For each species, several samples were collected to ensure that all target species were represented across the study plots. The eight sample plots share similar stand density (1200–1500 stems per hectare) and tree species composition (mainly broad-leaf trees), and the main soil type across all plots is dark brown forest soil. Based on the team's previous tree-by-tree survey results of multiple plots, we determined the dominant broad-leaf species by counting the number of trees with DBH (diameter at breast height) > 5 cm in the inventory data. Then, two~three upper, healthy dominant trees were selected from the dominant species as the test trees: those with symmetrical growth (leaf direction balanced distribution), healthy growth, and those free of diseases and pests. For eight species, we collected and analyzed two or three samples per species and 10 leaves at each of the 12 sampling positions to allocate the data. In order to reduce the error caused by the relief of the terrain, the sample plots were all selected in an area with a small slope, and the four corners of each plot were located and recorded by the SOUTH RTK (SOUTH Inc., Guangzhou, China) instrument.

Eight species of broad-leaved tree species in Northeast China were selected for the experiment: *Betula platyphylla* Suk., *Tilia amurensis* Rupr., *Juglans mandshurica* Maxim., *Phellodendron amurense* Rupr., *Quercus mongolica* Fischer. ex Ledebour., *Fraxinus mandschurica* Rupr., *Populus davidiana* Dode, and *Ulmus davidiana* Planch var. *japonica* (Rehd.) Nakai. These samples were kindly provided by Maoershan Forest Farm.

During sampling, the canopy of each sample tree was divided into three equal layers according to the canopy thickness, and a number of healthy and pest-free leaves were selected from four directions of east, south, west, and north in each layer. We used an averruncator to collect the branch; filled a form with the name of the plot, tree species, the height of the tree, the height of the sample, the height of the lowest branch, the horizontal distance from the sample point to the trunk, the diameter at breast height, the East–West crown diameter, the North–South crown diameter, and the coordinates of the tree; and named them according to the species–number–layer–orientation format. The twelve samples were collected from each tree and the branches were placed in plastic bags marked according to the species–number–layer–orientation format (Figure 2).

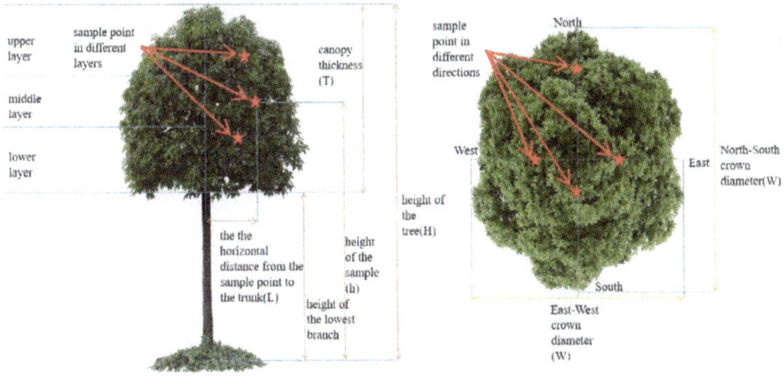

Figure 2. The diagram of 3D sampling in tree canopy. The red star-shaped symbol shows an example of sampling position in the canopy.

After finishing the collection and marking work, we returned to the laboratory as quickly as possible. We dipped the ends of the collected branches into water, chose

10 randomly sized leaves from the collected branches and placed them into a plastic bag, and then placed them in an ice-chilled styrofoam container after naming them in the same format. Subsequently, spectral measurements of these leaves were conducted using the SVC HR-1024i spectrometer (Spectra Vista Corporation, Poughkeepsie, NY, USA) (Figure 3).

Figure 3. SVC HR-1024i spectrometer.

During spectral measurements, the indoor environment was carefully controlled to ensure the absence of ambient light sources and the HLG-150 fiber optic light source (Nanjing Chunhui Science Technology Industrial Co., Ltd., Nanjing, China) was fixed as the unique incident light source. Each leaf's reflection spectrum was measured after using a whiteboard placed on a black substrate to reference the radiation (Figure 4). The leaf reflectance spectrum was measured within the wavelength range of 400 nm to 2500 nm. Five measurements were taken for each leaf and an average was taken, with 5 to 10 repeated experiments conducted for each combination of layer and direction. Following spectral measurement, data were exported and resampled to a 10 nm spectral resolution using the built-in software of SVC HR1024i (version 1.21.14, Spectra Vista Corporation, Poughkeepsie, NY, USA). The output was formatted as '*.csv' files for subsequent statistical analysis.

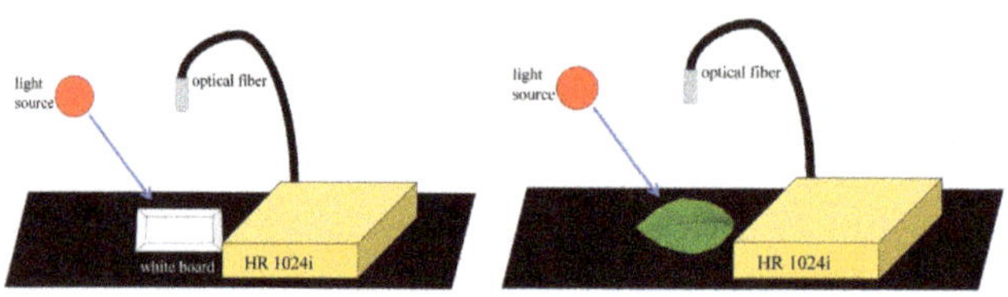

Figure 4. A schematic diagram of leaf reflectance spectrum measurement.

After the steps mentioned above, the Li-3000c was used to count leaf area and record the leaf area of the samples. Then, the leaves were put in envelopes and named according to the species–number–layer–orientation format, before being put in a 70 °C oven for at least 24 h until they reached a constant weight.

$$\text{LMA} = \frac{DW}{A} \tag{1}$$

where DW is the weight of the dry leaf (g), and A is the leaf area (m^2).

The weight of the dry leaf of each envelope was measured by electronic scales, and the LMA (Leaf Mass per Area, g/m^2) was obtained according to the ratio of the dry weight to the leaf area.

2.3. Data Analysis

To delineate the spatial distribution of leaf blades within the canopy, we employed the mean relative depth into the crown (RDINC) with 10% relative depth increments for every layer (the 80%~100% range was allocated to one piece to balance the data volume) and divided the mean relative horizontal distance from the sample point to the trunk (mHD) into three equally sized segments.

To estimate LMA (Leaf Mass per Area), we chose RDINC, mHD, DBH, and vegetation index as the candidates; we chose tree size and layer as the dummy variables; and four methods were chosen to estimate LMA: PLS (Partial Least Squares Regression), LMM (Linear Mixed Model), SVM (Support Vector Machine), and RF (Random Forest).

The Partial Least Squares (PLS) Regression method represents an advanced approach in multivariate statistical analysis, adept at performing both multivariate linear regression and principal component analysis concurrently. When confronted with challenges such as multicollinearity among variables, a large number of variables with a limited sample size, or heteroscedasticity, using PLS regression confers unmatched advantages over traditional linear regression methodologies. However, the PLS regression model cannot perform variable selection, resulting in an overly complex model when incorporating all variables. Therefore, this study performed a preliminary variable-importance ranking using the random forest algorithm before model construction. By including only the variables with higher importance in the model, the study aimed to enhance the usability and accuracy of the model.

Linear Mixed Model (LMM): This extension of Generalized Linear Models encompasses both fixed and random effects. By modeling fixed effects across different levels of random effects, an LMM can offer improved explanatory power for analyzing data. In this study, diameter at breast height (DBH) was classified into 5 cm intervals as a random effect, while vegetation indices and the three-dimensional positions of leaves within the canopy were used as fixed effects in constructing the Linear Mixed Model (LMM). This approach aimed to capture the differences in three-dimensional LMA prediction models across different DBH classes, thereby enhancing the predictive capability of the model.

Support Vector Machine Regression (SVM): The Support Vector Machine (SVM) model represents a relatively recent machine learning technique that has been extensively employed to address classification and regression problems. The fundamental principle of SVM involves the use of nonlinear mapping to project the original data into a high-dimensional space, wherein a hyperplane is identified. This hyperplane, supported by several vectors known as support vectors, maximizes the distance from the data points in the training set. By fine-tuning various hyperparameters, SVM achieves a balance between model generalization and overfitting. In the SVM model constructed for this experiment, R was utilized to perform a grid search for the cost and epsilon hyperparameters, aiming to minimize the Mean Absolute Error (MAE). The optimal hyperparameters, corresponding to the grid points where the MAE is lowest, were selected. The cost parameter controls the penalty factor, determining the tolerance for misclassification in the SVM model; a high cost may lead to overfitting by overemphasizing the correct classification of each sample, whereas a low cost may reduce model accuracy. The epsilon parameter defines the tolerance margin within which prediction errors are permitted; a large epsilon includes more points within the tolerance margin, potentially decreasing model accuracy, while a small epsilon reduces tolerance for errors, which may also result in overfitting. Thus, these hyperparameters are crucial for adjusting the complexity and generalization ability of the model to achieve optimal performance.

Random Forest Model (RF): The core concept of random forests involves aggregating the classification results of several weak classifiers through voting to generate a robust classifier. By employing bootstrapped random sampling with replacement for training data selection, each tree within the forest is trained on a different subset of the data. This ensemble machine learning algorithm is suitable for analyzing nonlinear relationships among complex variables. The aforementioned four methods were all subjected to cross-validation for accuracy assessment, aiming to identify the optimal model for predicting the three-dimensional distribution of LMA.

In this study, we constructed models for Leaf Mass per Area (LMA) using H (sampling height), HD (horizontal distance from leaf to trunk), and DBH (diameter at breast height), along with vegetation indices, which are listed in the variable ranking for correlation with LMA (importance ranking was integrated using random forest). The top 15 vegetation indices were selected for model construction (Table 1).

Table 1. The vegetation index used in this study.

Name	Formula *	Ref
Red Edge	Max first derivative in (R_{680}~R_{760})	[30]
Yellow Edge	Max first derivative in (R_{560}~R_{640})	[30]
Blue Edge	Max first derivative in (R_{490}~R_{530})	[30]
ReMSR	$\frac{\frac{R_{750}}{R_{705}}-1}{\frac{R_{750}}{R_{705}}+1}$	[31]
RDVI	$\frac{R_{800}-R_{670}}{\sqrt{R_{800}+R_{670}}}$	[32]
TVI	$\frac{120\times(R_{750}-R_{550})-200\times(R_{670}-R_{550})}{2}$	[33]
RVSI	$\frac{(R_{750}+R_{710})}{2}-R_{730}$	[34]
MCARI1	$1.5\times[1.2\times(R_{712}-R_{670})-0.5\times(R_{712}-R_{550})]\times\frac{R_{712}}{R_{670}}$	[35]
MCARI2	$\frac{1.5\times[1.2\times(R_{712}-R_{670})-0.5\times(R_{712}-R_{550})]}{0.5\times\left[2\times R_{800}+1-\sqrt{(2\times R_{800}+1)^2-8\times(R_{800}-R_{670})}\right]}\times\frac{R_{712}}{R_{670}}$	[35]
MTVI1	$1.5\times[1.2\times(R_{712}-R_{550})-2.1\times(R_{670}-R_{550})]$	[35]
MTVI2	$\frac{1.5\times[1.2\times(R_{712}-R_{550})-0.5\times(R_{670}-R_{550})]}{0.5\times\left[2\times R_{800}+1-\sqrt{(2\times R_{800}+1)^2-8\times(R_{800}-R_{670})}\right]}$	[35]
ARSIs	$\frac{R_{660}-R_{650}}{R_{520}-R_{650}}$	[36]
NDVI(a,b)	$\frac{R_a-R_b}{R_a+R_b}$	/
RVI(a,b)	$\frac{R_a}{R_b}$	/
SR(a,b)	R_a-R_b	/

* The result based on SVC HR 1024i, Excel 2016 (Microsoft Corporation, Redmond, WA, USA), R (version 4.2.2, R Foundation for Statistical Computing, Vienna, Austria).

The ARSIs vegetation index is extracted based on Figure 5. The bands with a correlation value more than 0.1 are taken as the biochemical significant wavelength (Figure 5a blue shadow), the highest R^2 increment in the biochemical insignificant wavelength is taken as R_{spec} (Figure 5b red dashed line), the lowest R^2 in the biochemical insignificant wavelength is taken as the reference wavelength R_{ref} (Figure 5c red dashed line), and the wavelength corresponding to the maximum value of ${R_{index}}^2$ in the biochemical significant wavelength (Figure 5d red dashed line) is calculated according to the formula as the reference wavelength R_{indi} [36].

$$ {R_{Index}}^2 = \left[Correlation\left(\frac{R_{ref}-R_{spec}}{R_{indi}-R_{spec}}, Y\right)\right]^2 \tag{2} $$

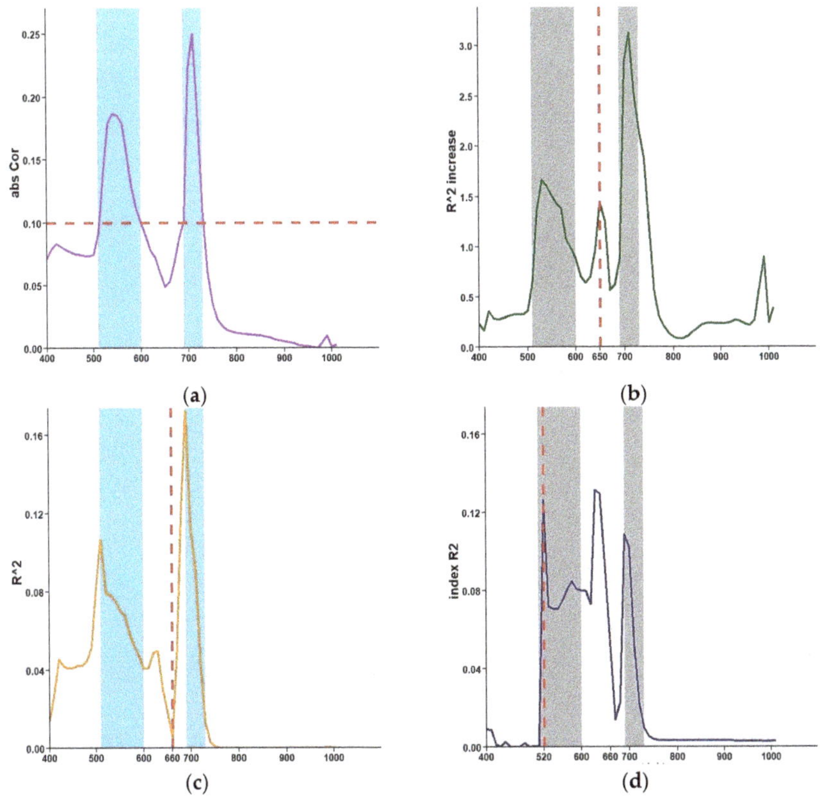

Figure 5. (**a**) Biochemical significant wavelength; (**b**) R_{spec}, the highest R^2 increment in the biochemical insignificant wavelength; (**c**) R_{ref}, the lowest R^2 in the biochemical insignificant wavelength; (**d**) R_{indi}, and the wavelength of the maximum index R^2 in the biochemical significant wavelength.

Due to the spectral range of the hyperspectral data, a range of 400–1000 nm was selected in this study. A value of 660 nm was chosen as the R_{ref}, 520 nm was chosen as the R_{indi}, and 650 nm was chosen as the R_{spec}.

$$ARSIs = \frac{\mathrm{R_{ref}} - \mathrm{R_{spec}}}{\mathrm{R_{indi}} - \mathrm{R_{spec}}} = \frac{\mathrm{R_{660}} - \mathrm{R_{650}}}{\mathrm{R_{520}} - \mathrm{R_{650}}} \tag{3}$$

For the purpose of the article, an additional attribute of 'tree size' was created based on the package 'mclust' in R to cluster the original trees into 'big', 'medium', and 'small' by Cluster.

$$\text{Cluster} = \frac{DBH^2 \times height_{tree}}{4} \tag{4}$$

2.4. Validation

The accuracy of the model was evaluated by the parameters below: determination coefficient (R^2), root mean squared error ($RMSE$), and mean absolute error (MAE).

Determination coefficient (R^2)

$$R^2 = \frac{\sum_{i=1}^{n}(\hat{y}_i - \overline{y})^2}{\sum_{i=1}^{n}(y_i - \overline{y})^2} \tag{5}$$

Root mean squared error (*RMSE*)

$$RMSE = \sqrt{\frac{1}{n}\sum_{i=1}^{n}(y_i - \hat{y}_i)^2} \quad (6)$$

Mean absolute error (*MAE*)

$$MAE = \frac{\sum_{i=1}^{n}|y_i - \hat{y}_i|}{n} \quad (7)$$

where $\hat{y}_i$ is the predicted value, y_i is the observed value, $\overline{y}_i$ is the average of observed value, and n is the number of samples.

3. Results

3.1. One-Way ANOVA of LMA for Layers and Directions in Canopy

3.1.1. One-Way ANOVA Between Different Layers Inside Canopy

This research analyzed the spatial distribution features of LMA (Leaf Mass per Area) in the canopy and the results are shown the in table and figures. Table 2 shows the range of LMA for different layers and Figure 6A shows the distribution in different layers of leaf mass per area (LMA); the pairwise comparison (Figure 6B) shows that the mean relative LMA has a significant difference between the high group and the low group, and between the middle group and the low group at the level of $p < 0.001$ ($p = 1.28 \times 10^{-8} < 0.001$).

Table 2. The distribution of original LMA data for the layers inside the canopy.

	Layer	Max	Min	Average	Std. *
LMA (g/m^2)	High	193.72	41.17	87.18	28.17
	Middle	148.75	41.24	78.60	23.71
	Low	135.72	24.75	75.17	22.67

* Standard deviation.

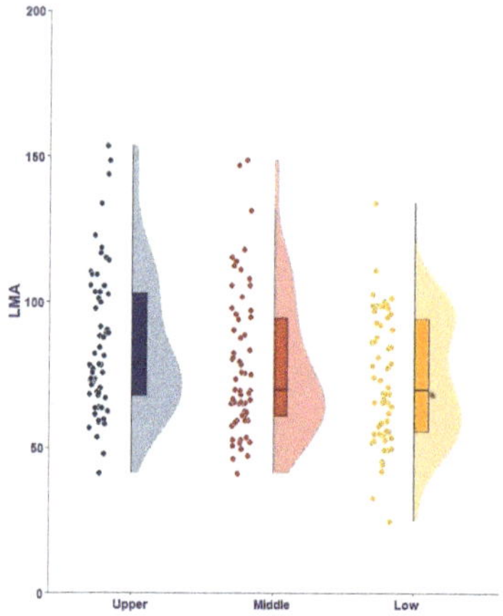

(A)

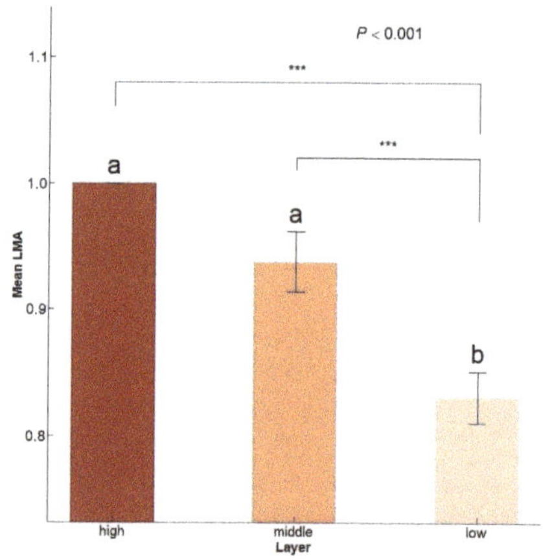

(B)

Figure 6. The one-way ANOVA for LMA of different layers. (**A**) The boxplot of LMA in different layers. (**B**) The letters (a, b) are the group of Post-hoc test, different letters represent significant difference of different group. The significance marker '***' indicates the p-value < 0.001.

3.1.2. One-Way ANOVA for Different Directions Inside Canopy

Table 3 shows the range of LMA for different directions and Figure 7A shows the distribution for different directions of leaf mass per area (LMA); the pairwise comparison (Figure 7B) of the mean relative LMA shows an insignificant difference between the directions ($p = 0.272 > 0.05$). After one-way analysis of variance with the direction and layer data, we concluded that the direction has little relationship with the leaf mLMA.

Table 3. The distribution of original LMA data in the directions inside the canopy.

	Direction	Max	Min	Average	Std. *
LMA (g/m^2)	E	153.63	41.17	79.63	23.49
	S	157.30	41.24	80.61	25.56
	W	193.72	24.75	83.60	30.57
	N	143.93	32.92	78.23	22.89

* Standard deviation.

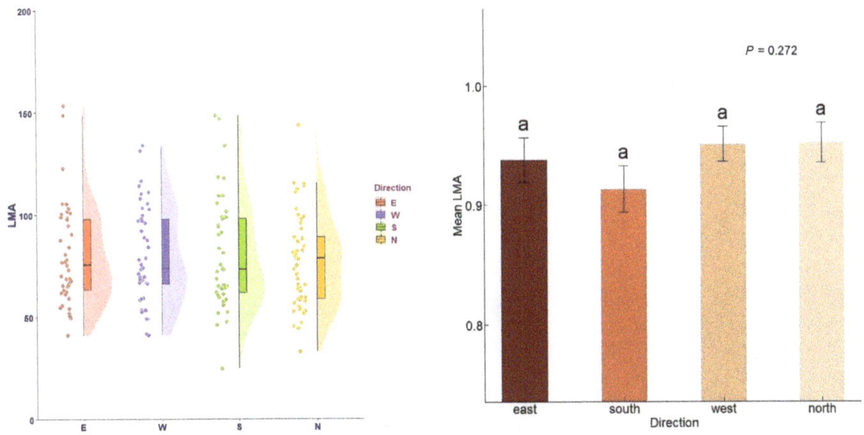

(A) (B)

Figure 7. The one-way ANOVA for LMA and directions. (**A**) The boxplot of LMA in different directions. (**B**) The letter (a) is the group of Post-hoc test, same letter represents insignificant difference of different group.

This demonstrates that leaf mass per area (LMA) at a given canopy position can be sufficiently characterized by its three-dimensional coordinates alone, without requiring consideration of directional effects.

3.2. Frequency Distribution of Data

We divided the canopy vertically into nine sections and horizontally into several segments to describe the trends of LMA (leaf mass per area) with positional blocks. The original data were divided into eight equal parts for the 0%~80% section of RDINC in 10% increments, and into one bigger part for the 80%~100% section (because the points in 80%~90% and the 90%~100% are both far less than other equal parts respectively), after removing errors and fewer than three observations in each group from the box-plot (Figure 8).

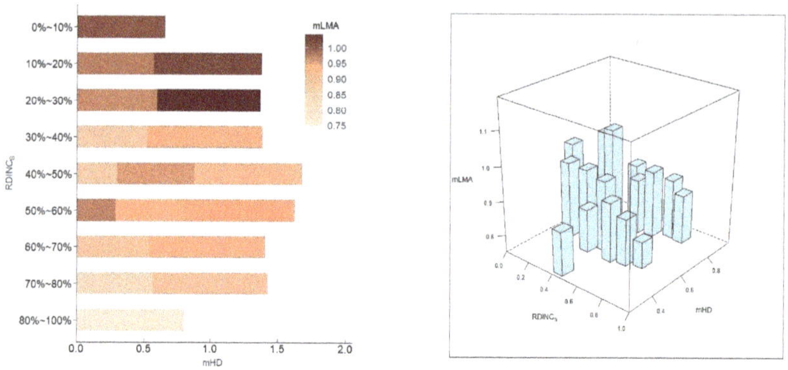

(a) (b)

Figure 8. The three-dimensional distribution of the mLMA in the canopy: (**a**) the mLMA of the different position sections; (**b**) the trend in mLMA against increases in mHD and RDINC.

Figure 9 shows that in the same layer of RDINC, the mLMA tends to increase with the increase in mHD; in the same layer of mHD, the mLMA tends to decrease with the increase in RDINC. Detail are shown in Table 4. Each 'Group' is divided into 3 parts by mHD in 0%~33%, 33%~66%, and 66%~100%, and remove the fewer than three observations parts.

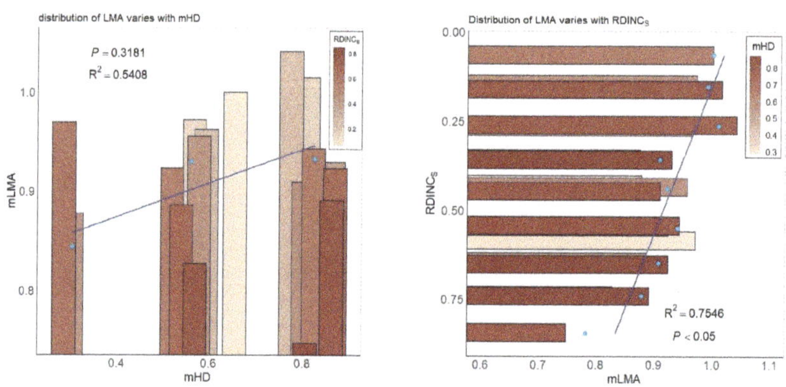

(a) (b)

Figure 9. The linear regression p range and R^2 for the mean value of (**a**) mLMA with mHD; (**b**) mLMA with RDINC.

Figure 9 illustrates that mLMA tends to increase with the increase in mHD, and mLMA tends to decrease with the increase in RDINC. In addition to this, the line in Figure 9a shows a low slope, and $p > 0.05$ for the relation between mHD and mLMA; the simple linear regression is not significant, so we attempt to use machine learning models and nonlinear models. Nevertheless, the R^2 value and fact that $p < 0.05$ show a good result in linear regression in Figure 9b. The blue points and lines correspond to the mean value of each RDINC and mHD group. Therefore, we speculate that mLMA might have some functional relationship with RDINC and mHD.

Table 4. The mLMA and mHD distribution for the values of RDINC.

Group	RDINC *	mLMA *	mHD *
0%~10%	0.06417	1.0000	0.6544
	0.1458	0.9724	0.5679
10%~20%	0.1613	1.0149	0.8098
	0.2653	0.9625	0.5926
20%~30%	0.2607	1.0413	0.7740
	0.3550	0.8748	0.5206
30%~40%	0.3587	0.9295	0.8634
	0.4278	0.8773	0.3010
	0.4343	0.9559	0.5771
40%~50%	0.4465	0.9096	0.7999
	0.5841	0.9697	0.2849
	0.5462	0.9231	0.5189
50%~60%	0.5432	0.9435	0.8210
	0.6440	0.8861	0.5390
60%~70%	0.6518	0.9230	0.8672
	0.7387	0.8270	0.5687
70%~80%	0.7401	0.8910	0.8596
80%~100%	0.8436	0.7469	0.8045

* RDINC: Relative depth into crown; mLMA: mean relative LMA; mHD: mean relative horizontal distance from point to trunk.

3.3. Model Establish and Evaluation

In this experiment, four modeling methods were used. Variable selection was performed by conducting correlation tests to rank the independent variables (Figure 10).

Figure 10. Top 15 importance ranking using the random forest algorithm before model construction.

The top 15 vegetation indices, along with sampling height (Hs), horizontal distance from sample leaf to the trunk (HD), and diameter at breast height (DBH), were chosen for modeling.

Two models are established in the data preprocessing step: first, we trained a layer categorizer based on tree height, DBH, and sample height; second, we trained a subbranch position prediction model based on tree height and DBH. The two model accuracies are 98.40% (Figure 11a) and 96.39% (Figure 11b).

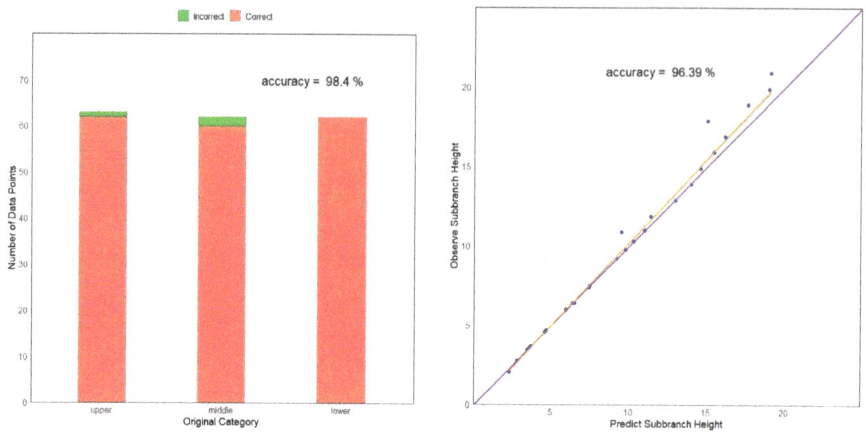

(a) (b)

Figure 11. (**a**) The accuracy of the layer categorizer; (**b**) The accuracy of subbranch position prediction. The purple line is the $y = x$ and the yellow line is the predict result of RF model, which shows the accuracy = 96.39%.

The Random Forest model consists of two sub-models: 1. a model to estimate the top-layer LMA (leaf mass per area) with the vegetation index and tree height, tree size, and DBH; 2. a model to estimate the distribution of nLMA which is normalized by the top-layer LMA. The final LMA prediction is the product of 1 and 2.

Model accuracy evaluation was conducted using 10-fold cross-validation to find the highest prediction accuracy and the best model was selected for constructing the three-dimensional distribution of LMA.

In this experiment, we are going to select a coefficient of determination R^2, root mean square error RMSE, and residual analysis p-Value as indicators to evaluate the fitting effect and prediction accuracy of the model. The significance of the RF model in Table 5 shows the best test with a cross-validation R^2 between 0.716 and 0.939 and the lowest RMSE. The test min R^2 of the LMM model (cross-validation R^2 0.316~0.649) is better than that of the PLS model (cross-validation R^2 0.261~0.448), and SVM has the lowest test min R^2 (cross-validation R^2 0.238~0.716). The result shows that the LMA prediction based on 3D position and the vegetation index with Random Forest model can describe LMA with a higher accuracy (Figure 12).

Table 5. The results of each model.

Method *	Test Max R^2	Test Min R^2	Test Min RMSE (g/m^2)	p-Value
LMM ***	0.649	0.316	11.249	2.544×10^{-6}
PLS **	0.448	0.261	17.387	4.036×10^{-3}
SVM ***	0.716	0.238	11.689	3.322×10^{-5}
RF ***	**0.939**	**0.716**	**6.325**	$\mathbf{7.889 \times 10^{-12}}$

* Significance values: '***' $p < 0.001$; '**' $p < 0.01$; '*' $p < 0.05$.

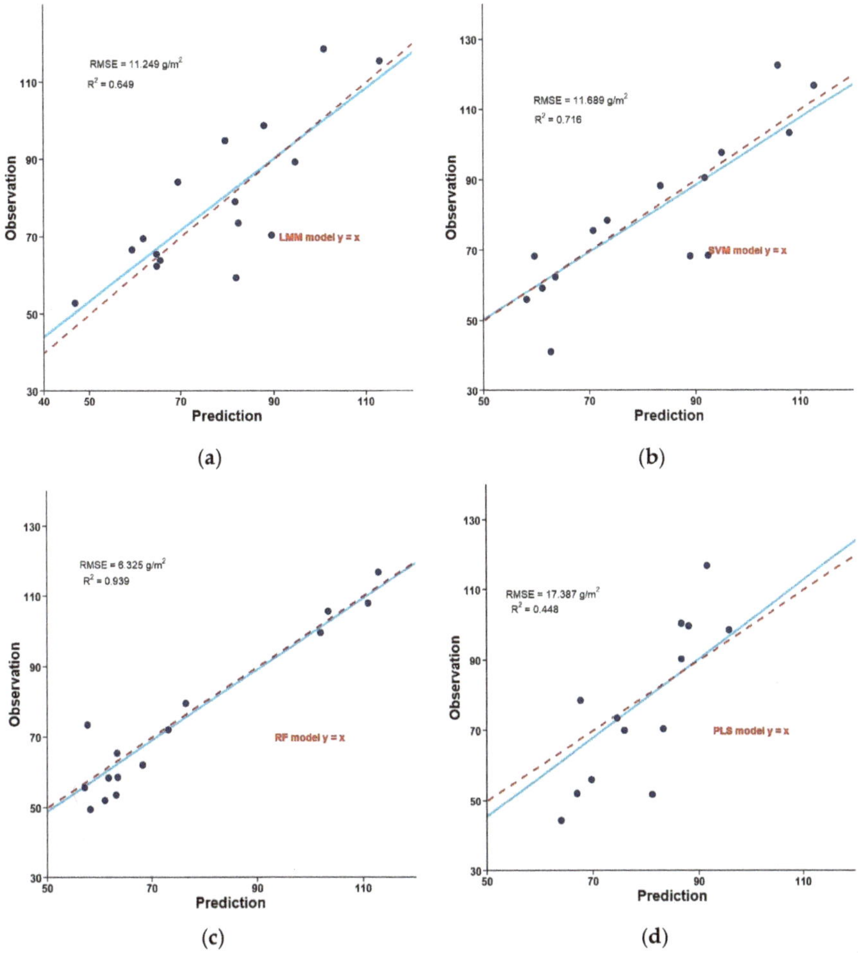

Figure 12. The results of four models: (**a**) Linear Mixed Model; (**b**) Support Vector Machine; (**c**) Random Forest; (**d**) Partial Least Squares.

3.4. The Result of Prediction

In this study, a crown-like scatter plot was simulated, and interpolation based on the corresponding LMA (Leaf Mass per Area) values of the collected sample points' relative spatial positions was performed to describe the LMA distribution within the canopy (Figure 13).

We employed a random forest model to predict the three-dimensional LMA of the canopy. The results indicated that relative spatial position could explain at least 30% of the within-canopy LMA variability. By estimating the three-dimensional distribution and multiplying it with the estimated top LMA, we obtained an estimate of the actual LMA, with a ten-fold cross-validated R^2 value of 0.7387 for the test set. The images of the LMA interpolation and predicted values are shown in Figure 14.

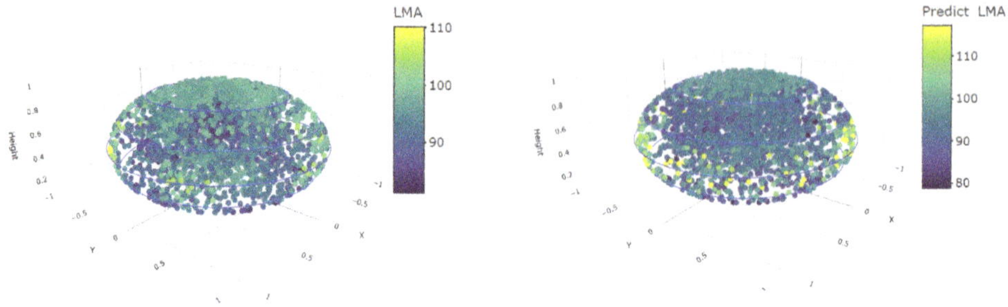

(a) (b)

Figure 13. The simulated point cloud: (**a**) observed point cloud; (**b**) prediction point cloud.

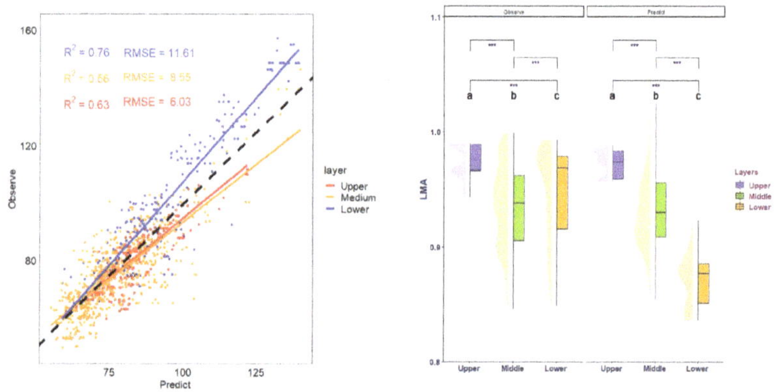

(a) (b)

Figure 14. (**a**) Validation of stratified LMA predictions against observed values in simulation data; (**b**) one-way ANOVA and multiple comparisons of stratified LMA predictions in simulation data. The letters (a, b, c) are the group of Post-hoc test, different letters represent significant difference of different group. The significance marker '***' indicates the p-value < 0.001.

The results of the stratified accuracy validation are presented in Figure 14. The model predictions slightly overestimate LMA for the upper and middle layer, while slightly underestimating it for the lower layer. Overall, the prediction accuracy of the model is highest for the upper layer (R^2 = 0.888), followed by the lower layer (R^2 = 0.789) and the middle layer (R^2 = 0.560). The mean differences between layers are significant, following the distribution pattern of upper > middle > lower, consistent with trends reported in previous studies, thereby lending credibility to the results. To represent the distribution patterns of predicted LMA across different layers visually, each layer was sliced and the corresponding LMA values were normalized to three ranges of 0–0.33, 0.33–0.66, and 0.66–1. This normalization facilitated the clear visualization of the three layers within the z-axis range of 0 to 1. As shown in Figure 15, there is a noticeable increase in LMA with increasing HD ($|x|, |y|$) across all three layers, with the upper and middle layer exhibiting more pronounced trends. When the predicted LMA at the top layer is high, the model's estimation accuracy fluctuates, the impact of random errors in the observed values increases, and the confidence in the estimated results decreases.

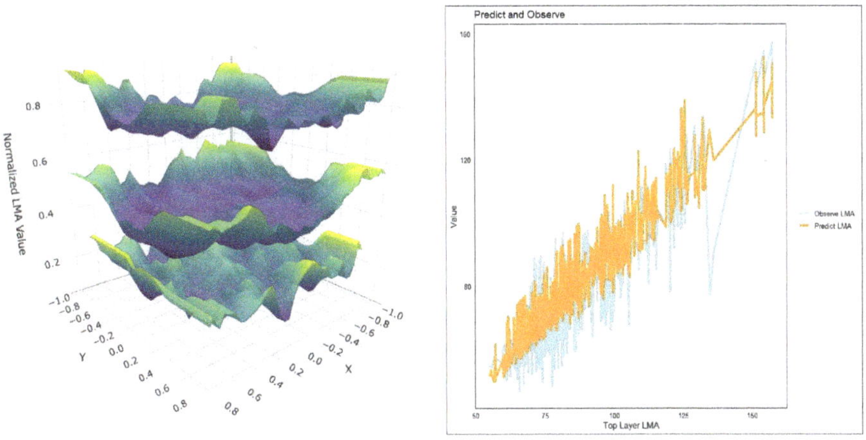

(**a**) (**b**)

Figure 15. (**a**) LMA distribution trends for each layer; (**b**) changes in observed and predicted values with increasing top-layer LMA.

When the top-layer LMA is less than 125 g/m^2, the model exhibits relatively stable prediction variability. Additionally, the predicted LMA values for the lower layer are generally lower than the observed values. The overall trend aligns with findings from previous studies, indicating that the model can be effectively used for large-scale predictions.

4. Discussion

In this study, a three-dimension estimation of canopy LMA (Leaf Mass per Area) for broad-leaved tree species in the study area was developed, which offered a possibility for improving the three-dimension accuracy of various vegetation biophysical parameter models with LMA as the parameter in this area. The number of articles on stratification analysis of vertical changes in tree canopy has been increasing in recent years. This study shows a trend that mLMA inside the canopy follows the following trend: upper layer > middle layer > lower layer, which has been found by previous studies [37]. Some studies have shown that considering the three-dimensional differences inside the canopy can offer a better prediction of the canopy NPP [38]. There are many studies on the vertical gradient changes in vegetation canopy parameters. For example, the vertical distribution of vegetation canopy leaves [37], biomass [39], the distribution of forest canopy leaf area [40], the change trend of leaf traits between different layers in the canopy [18], and the distribution of canopy nitrogen and LMA, as affected by the change in vertical ecological factors in the canopy, have been studied [41,42]. Scholars have gradually realized that the spatial structure of the canopy can be an important factor affecting canopy vegetation parameters. In order to improve the utilization rate of light energy, plants will enhance the complexity of canopy structure spontaneously to achieve more light interception and maintain canopy NPP [43–45]. The allocation of biomass can also be determined by leaf area and LMA [5]. Obviously, LMA plays an important role and is linked to other vegetation parameters. And its three-dimensional prediction has gradually become more important in the study of the three-dimensional structure of tree canopy; it shows great significance for improving the accuracy of canopy photosynthesis capacity prediction.

It is now generally believed that the photosynthetic capacity of trees is related to the LMA and to the nitrogen distribution of leaves, which are considered important parameters in vegetation photosynthesis [21,46]. Leaf nitrogen content is a probably function of LMA [42]. However, some studies have shown that, unlike LMA, which is simple to measure and has significant changes in canopy position, there is no definite relationship between leaf nitrogen content and changes in canopy position [47]. Moreover, LMA shows a significant light dependence in the canopy, which has also been proved to be one of the important reasons for the three-dimensional variation in LMA in the canopy [48]. The reason for selecting LMA as the prediction target in this study is that LMA can better reflect leaf structure across variations in the plant canopy environment such as light and water gradients, and it is possible for us to predict with a simplified three-dimensional position. The model in this study may significantly streamline the LMA measurement process in regions with intricate terrain features, including those with varied exposure and undulating surfaces.

In a previous study, people used LVA (leaf volume per area) × LD (leaf density) to calculate LMA, but these are leaf structure parameters based on leaf scale, which is difficult to apply when predicting large quantities. Hence, our model uses simple positional parameters and can be used for the estimation of large quantities of LMA with a high accuracy.

However, for the same vertical layer of the canopy, the light and water fluxes at different horizontal positions can be totally different, which is likely to lead to variation in vegetation parameters within the same layer, but there are relatively few relevant studies on this. In our experiment, not only the vertical canopy and environmental changes, but also the horizontal changes in each canopy layer were considered as influencing factors. Leaves in different layers, or even leaves at different horizontal positions in the same layer, have different coping strategies against the changes in light and water flux received. Leaves change their resource allocation strategy, which is then reflected in the vegetation parameters. Although different species respond differently to different environmental factors [11], the accuracy of tree species identification for individual trees over larger areas needs to improve; this paper has constructed a distribution logic that is applicable to a larger range and applicable to all broad-leaved tree species in the area, and the model has a decent prediction accuracy (min 10-fold R^2 = 0.716).

Recent studies have been conducted on the spatial differences in photosynthetic parameters in different layers and their contribution to the net photosynthetic rate at different positions in the canopy (the canopy has been divided into three layers in the vertical direction—upper, middle, and lower—and two parts in the horizontal direction—inner and outer). After qualitative comparison, the photosynthetic parameters show clear vertical and horizontal gradient distribution [49]. Similar results are shown with this study.

The model obtained in this paper describes the changes in different 3D positions as follows: The LMA increases with the increase in vertical height of the sample point in the canopy and the horizontal distance from the sample point to the trunk. Due to the stronger illumination at higher positions within the canopy and further away from the trunk in the same layer, plants allocate more resources towards building the mechanical tissue of leaves to cope with the water stress caused by intense light at the top of the canopy. The height of the top canopy and the distance from the trunk causes it greater resistance to get water to these places. Therefore, the transpiration pull required by plants to transport water to these places is relatively high. In order to prevent the water potential becoming too low, which may also cause high xylem tension, the plant closes its stomata in advance, which leads to a decrease in photosynthetic capacity in these positions. Previous studies have suggested that the photosynthetic capacity of leaves with a larger LMA tends to be lower [50].

The results obtained in this study are consistent with previous research hypotheses: light and hydraulic gradient affect the three-dimensional distribution of LMA; light and hydraulic power restrict the change in LMA. The higher the sample point is and the farther it is from the trunk, the higher the LMA will be [28]. Previous review articles have summarized plant LMA under environmental stress, showing that plants with a high LMA are more tolerant to stress, which is similar to this study.

It has been proposed that a combination of accurate LiDAR and optical remote sensing can be used for three-dimension estimation of vegetation parameters [51]. The three-dimensional structure of the tree crown was accurately extracted by laser point cloud data in a recent paper, enabling the height of the lowest branches to be located and the layer-wise extraction of the crown radius inside the canopy within different RDINC parts. Notably, an impressive prediction accuracy of 91% was achieved within the RDINC range of 0.15~1, suggesting its potential as a viable alternative method of traditional field measurement [52]. This paper quantitatively estimates the three-dimensional variations in LMA at different canopy positions for dominant broad-leaved tree species in Maoershan Mountain. Combining the LMA of a certain position at the top of the canopy obtained by remote sensing image, the sample point height and distance from the trunk were obtained by LiDAR. Then, the three-dimensional distribution of LMA of the canopy can be estimated by the model obtained in this study. We use the following parameters: sample height and sample distance from the trunk, which could be measured easily to estimate the three-dimensional LMA distribution as accurately and conveniently as possible. In this way, it can be possible to obtain the three-dimensional changes in other important vegetation parameters coupled with LMA, such as light and efficiency, biomass estimation, etc., so as to improve the estimation accuracy of the corresponding parameters.

In addition, some studies have considered the seasonal variation in leaf parameters [24]. Temperature changes also have a great effect on the LMA of plants because of the variation in leaf LVA and LD in response to the transformation of the environment temperature [53]. And only the spatial dimension of LMA was analyzed in this study. However, the level of the response of leaves at different locations to season, age [54], and temperature is not always consistent, which is also a challenge for us in the future.

5. Conclusions

In this study, the 3D LMA (Leaf Mass per Area) estimating method was mentioned. The results showed that LMA decreases with increasing depth and increases with increasing distance between the sample point and the trunk. This research established four methods to estimate LAM and the Random Forest model (RF) achieved the highest accuracy. The R^2 of 10-fold cross-validation ranged from 0.716 to 0.939. Then, a point cloud simulating the canopy shape was created and interpolated the LMA observations and top-layer LMA to the corresponding points. The LMA distribution of the point cloud was predicted based on its position by the RF model and the corresponding top-layer LMA with an R^2 of 0.7387. This model effectively describes the LMA distribution of an unknown broad-leaf tree canopy and can be extended to larger scales using LiDAR and optical remote sensing for broader data acquisition, enabling the observation of the three-dimensional distribution of canopy LMA on a larger scale.

The 3D model has a decent accuracy and can be used in more cases of data prediction. However, the tree age, temperature, and the season variation might be variable in different canopy locations; the model might be optimized for this in the future studies.

Author Contributions: X.Y. and D.Z. designed and proofread this article and collected the data. D.Z. analyzed the data and wrote the article. Y.W., S.Y., Y.L., Z.Y. and X.Z. revised the article. All authors have read and agreed to the published version of the manuscript.

Funding: This research was funded by the National Natural Science Foundation of China, grant number 31971580, and the China Postdoctoral Science Foundation, grant number 2019M661239.

Data Availability Statement: The raw data supporting the conclusions of this article will be made available by the authors on request.

Acknowledgments: Ying Yu, Zhiguo Liang, Yuqi Liu, Wenrui Zheng, Jie Chen, Yuxuan Li, Guanran Wang, Jinxin Piao, and Hang Xu help collected the data. Delvina A. Meshili edited the language.

Conflicts of Interest: The authors declare no conflicts of interest.

Abbreviations

The following abbreviations are used in this manuscript:

LMA	Leaf Mass per Area
mLMA	Mean relative Leaf Mass Per Area
mHD	Mean relative Horizontal Distance from sample point to the trunk
RDINC	Relative depth Into the Crown
LMM	Leaf Mixed Models
SVM	Support Vector Machine Regression
RF	Random Forest
PLS	Partial Least Squares

References

1. Li, H.; Luan, C.; Xia, X.; Chen, C.; Sun, R.; Wang, H.; Li, B. Prediction of future extreme precipitation scenarios in China based on CMIP6 climate model. *Water Resources and Hydropower Engineering.* **2023**, *54*, 16–29. [CrossRef]
2. Filonchyk, M.; Peterson, M.P.; Zhang, L.; Hurynovich, V.; He, Y. Greenhouse gases emissions and global climate change: Examining the influence of CO_2, CH_4, and N_2O. *Sci. Total Envir.* **2024**, *935*, 173359. [CrossRef] [PubMed]
3. Onoda, Y.; Wright, I.J.; Evans, J.R.; Hikosaka, K.; Kitajima, K.; Niinemets, U.; Poorter, H.; Tosens, T.; Westoby, M. Physiological and structural tradeoffs underlying the leaf economics spectrum. *New Phytol.* **2017**, *214*, 1447–1463. [CrossRef]
4. Liu, M.; Liang, G. Research progress on leaf mass per area. *Chin. J. Plant Ecol.* **2016**, *40*, 847–860.
5. Duursma, R.A.; Falster, D.S. Leaf mass per area, not total leaf area, drives differences in above-ground biomass distribution among woody plant functional types. *New Phytol* **2016**, *212*, 368–376. [CrossRef] [PubMed]
6. McGill, B.J.; Enquist, B.J.; Weiher, E.; Westoby, M. Rebuilding community ecology from functional traits. *Trends Ecol. Evol.* **2006**, *21*, 178–185. [CrossRef]
7. Wright, I.J.; Reich, P.B.; Westoby, M.; Ackerly, D.D.; Baruch, Z.; Bongers, F.; Cavender-Bares, J.; Chapin, T.; Cornelissen, J.H.C.; Diemer, M.; et al. The worldwide leaf economics spectrum. *Nature* **2004**, *428*, 821–827. [CrossRef]
8. Oren, R.; Schulze, E.D.; Matyssek, R.; Zimmermann, R. Estimating photosynthetic rate and annual carbon gain in conifers from specific leaf weight and leaf biomass. *Oecologia* **1986**, *70*, 187–193. [CrossRef]
9. Villar, R.; Ruiz-Robleto, J.; Luis Ubera, J.; Poorter, H. Exploring variation in leaf mass per area (LMA) from leaf to cell: An anatomical analysis of 26 woody species. *Am. J. Bot.* **2013**, *100*, 1969–1980. [CrossRef]
10. Westoby, M.; Warton, D.; Reich, P.B. The time value of leaf area. *Am. Nat.* **2000**, *155*, 649–656. [CrossRef]
11. Poorter, H.; Niinemets, Ü.; Poorter, L.; Wright, I.J.; Villar, R. Causes and consequences of variation in leaf mass per area (LMA): A meta-analysis. *New Phytol.* **2009**, *182*, 565–588. [CrossRef] [PubMed]
12. Bowes, G.; Ogren, W.; Hageman, R. Light Saturation, photosynthesis rate, RuDP carboxylase activity, and specific leaf weight in soybeans grown under different light intensities 1. *Crop Sci.* **1972**, *12*, 77–79. [CrossRef]
13. Björkman, O. Responses to different quantum flux densities. In *Physiological Plant Ecology I: Responses to the Physical Environment*; Springer: Berlin/Heidelberg, Germany, 1981; pp. 57–107.
14. Li, L.; Wang, X.; Niu, J.; Cui, J.; Zhang, Q.; Wan, W.; Liu, B. Effects of elevated atmospheric O3 concentrations on early and late leaf growth and elemental contents of Acer truncatum Bung under mild drought. *Acta Ecol. Sin.* **2017**, *37*, 31–34. [CrossRef]

15. Chaves, M.M.; Maroco, J.P.; Pereira, J.S. Understanding plant responses to drought—From genes to the whole plant. *Funct. Plant Biol.* **2003**, *30*, 239–264. [CrossRef] [PubMed]
16. Leuning, R.; Kelliher, F.M.; De Pury, D.; Schulze, E.D. Leaf nitrogen, photosynthesis, conductance and transpiration: Scaling from leaves to canopies. *Plant Cell Environ.* **1995**, *18*, 1183–1200. [CrossRef]
17. De Pury, D.; Farquhar, G. Simple scaling of photosynthesis from leaves to canopies without the errors of big-leaf models. *Plant Cell Environ.* **1997**, *20*, 537–557. [CrossRef]
18. Tian, J.; Wei, L.; He, N.; Xu, L.; Chen, Z.; Hou, J. Vertical variation of leaf functional traits in temperate forest canopies in China. *Acta Ecol. Sin.* **2018**, *38*, 8383–8391.
19. Yao, H.; Zhang, Y.; Yi, X.; Zhang, X.; Zhang, W. Cotton responds to different plant population densities by adjusting specific leaf area to optimize canopy photosynthetic use efficiency of light and nitrogen. *Field Crops Res.* **2016**, *188*, 10–16. [CrossRef]
20. Reich, P.B.; Walters, M.B.; Ellsworth, D.S. From tropics to tundra: Global convergence in plant functioning. *Proc. Natl. Acad. Sci. USA* **1997**, *94*, 13730–13734. [CrossRef]
21. Meir, P.; Kruijt, B.; Broadmeadow, M.; Barbosa, E.; Kull, O.; Carswell, F.; Nobre, A.; Jarvis, P.G. Acclimation of photosynthetic capacity to irradiance in tree canopies in relation to leaf nitrogen concentration and leaf mass per unit area. *Plant Cell Environ.* **2002**, *25*, 343–357. [CrossRef]
22. Ellsworth, D.; Reich, P. Canopy structure and vertical patterns of photosynthesis and related leaf traits in a deciduous forest. *Oecologia* **1993**, *96*, 169–178. [CrossRef]
23. Cavaleri, M.A.; Oberbauer, S.F.; Clark, D.B.; Clark, D.A.; Ryan, M.G. Height is more important than light in determining leaf morphology in a tropical forest. *Ecology* **2010**, *91*, 1730–1739. [CrossRef]
24. Martinez, K.A.; Fridley, J.D. Acclimation of leaf traits in seasonal light environments: Are non-native species more plastic? *J. Ecol.* **2018**, *106*, 2019–2030. [CrossRef]
25. Coble, A.P.; VanderWall, B.; Mau, A.; Cavaleri, M.A. How vertical patterns in leaf traits shift seasonally and the implications for modeling canopy photosynthesis in a temperate deciduous forest. *Tree Physiol.* **2016**, *36*, 1077–1091. [CrossRef] [PubMed]
26. Abdollahnejad, A.; Panagiotidis, D.; Surový, P.; Ulbrichová, I. UAV Capability to Detect and Interpret Solar Radiation as a Potential Replacement Method to Hemispherical Photography. *Remote Sens.* **2018**, *10*, 423. [CrossRef]
27. Chlus, A.; Kruger, E.L.; Townsend, P.A. Mapping three-dimensional variation in leaf mass per area with imaging spectroscopy and lidar in a temperate broadleaf forest. *Remote Sens. Environ.* **2020**, *250*, 112043. [CrossRef]
28. Williams, G.; Nelson, A. Spatial variation in specific leaf area and horizontal distribution of leaf area in juvenile western larch (Larix occidentalis Nutt.). *Trees* **2018**, *32*, 1621–1631. [CrossRef]
29. Aber, J.D.; Reich, P.B.; Goulden, M.L. Extrapolating leaf CO_2 exchange to the canopy: A generalized model of forest photosynthesis compared with measurements by eddy correlation. *Oecologia* **1996**, *106*, 257–265. [CrossRef]
30. Yang, X.; Fan, W.; Yu, Y. Leaf Chlorophyll Content Retrieval from Hyperspectral Remote Sensing Images. *J. Northeast. For. Univ.* **2010**, *38*, 123–124+135. [CrossRef]
31. Wu, C.; Niu, Z.; Tang, Q.; Huang, W. Estimating chlorophyll content from hyperspectral vegetation indices: Modeling and validation. *Agric. For. Meteorol.* **2008**, *148*, 1230–1241. [CrossRef]
32. Jordan, C.F. Derivation of leaf-area index from quality of light on the forest floor. *Ecology* **1969**, *50*, 663–666. [CrossRef]
33. Broge, N.H.; Leblanc, E. Comparing prediction power and stability of broadband and hyperspectral vegetation indices for estimation of green leaf area index and canopy chlorophyll density. *Remote Sens. Environ.* **2001**, *76*, 156–172. [CrossRef]
34. Naidu, R.A.; Perry, E.M.; Pierce, F.J.; Mekuria, T. The potential of spectral reflectance technique for the detection of Grapevine leafroll-associated virus-3 in two red-berried wine grape cultivars. *Comput. Electron. Agric.* **2009**, *66*, 38–45. [CrossRef]
35. Guan, L.; Liu, X. Two kinds of modified spectral indices for retrieval of crop canopy chlorophyll content. *Adv. Earth Sci.* **2009**, *24*, 548–554.
36. Ran, D.; Sun, Z.; Lu, S.; Omasa, K. Optimizing angular resistant spectral indices to estimate leaf biochemical parameters from multi-angular spectral reflection. *Agric. For. Meteorol.* **2024**, *348*, 109916. [CrossRef]
37. Lin, Z.; Tianxiang, L.U.O.; Kunmei, D.; Wenhua, L.I. Vertical variations in specific leaf area and leaf dry matter content with canopy height in Pinus yunnanensis. *J. Beijing For. Univ.* **2008**, *30*, 40–44.
38. Parker, G.G.; Harmon, M.E.; Lefsky, M.A.; Chen, J.; Pelt, R.V.; Weis, S.B.; Thomas, S.C.; Winner, W.E.; Shaw, D.C.; Frankling, J.F. Three-dimensional structure of an old-growth Pseudotsuga-Tsuga canopy and its implications for radiation balance, microclimate, and gas exchange. *Ecosystems* **2004**, *7*, 440–453. [CrossRef]
39. Williams, G.M.; Nelson, A.S.; Affleck, D.L. Vertical distribution of foliar biomass in western larch (Larix occidentalis). *Can. J. For. Res.* **2018**, *48*, 42–57. [CrossRef]
40. Sterba, H.; Dirnberger, G.; Ritter, T. Vertical Distribution of Leaf Area of European Larch (*Larix decidua* Mill.) and Norway Spruce (*Picea abies* (L.) Karst.) in Pure and Mixed Stands. *Forests* **2019**, *10*, 570. [CrossRef]

41. Chmura, D.J.; Tjoelker, M.G. Leaf traits in relation to crown development, light interception and growth of elite families of loblolly and slash pine. *Tree Physiol.* **2008**, *28*, 729–742. [CrossRef]
42. White, J.D.; Scott, N.A. Specific leaf area and nitrogen distribution in New Zealand forests: Species independently respond to intercepted light. *For. Ecol. Manag.* **2006**, *226*, 319–329. [CrossRef]
43. Ishii, H.; Asano, S. The role of crown architecture, leaf phenology and photosynthetic activity in promoting complementary use of light among coexisting species in temperate forests. *Ecol. Res.* **2010**, *25*, 715–722. [CrossRef]
44. Hardiman, B.S.; Bohrer, G.; Gough, C.M.; Vogel, C.S.; Curtis, P.S. The role of canopy structural complexity in wood net primary production of a maturing northern deciduous forest. *Ecology* **2011**, *92*, 1818–1827. [CrossRef] [PubMed]
45. Gough, C.M.; Vogel, C.S.; Hardiman, B.; Curtis, P.S. Wood net primary production resilience in an unmanaged forest transitioning from early to middle succession. *For. Ecol. Manag.* **2010**, *260*, 36–41. [CrossRef]
46. Chiang, J.-M.; Brown, K.J. The effects of thinning and burning treatments on within-canopy variation of leaf traits in hardwood forests of southern Ohio. *For. Ecol. Manag.* **2010**, *260*, 1065–1075. [CrossRef]
47. Hardiman, B.S.; Gough, C.M.; Halperin, A.; Hofmeister, K.L.; Nave, L.E.; Bohrer, G.; Curtis, P.S. Maintaining high rates of carbon storage in old forests: A mechanism linking canopy structure to forest function. *For. Ecol. Manag.* **2013**, *298*, 111–119. [CrossRef]
48. Niinemets, Ü.; Keenan, T.F.; Hallik, L. A worldwide analysis of within-canopy variations in leaf structural, chemical and physiological traits across plant functional types. *New Phytol.* **2015**, *205*, 973–993. [CrossRef]
49. Liu, Q.; Dong, L.H.; Li, F.R.; Li, X. Spatial heterogeneity of canopy photosynthesis for Larix olgensis. *J. Appl. Ecol.* **2016**, *27*, 2789–2796. [CrossRef]
50. Niinemets, Ü.; Ellsworth, D.S.; Lukjanova, A.; Tobias, M. Site fertility and the morphological and photosynthetic acclimation of Pinus sylvestris needles to light. *Tree Physiol.* **2001**, *21*, 1231–1244. [CrossRef]
51. Swatantran, A.; Dubayah, R.; Roberts, D.; Hofton, M.; Blair, J.B. Mapping biomass and stress in the Sierra Nevada using lidar and hyperspectral data fusion. *Remote Sens. Environ.* **2011**, *115*, 2917–2930. [CrossRef]
52. Wang, F.; Sun, Y.; Jia, W.; Li, D.; Zhang, X.; Tang, Y.; Guo, H. A Novel Approach to Characterizing Crown Vertical Profile Shapes Using Terrestrial Laser Scanning (TLS). *Remote Sens.* **2023**, *15*, 3272. [CrossRef]
53. Atkin, O.K.; Loveys, B.; Atkinson, L.J.; Pons, T. Phenotypic plasticity and growth temperature: Understanding interspecific variability. *J. Exp. Bot.* **2006**, *57*, 267–281. [CrossRef] [PubMed]
54. Osnas, J.L.; Lichstein, J.W.; Reich, P.B.; Pacala, S.W. Global leaf trait relationships: Mass, area, and the leaf economics spectrum. *Science* **2013**, *340*, 741–744. [CrossRef] [PubMed]

forests

MDPI

Article

TreeDBH: Dual Enhancement Strategies for Tree Point Cloud Completion in Medium–Low Density UAV Data

Yunlian Su [1], Zhibo Chen [1,2,*] and Xiaojing Xue [1]

1 School of Information Science and Technology, Beijing Forestry University, Beijing 100083, China; syl3220299@bjfu.edu.cn (Y.S.); xuexiaojing@bjfu.edu.cn (X.X.)
2 Engineering Research Center for Forestry-Oriented Intelligent Information Processing, National Forestry and Grassland Administration, Beijing 100083, China
* Correspondence: zhibo@bjfu.edu.cn

Abstract: Medium–low density UAV point clouds often suffer from incomplete lower canopy structures and sparse distributions due to self-occlusion. While existing point cloud completion models achieve high metric accuracy, they inadequately address missing regions in trunks and lower canopy areas. To resolve these issues, this paper proposes a hierarchical random sampling strategy and a spatially constrained loss function. First, we dynamically stratify point clouds based on density distribution characteristics, employing hierarchical random sampling to preserve proportional representation of lower-level points, thereby effectively retaining basal tree structure information. Second, we introduce a distance constraint term for mid-lower point clouds into the symmetrical Chamfer distance (CD) loss, compelling models to prioritize completion quality in trunk base regions. Experiments on the FOR-instance-created completion dataset and Xiong'an dataset demonstrate that our method significantly enhances structural recovery capability at tree trunk bases, with visual results outperforming the baseline SeedFormer model. Additionally, we refer to existing point cloud-based diameter at breast height (DBH) calculation methods to measure the completed trees and compare the computed results with the measured values to evaluate the accuracy of the completion effect. Experimental results show that, after integrating our proposed strategies with existing completion methods, the accuracy of DBH measurement from point clouds is significantly improved. This study provides novel insights for addressing structural bias in tree point cloud completion and offers valuable references for digital forestry resource management.

Keywords: UAV point clouds; point cloud completion; deep learning

1. Introduction

Point cloud technology, as a three-dimensional coordinate-based digitalization method, enables rapid acquisition of tree morphological features, structural characteristics, and spatial distribution patterns through sensor-captured arboreal point cloud data [1,2]. Unmanned Aerial Vehicles (UAVs) equipped with LiDAR systems demonstrate exceptional capability in large-scale forest point cloud data collection [3–5]. Compared to conventional ground surveys, UAV platforms can rapidly survey extensive areas while overcoming accessibility limitations in remote regions and complex terrains. However, the inherent geometric complexity of trees and environmental constraints during scanning, including equipment precision limitations, flight altitude restrictions, self-occlusion within tree canopies, and occlusion from surrounding vegetation, frequently result in data incompleteness, particularly in lower canopy regions, inter-tree spaces, and dense forest stands, point clouds often exhibit sparsity issues. To address data incompleteness and sparsity in medium–low density UAV point clouds, deep learning-based point cloud completion methods [6–8] offer promising solutions for reconstructing missing segments, thereby delivering more comprehensive and precise 3D tree representations. Enhanced point cloud integrity significantly improves the accuracy of extracting critical biophysical parameters [9–11] including tree height, diameter at breast height (DBH), and crown projection area. These parameters hold substantial application value for forest resource monitoring, ecological environment studies, and sustainable forestry management.

Point cloud completion methodologies can be categorized into traditional approaches and deep learning-based techniques [12]. Traditional methods [13–17] infer missing regions by leveraging the local geometric features of point clouds. Common geometric reconstruction strategies include neighborhood-based interpolation, Poisson reconstruction, and triangular mesh generation. These approaches necessitate inherent symmetry and regularity in raw point clouds, often yielding significant errors when handling large missing areas or complex point distributions. In recent years, deep learning-based point cloud completion has emerged as a prominent research focus. Predominant frameworks encompass voxel-based [18–20], view-based [21–24], and point-based [25–27] paradigms. Voxel-based methods reconstruct missing regions by analyzing structural relationships among volumetric pixels, while view-based methods infer complete models through multi-view projections generated from varying perspectives. The advent of PointNet [28] enabled direct processing of raw point clouds via deep neural networks, marking a paradigm shift. The PCN [29] network pioneered point-based completion frameworks by integrating the architectural principles of FoldingNet [30] and PointNet, employing a coarse-to-fine decoder to predict complete point clouds. Subsequent advancements include SANet [31], which enhanced completion performance through skip-attention mechanisms, and PF-Net [32], which improved precision via multi-resolution feature extraction, hierarchical generation, and adversarial discriminators for quality assessment. Transformer architectures [33] demonstrate inherent suitability for processing unordered point cloud data through permutation-invariant attention mechanisms, effectively modeling global point relationships. The Pointr [34] framework represents incomplete point clouds as point proxies, utilizing transformer-based [35] encoders–decoders for structural restoration. Recent innovations like SnowflakeNet [36] implement deconvolutional upsampling with skip-Transformer modules across decoding layers to progressively refine missing regions while preserving local geometric features.

While existing point cloud completion methods have achieved notable progress in global structure modeling and local detail recovery through ongoing research, most approaches predominantly rely on synthetic datasets characterized by planar surfaces and uniform point density. Furthermore, the simulation of missing regions typically employs random spherical elimination [32,34], where points within a randomly positioned sphere are removed. However, in real-world scenarios, missing regions in UAV-captured tree point clouds predominantly result from occlusion by foliage and branches. Limited by LiDAR penetration capabilities, point density gradually diminishes near ground surfaces, exacerbating density disparities between upper and lower canopy layers when applying current completion methods to practical cases, ultimately compromising completion quality. To address this challenge, this study proposes a hierarchical random sampling strategy tailored to authentic UAV point cloud characteristics, with SeedFormer [37] as the baseline framework. The strategy initiates with statistical analysis of lower-trunk point density, calculating its proportional representation within the complete point cloud and determining the median proportion threshold. During sampling, it prioritizes retaining lower-level points above this median threshold, ensuring moderate input density for trunk base regions at the data level and mitigating adverse effects from excessive sparsity on completion performance. Additionally, we enhance model attention to lower canopy structures by incorporating the Chamfer distance [38] between completed point clouds and ground-truth lower-trunk regions into the loss function. Through these dual strategies, the model achieves effective trunk structure completion even under low-density lower canopy conditions, significantly improving point cloud integrity. This advancement enables more structurally coherent and ecologically plausible tree reconstructions.

Terrestrial laser scanning (TLS)-acquired point clouds exhibit high quality with distinct trunk profiles, enabling accurate DBH estimation. In contrast, most UAV-derived tree point clouds show degraded quality at DBH measurement heights (1.3–1.5 m), which is often insufficient to support reliable DBH computation. In traditional forestry research, DBH estimation methods based on allometric equations [39] are widely used. These methods rely on tree growth models and empirical parameters, providing relatively stable DBH estimates to a certain extent. To reduce the workload of manual measurements, we aim to improve the automation level of DBH measurement using UAV point cloud data. Jie Xu et al. [40] employed the BlendMask algorithm to accurately delineate tree crown contours in UAV imagery and introduced a Bayesian neural network to establish a correlation model between individual tree crown size and DBH. The CaR3DMIC method [41] directly extracts key tree features, such as crown height and crown area, from image data. However, in environments with medium- and low-density point cloud data, the complexity of tree morphology and occlusion effects make it challenging to extract other critical parameters directly, leading to difficulties in ensuring the accuracy of DBH measurement. Compared to traditional allometric equations, point cloud-based methods can reconstruct tree trunk structures in a data-driven manner, reducing reliance on empirical formulas and minimizing systematic errors. Furthermore, our approach focuses on improving the overall quality of tree structure completion, particularly the integrity of lower regions. DBH measurement based on point clouds provides a more intuitive reflection of the effectiveness of our completion method. Therefore, we measure DBH using point clouds [42] and introduce quality evaluation metrics to qualitatively analyze the accuracy of trunk completion. We conducted single-tree point cloud completion experiments on both our proprietary dataset and the adjusted FOR-instance dataset [43]. Experimental results demonstrate that the adjusted model maintains completion performance while more accurately restoring tree contours, significantly enhancing DBH measurement accuracy.

Our primary contributions are threefold:

(1) This study established a proprietary dataset through UAV-mounted LiDAR and terrestrial laser scanning (TLS) data acquisition, while constructing a dedicated point cloud completion benchmark based on the FOR-instance dataset;
(2) To address self-occlusion-induced incompleteness and canopy layer density disparities in medium–low density UAV point clouds, we proposed two enhancement strategies that significantly improve the integrity and precision of tree point cloud completion;
(3) In this study, we applied a DBH measurement method to fit the DBH of the completed point clouds and compare the results with ground-truth measurements, verifying the impact and value of point cloud completion on single-tree parameter extraction.

2. Materials

2.1. Point Cloud Completion Dataset Construction Using the FOR-Instance Benchmark

2.1.1. Overview of the FOR-Instance Dataset

The FOR-instance dataset [43] (available at https://doi.org/10.5281/zenodo.8287792 accessed on 27 August 2024) comprises UAV LiDAR-derived point clouds collected from

30 forest plots across five countries. It features manually annotated semantic class labels and tree IDs for individual points, with seven semantic categories: unclassified points, low vegetation, ground surface, incomplete tree points, trunks, live branches, and dead branches. Field-measured DBH values are also included for selected trees.

The dataset encompasses five forest types:

- NIBIO: Boreal conifer-dominated forests (Norway, 42% of annotated data) [44];
- CULS: Temperate conifer-dominated forests (Czech Republic, 5%) [45];
- TUWIEN: Deciduous floodplain forests (Austria, 29%) [46];
- RMIT: White peppermint eucalyptus-dominated stands (Australia, 11%);
- SCION: Monocultural radiata pine plantations (New Zealand, 13%).

FOR-instance captures realistic structural complexities, including curved trunks, occluded or sparsely sampled trunks, fallen trees, and understory vegetation. Notably, 89% of trees are visible from the upper canopy layer, while 11% reside in the sub-canopy stratum with obscured crowns. Geographical disparities manifest through distinct species compositions and terrain variations across regions. Significant interspecific differences in crown architecture are observed. Point cloud density varies substantially, ranging from 9629 pts/m^2 in NIBIO regions to 498 pts/m^2 in RMIT areas, collectively positioning FOR-instance as a high-density point cloud benchmark.

2.1.2. FOR-Instance Dataset Processing

We initially extracted all individual trees from the FOR-instance dataset using tree ID annotations, retaining specimens with over 16,384 points per tree as labeled point clouds to ensure high-density characteristics and model input compatibility. Building upon this foundation, we synthesized medium–low density UAV-like point clouds as model inputs by simulating real-world sensor limitations. Specifically, we simulated the self-occlusion phenomenon of LiDAR sensors and the uneven point cloud density between the upper and lower layers of the tree canopy by applying a weighted random downsampling method to process the point cloud data. Specifically, we first generated a viewpoint based on the center, diameter, and maximum height of the point cloud's bounding box. Then, using this viewpoint and a defined radius, we classified the points into visible and occluded categories. During the random downsampling process, we stochastically removed points based on deletion weights. Visible points were assigned lower deletion weights to preserve as much information as possible, while occluded points were given higher deletion weights to increase their probability of removal, thereby enhancing the missing rate of point clouds in occluded areas. This process constrained final point densities to 150–360 pts/m^2 across all plots, representative of practical UAV LiDAR survey conditions. Ultimately, input point clouds were rigorously paired with their corresponding labeled counterparts to ensure spatial correspondence. Figure 1 illustrates the visual comparison of downsampling effects across five representative regions.

Following rigorous screening, we obtained 607 qualified individual trees, with 80% (486 trees) allocated to the training set and 20% (121 trees) to the test set. The final dataset composition per region is detailed in Table 1. Due to the low initial point cloud acquisition density in the RMIT region, this area yielded the minimal dataset contribution, containing only four trees exceeding 16,384 points. Three RMIT samples were assigned to the training set and one to the test set through stratified random sampling across five geographic regions. To enhance model generalizability, we implemented data augmentation on the training set via geometric transformations including translation and random rotation about the centroid axis. Post-augmentation, the training set expanded to 3888 input samples while maintaining the test set size. The spatial distribution is visualized in Figure 2.

Figure 1. Downsampling effects visualization on the FOR-instance dataset.

Table 1. Training–testing split statistics of the FOR-instance point cloud completion dataset. In the "Training Set" column, the number before the multiplication sign represents the original number of trees before data augmentation. We applied an 8× data augmentation to the trees in the training set, resulting in a final training sample size that is eight times the original number. The number of trees in the test set remains unchanged and has not undergone data augmentation.

Region	TrainSet	TestSet
CULS	37 × 8 = 296	10
NIBIO	315 × 8 = 2520	81
RMIT	3 × 8 = 24	1
SCION	78 × 8 = 624	18
TEWIEN	53 × 8 = 424	11
Total	486 × 8 = 3888	121

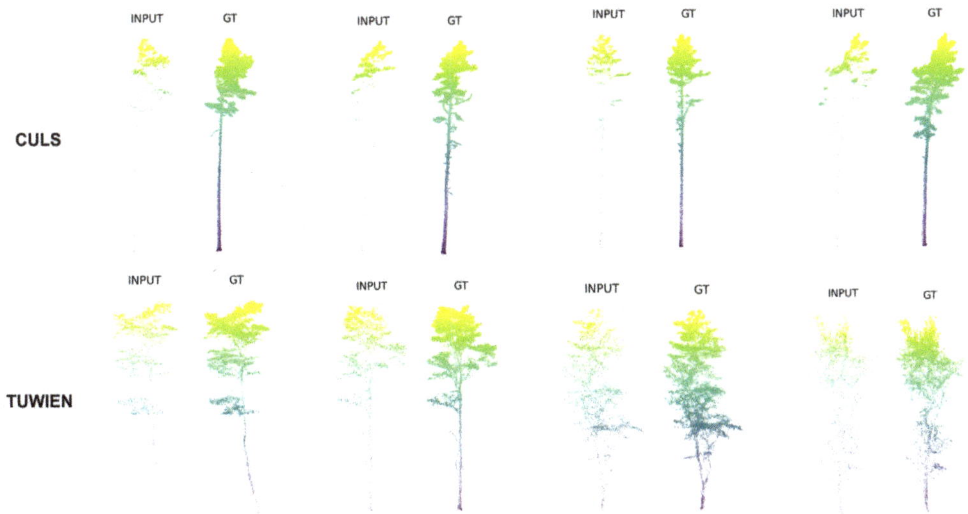

Figure 2. *Cont.*

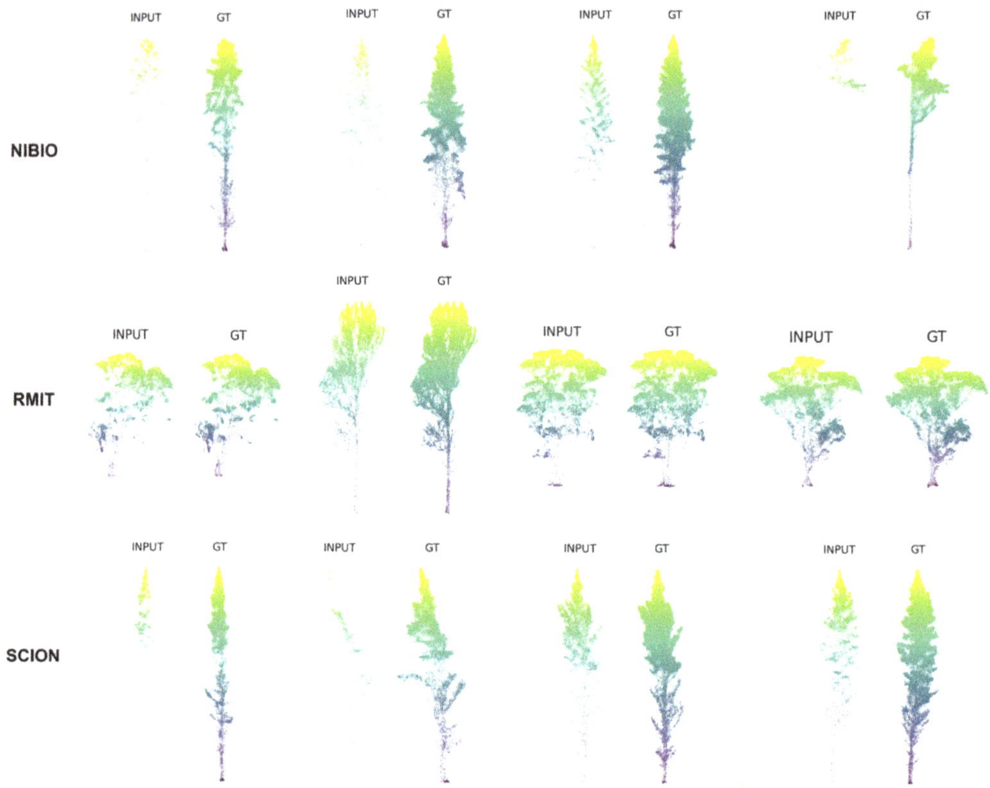

Figure 2. Representative samples from the FOR-instance completion dataset. "INPUT" represents the input point cloud from the dataset, while "GT" (ground truth) denotes the corresponding real tree point cloud. It is primarily used for model training, performance evaluation, and result validation. The image assigns colors to the trees based on their height.

2.2. Xiong'an Dataset

2.2.1. Study Area

The study area of this dataset is located in central Hebei Province, China (38°43′–39°10′ N, 115°38′–116°20′ E), encompassing Xiong County, Rongcheng County, Anxin County, and surrounding regions, as shown in Figure 3. The Xiong'an New Area lies in the mid-latitude zone and features a warm temperate monsoon continental climate with four distinct seasons. The annual average temperature is 11.7 °C, with the highest monthly average temperature of 26 °C in July and the lowest monthly average temperature of −4.9 °C in January. The region receives an average of 2685 h of sunshine per year and an annual precipitation of 551.5 mm, 80% of which falls between June and September. The frost-free period lasts approximately 185 days. The dominant tree species in this area include 13 major species, such as *Robinia pseudoacacia*, *Fraxinus chinensis*, *Ulmus pumila*, *Sophora japonica*, *Acer truncatum*, *Catalpa bungei*, *Ailanthus altissima*, *Juniperus chinensis*, *Pinus armandii*, *Prunus davidiana*, *Populus* spp., *Ginkgo biloba*, *Pinus tabuliformis*, *Platycladus orientalis*, *Platanus acerifolia*, *Catalpa ovata*, *Malus micromalus*, *Salix matsudana*, *Koelreuteria paniculata*, *Toona sinensis*, *Acer pictum*, among others. In this study, 13 experimental plots were selected within the Xiong'an region, covering eight dominant tree species: *Ginkgo biloba*, *Sophora japonica*, *Eucommia ulmoides*, *Fraxinus chinensis*, *Koelreuteria paniculata*, *Catalpa ovata*, *Ulmus* spp., and *Acer pictum*.

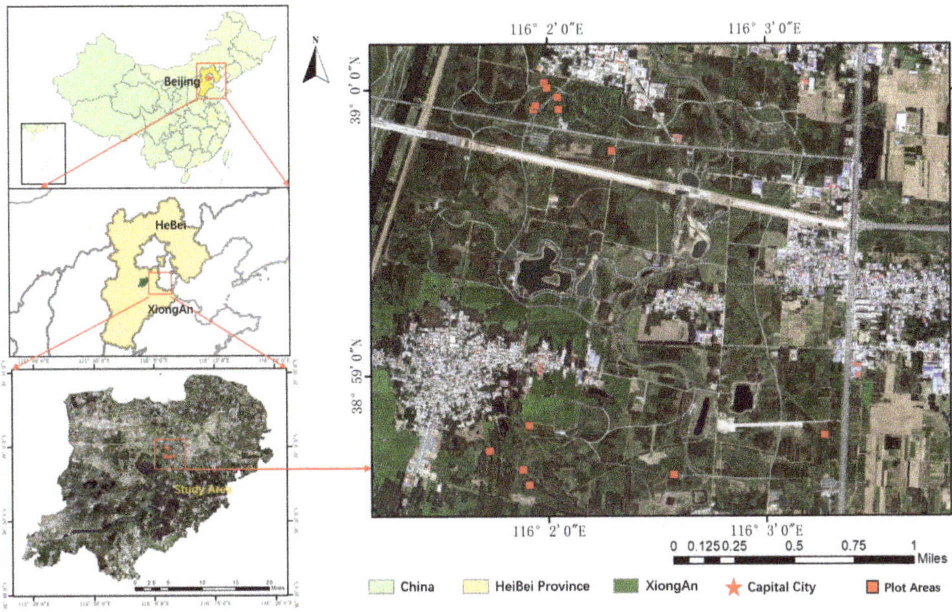

Figure 3. General overview of the study area. The study area is located within Xiong'an New Area, Hebei Province, China. The red squares indicate the locations of the experimental plots.

2.2.2. ULS Data

This study utilized the DJI Zenmuse L2 LiDAR module mounted on the DJI Matrice 350 UAV platform to acquire unmanned airborne laser scanning (ULS) data. The LiDAR system operates at a wavelength of 905 nm, with a ranging accuracy of 2 cm at a distance of 150 m. It features a laser divergence angle of 0.6 mrad (horizontal) × 0.2 mrad (vertical) and a pulse repetition frequency of 240 kHz. In multiple return mode, the system can acquire data at a maximum rate of 1.2 million points per second. Data collection was conducted from August to September 2024, during the morning hours under optimal lighting conditions. The flight speed was set to 18 m/s, and the flight altitude was maintained at 80 m above the takeoff point. The surveyed area covered 13 experimental plots included in the dataset, with point cloud densities ranging from 148.16 to 462.87 points per square meter (pts/m^2), and an average density of 260.28 pts/m^2 across all plots. After data acquisition, noise in the raw scan data was removed using Gaussian filtering. The point cloud was then classified into ground and non-ground points using the cloth simulation filtering algorithm. Finally, the point cloud data were normalized based on ground elevation.

2.2.3. TLS Data

We utilized the RIEGL VZ-600i laser scanner (RIEGL Laser Measurement Systems GmbH, Horn, Austria) for terrestrial laser scanning (TLS) data acquisition. This instrument features a scanning field of view of ≥360° × 105° (horizontal × vertical) and achieves a scanning accuracy of ≥5 mm at a distance of 100 m, with a laser pulse repetition rate of 2.2 million points per second. The TLS data collection was conducted concurrently with UAV-based data acquisition, covering 13 experimental plots, each measuring 20 m × 20 m. For each plot, three independent scans were performed. The acquired scans were subsequently processed using the accompanying software for automatic registration, noise removal, and point cloud normalization. To ensure data quality while improving processing efficiency, we employed CloudCompare software (version 2.13.0) [47] and applied the octree-based sampling method to downsample the point cloud. After downsampling, the density of the ground point cloud ranged from 10,687.20 to 52,323.46 points per square meter (pts/m^2), with an average density of 36,771.18 pts/m^2 across the 13 plots. The acquired point cloud data are illustrated in the corresponding Figure 4.

Figure 4. Partial visualization of sample plots in the Xiong'an collected dataset. The image assigns colors to the trees based on their height.

2.2.4. Field Data

The sample plot data collection was conducted concurrently with UAV-based LiDAR and terrestrial LiDAR data acquisition. The selection of sample plots considered multiple factors, including dominant local tree species, growth season, accessibility, and terrain, to ensure representativeness. An RTK device was used to ensure precise plot positioning and alignment, while standard measuring tapes were employed to measure distances within the plots. Each sample plot was set to a size of 20 m $\times$ 20 m, with its center coordinates recorded. Within each plot, we measured the DBH and geographic coordinates of individual trees. Following standard field measurement protocols, trees with a DBH greater than 5 cm were recorded, with DBH measurements taken at approximately 1.3 m above ground level. The actual data collection situation is shown in Figure 5.

Figure 5. Display of the actual situation of partial plots in the Xiong'an dataset.

2.2.5. Dataset Creation

We first performed registration of the ground-based LiDAR point cloud data and UAV-derived point cloud data using CloudCompare. Subsequently, we conducted manual annotation for both datasets, classifying the sample plots into three categories: grassland, trees, and unclassified points. Each individual tree was assigned a unique label.

After completing the data annotation, we generated rectangular bounding boxes for each tree's terrestrial and UAV-derived point clouds and determined the tree centers based on these bounding boxes. Each terrestrial tree point cloud was then matched to its nearest neighboring UAV-derived tree point cloud. To establish tree correspondences, we calculated the overlap ratio between the bounding boxes of each terrestrial tree and its nearest UAV-derived tree. Pairs with the highest overlap ratio exceeding 50% were identified as the same tree. In this process, the UAV-derived point cloud was designated as the input point cloud, while the terrestrial point cloud served as the ground truth.

After matching the UAV-derived tree point clouds with the terrestrial tree point clouds, we further associated them with field-measured tree data to determine the DBH of each tree. Specifically, a polygonal bounding box was generated for each UAV-derived tree. If a field-measured tree was located within the bounding box and was closest to its center point, it was identified as the corresponding field-measured tree for that UAV-derived point cloud. Through this two-step matching process, we established the DBH, terrestrial point cloud, and UAV-derived point cloud for each tree. Finally, we conducted a manual verification of the matching results to ensure accuracy.

In the end, we successfully matched 430 trees, with about 80% (351 trees) used as the training set and the remaining 20% (79 trees) as the test set. In accordance with the data processing method used in the FOR-instance dataset, the trees in the test set were obtained through stratified random sampling within the 13 sample plots. For each sample in the training set, we performed data augmentation, including operations such as translation and random rotation around the central axis. After augmentation, the training set contained 2808 input samples. No data augmentation was applied to the test set samples. Figure 6 shows the visualization of some of the point clouds from our training dataset. And the final dataset composition per region is detailed in Table 2.

Figure 6. Partial visualization of the completed dataset from the Xiong'an dataset creation. The image assigns colors to the trees based on their height.

Table 2. Training–testing split statistics of the Xiong'an point cloud completion dataset. In the "Training Set" column, the number before the multiplication sign represents the original number of trees before data augmentation. We applied an 8× data augmentation to the trees in the training set, resulting in a final training sample size that is eight times the original number. The number of trees in the test set remains unchanged and has not undergone data augmentation.

PlotID	TrainSet	TestSet
150	35 × 8 = 280	8
151	42 × 8 = 336	10
152	27 × 8 = 216	6
153	31 × 8 = 248	7
155	28 × 8 = 224	7
157	31 × 8 = 248	7
158	21 × 8 = 168	5
159	23 × 8 = 184	5
160	32 × 8 = 256	7
161	24 × 8 = 192	5
162	16 × 8 = 128	3
163	24 × 8 = 192	5
164	17 × 8 = 136	4
Total	351 × 8 = 2808	79

3. Methods

3.1. SeedFormer Model

Based on the performance of the completion model and the practical results of tree completion, we selected SeedFormer [37] as the foundational model for our strategy. SeedFormer is an advanced deep learning neural network for point cloud completion. As shown in Figure 7, SeedFormer primarily consists of an encoder, a seed generator, and a coarse-to-fine generation module.

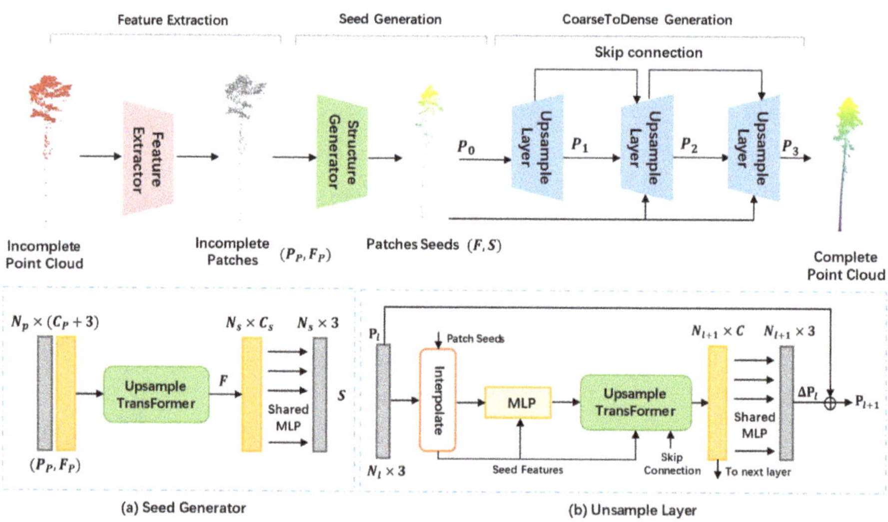

Figure 7. Structure of the SeedFormer model. (**a**) The seed generator is used to obtain Patch Seeds, which are then propagated to subsequent layers through interpolation of seed features. (**b**) In the coarse-to-fine generation process, multiple upsampling layers are used, with the Upsample Transformer applying skip connections and seed feature encoding to generate new points.

The encoder section uses Point Transformer and Set Abstraction layers to extract features from the input point cloud, outputting the center coordinates of Patch Seeds and the feature representations of the seed points. These coordinates and seed point features represent the local set characteristics of the input point cloud and provide support for subsequent point cloud completion. The seed generator aims to output a rough structure of the complete shape, specifically by generating seed points. It first utilizes a Transformer structure to transform the Patch Seeds features F_p into new seed features F:

$$F = UpTrans(F_p, P_p) \tag{1}$$

Subsequently, a multilayer perceptron (MLP) is used to map the seed features F into the three-dimensional coordinate space, yielding N_s seed points S:

$$S = \{x_i\}_{i=1}^{N_s} \in \mathbb{R}^{N_s \times 3} \tag{2}$$

During the seed generation phase, the Softmax normalization in the Transformer structure is removed to prevent the distribution of seed points from being constrained, allowing for a more flexible coverage of the complete shape. This design enables SeedFormer to both restore existing structures and reasonably infer the missing parts. The combination of the seed points S and features F forms the Patch Seeds, which serve as the foundation for the point cloud completion. Finally, the Upsample Transformer progressively generates the point cloud layer by layer, recovering the missing parts and ultimately producing the complete point cloud shape. The core idea of the Upsample Transformer is local attention aggregation, where, during new point generation, the spatial and semantic information of neighboring points is integrated to ensure that the new points maintain consistent geometric relationships with the surrounding structure. The specific process is as follows: the feature f_i of each point cloud is mapped through a multilayer perceptron (MLP) into three vectors: query (q_i), key (k_j), and value (v_j), which are used to calculate the relationship between the current point and its neighboring points and to generate new points. The query vector q_i comes from the feature of the current point cloud, the key vector k_j comes from the features of the Patch Seeds, and the value vector v_j is used to generate the new points. The attention calculation is performed by examining the relationship between the query and key vectors, and the specific formula is as follows:

$$a_{ijm} = \alpha_m\left(\beta(q_i) - \gamma(k_j) + \delta\right), \tag{3}$$

In this, α_m, β, and γ are feature mapping functions used for the mapping of the query and key, while the positional encoding δ integrates the geometric relationships in the point cloud and the relationships between seed features to ensure that the generated new points account for both spatial positions and local patterns. The final point computation is then performed by aggregating neighborhood information through the weighted value vector, as expressed in the following formula:

$$h_{im} = \sum_{j \in \mathcal{N}(i)} a_{ijm} * \left(\psi(v_j) + \delta\right), \tag{4}$$

Here, h_{im} represents the aggregated point feature, ψ is a mapping function used to adjust the value vector features, and δ serves as the positional encoding, ensuring that the generated new points are positioned appropriately in space. Finally, the new points are predicted for their three-dimensional coordinates through a multilayer perceptron (MLP), completing the point cloud upsampling process. This process is refined step by step through multiple layers of Transformer structures, ultimately restoring the complete point cloud. The generated points effectively utilize the spatial distribution and semantic information of the existing points, thereby efficiently recovering the missing parts.

SeedFormer introduces a novel generative strategy combining local information enhancement with Transformer for the point cloud completion task. It preserves local geometric features through the Patch Seeds structure, aggregates neighborhood features using the Upsample Transformer, and employs a step-by-step refinement upsampling strategy. This approach ensures that the generated point cloud achieves optimal performance in both overall structural integrity and local detail fidelity. In real-world point cloud environments, factors such as noise, outliers, and occlusion can significantly impact the completion performance. However, the SeedFormer model structure is highly effective in addressing these challenges and improving completion performance. It leverages the Transformer mechanism to simultaneously learn global and local features, which allows it to effectively distinguish between noise points and true points, reducing the interference of outliers on the completion quality. Moreover, the model combines hierarchical feature extraction with global contextual information for shape reasoning, thereby alleviating the point cloud missing problem caused by occlusion to a large extent. Experimental results show that SeedFormer outperforms existing point cloud completion methods on multiple datasets, fully validating its exceptional capability in detail restoration and global shape reconstruction.

3.2. Hierarchical Random Sampling Method

To effectively process low- to medium-density UAV point cloud data and generate complete structure point clouds for individual trees, this paper proposes a height-based hierarchical random sampling strategy. The specific process is illustrated in Algorithm 1. First, the height information (z-coordinate) of the input point cloud is extracted, and the median value is calculated as the hierarchical threshold. Based on this threshold, the point cloud is divided into two parts: the low-height point cloud (below 50% height) and the high-height point cloud (above 50% height). Next, the proportion of the low-height point cloud, denoted as p_{low} (i.e., the ratio of low-height points to the total number of points), is calculated.

When $p\_low < 0.15$, to ensure the representativeness of the low-level point cloud, 15% of the points are randomly sampled from the low-level point cloud, and the remaining points are randomly selected from the high-level point cloud. If $p\_low \geq 0.15$, random sampling is performed directly across the entire point cloud. To ensure the stability and completeness of the sampling, if the number of points in one category of the point cloud is insufficient to meet the required sample size, random supplementation is performed to ensure the final number of points is met. Finally, the sampled points from both the low-level and high-level point clouds are merged to form the complete sampling result.

This sampling strategy ensures that the low-level point cloud accounts for at least 15% of the total points. This proportion is derived from a statistical analysis of the labeled point cloud, where points below 50% height account for approximately 15% of the total. For consistency, both the labeled point cloud and the input point cloud are sampled using this method, resulting in the labeled point cloud being uniformly sampled to 16,384 points, while the input point cloud is sampled to 2048 points.

Algorithm 1 Adaptive Point Cloud Sampling (APCS)

Input:
$\boldsymbol{P} \in \mathbb{R}^{N \times 3}$: Input point cloud,
$N \in N^{+}$: Target sample size
$\gamma \in [0,1]$: Median ratio of lower-layer points (pre-computed)
Output:
$\boldsymbol{P}_{samples} \in \mathbb{R}^{N \times 3}$: Sampled point cloud
Function APCS(P, n, γ)

Extract z-coordinates: $z \leftarrow P[:,2]$
Compute height threhold: $z_{th} \leftarrow \frac{\min(z)+\max(z)}{2}$
Partition point cloud: $P_{low} \leftarrow \{p \in P | p_z \leq z_{th}\}$; $P_{high} \leftarrow \{p \in P | p_z > z_{th}\}$;
Calculate actual ratio: $\alpha \leftarrow \frac{|P_{low}|}{|P|}$
if $\alpha < \gamma$ then
  $n_{low} \leftarrow \lfloor \gamma n \rfloor$;
  $n_{high} \leftarrow n - n_{low}$;
  if $|P_{low}| \geq n_{low}$ then
    $S_{low} \leftarrow UniformSubample(P_{low}, n_{low})$;
  else
    $S_{low} \leftarrow OverSubample(P_{low}, n_{low})$;
  if $\left|P_{high}\right| \geq n_{high}$ then
    $S_{high} \leftarrow UniformSubample\left(P_{high}, n_{high}\right)$;
  else
    $S_{high} \leftarrow OverSubample\left(P_{high}, n_{high}\right)$;
  $P_{sampled} \leftarrow Concatenate\left(S_{low}, S_{high}\right)$;
else
  if $|P| \geq n$ *then*
    $P_{sampled} \leftarrow UniformSubample(P, n)$;
  else
    $P_{sampled} \leftarrow OverSubample(P, n)$;
return $P_{sampled}$

3.3. Loss Function

The loss function is used to measure the discrepancy between the output point cloud and the ground truth point cloud. In this study, we employ the Chamfer distance [38] (CD), which is commonly used in most point cloud completion methods, as the loss function. The CD computes the average shortest distance between the output point cloud S_1 and the ground truth point cloud S_2, and is defined by the following formula:

$$CD(S_1, S_2) = \frac{1}{|S_1|} \sum_{x \in S_1} \min_{y \in S_2} \| x - y \|_2 + \frac{1}{|S_2|} \sum_{y \in S_2} \min_{x \in S_1} \| y - x \|_2 \tag{5}$$

The first term of this formula encourages the output point cloud to be as close as possible to the points in the ground truth point cloud, while the second term ensures that the output point cloud adequately covers the ground truth point cloud.

To further improve the accuracy of individual tree point cloud completion and the structural integrity, we proposed a loss function consisting of three components: d_1, d_2, and d_3, weighted by hyperparameters α and β. The formula for the loss function is as follows:

$$L\left(Y_{seed}, Y_{dense}, Y_{dense'}, Y_{gt}, Y_{gt'}\right) = d_1\left(Y_{seed}, \widetilde{Y}_{gt}\right) + \alpha d_2\left(Y_{dense}, Y_{gt}\right) + \beta d_3\left(Y_{dense'}, Y_{gt'}\right) \quad (6)$$

The first term d_1 represents the distance between the seed point cloud Y_{seed} and the downsampled label point cloud $\widetilde{Y}_{gt}$. The second term d_2 calculates the distance between the completed point cloud Y_{dnese} and the ground truth label point cloud Y_{gt}. The third term d_3 computes the distance between the completed point cloud's lower 25% height points $Y_{dense'}$ and the corresponding portion of the label point cloud $Y_{gt'}$. In addition to the common loss functions, we introduced d_3 to enhance the model's focus on the lower tree canopy during the completion process. In most trees, the lower 25% of their height is sufficient to cover the portion of the tree at or below breast height, which is critical for accurately reflecting important structural features of the tree.

3.4. Point Cloud-Based Diameter at Breast Height Measurement Method

For point cloud data, higher-quality point clouds are more conducive to accurately calculating the diameter at breast height (DBH). TLS data has higher point cloud density and uniform distribution, providing good data support for point cloud-based DBH calculation. In contrast, ULS data is affected by factors such as occlusion, resulting in lower data quality in the lower canopy, which compromises the reliability of DBH calculation when used directly. To address this issue, we apply a point cloud completion method to effectively restore the tree structure in UAV point clouds. However, due to the potential large areas of missing data in ULS, the complexity of tree completion is higher, and the completed point clouds may still fail to meet the requirements for DBH calculation. Therefore, during the calculation process, we introduce certain condition checks to ensure that only higher-quality data is used for DBH measurement, thus improving the reliability of the results.

This paper develops a DBH calculation method [42] based on trunk slicing, referring to existing methods. The calculation process is shown in Figure 8. The approach mainly includes the following steps:

(1) Point cloud slices are extracted from the trunk at a height between 1.25 m and 1.35 m. To address potential noise issues in the original data, denoising is performed based on the consistency of the point cloud normal vectors. Since the normal vectors of the trunk point cloud are relatively stable, while the normal vectors of noise points fluctuate significantly, this paper calculates the rate of change of the point cloud's normal vector direction and removes points with large gradient changes to improve data quality. Additionally, to ensure the reliability and stability of the slices, the number of valid points within a slice must be no less than 25. If the number of points is insufficient, the slice thickness is appropriately increased to ensure the data is adequate to support subsequent calculations.

(2) After the preprocessing of the sliced point cloud, it is projected onto the XOY plane, and its convex hull is calculated to generate a polygon representing the tree trunk boundary. To further optimize the boundary shape, Gaussian smoothing is applied to the boundary points to reduce the impact of local outliers on measurement accuracy, making the extracted tree trunk contour more stable.

(3) To improve the stability and robustness of DBH measurement, this paper randomly generates 10 sets of lines that vertically pass through the centroid of the point cloud's

outer circle and calculates the length of the intersection between the lines and the tree trunk boundary's convex hull.

(4) The quality is assessed by calculating the standard deviation and mean of the 10 measurement values. If the standard deviation is greater than the set threshold ($\sigma < 0.15 \times$ mean measurement value), the slice point cloud is considered unable to accurately represent the tree trunk, and the DBH measurement result for that tree is discarded. Otherwise, the average of the 10 measurement results is taken as the final DBH measurement value.

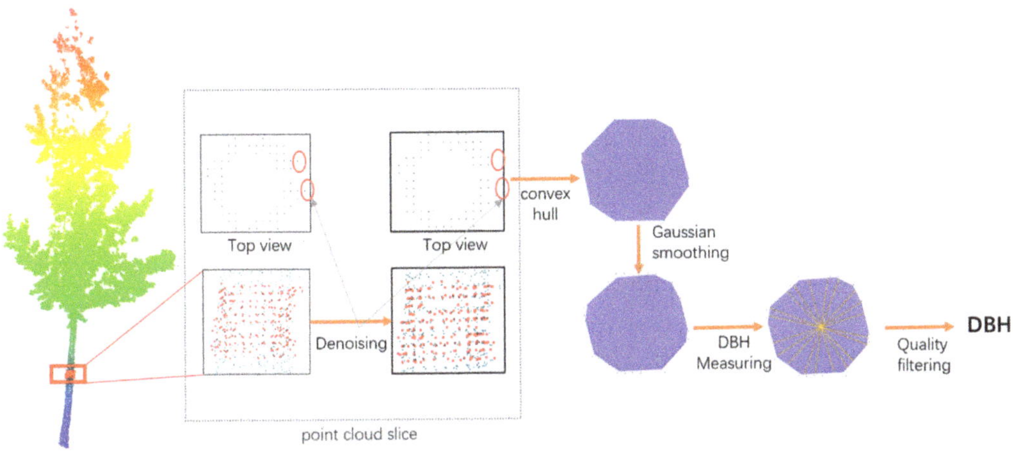

Figure 8. DBH measurement process. The blue points within the dashed box represent the point cloud slice of the tree trunk itself, while the red points represent the point cloud slice within the DBH range.

4. Results

4.1. Evaluation Metrics

Based on existing point cloud completion methods and commonly used evaluation metrics, this study selected L1 Chamfer distance (CD-L1), L2 Chamfer distance (CD-L2), and F1 score as the evaluation criteria for the point cloud completion performance. Specifically, CD-L1 and CD-L2 measure the difference between the completed point cloud and the ground truth point cloud. Smaller values indicate better completion performance by the network. The F1 score reflects the proportion of correctly reconstructed points in the completed point cloud; a higher score indicates better completion results. Furthermore, in conjunction with the point cloud-based diameter measurement method, the standard deviation between the completed and the actual tree diameter is quantitatively assessed. A smaller standard deviation suggests that the completed point cloud's diameter is closer to the actual measurement, indicating higher data consistency. This standard deviation provides an additional qualitative measure of the completion performance, particularly for low-to-medium density UAV point cloud data.

4.2. Experimental Setup

The hardware environment for the experiments in this study consists of an Intel(R) Xeon(R) E5-2650 v4 CPU, 128 GB of memory, and four 12 GB NVIDIA GeForce RTX 4090 GPUs; and one 24 GB NVIDIA RTX 4090 GPU; the operating system is Ubuntu 22.04. This overall hardware configuration provides a solid foundation for the experiments, ensuring the efficiency and stability of the training process. The experimental code was implemented using the PyTorch (version 1.11.0 + cu113) deep learning framework. The model was trained for 150 epochs using the AdamW optimizer (version from PyTorch, developer: Meta AI, Menlo Park, California, USA), with an initial learning rate of 0.0001, which decays by half every 50 epochs. The batch size was set to 12.

4.3. *Experient Results of Point Cloud Completion Models*

4.3.1. Experiments on FOR-Instance Dataset

To validate the effectiveness of the proposed method, we conducted comparative experiments on five typical forest scenarios from the FOR-instance dataset: CULS, NIBIO, RMIT, SCION, and TEWIEN. SeedFormer was used as the baseline model, and performance comparisons were made by incorporating the improvements introduced in this study. Table 3 presents the quantitative evaluation results.

Table 3. Comparison of CD-L1 before and after combining SeedFormer with the proposed strategy on the FOR-instance tree completion dataset. The metrics used are L1 CD $\times 1000$ (lower is better), L2 CD $\times 1000$ (lower is better) and F-Score@1% (higher is better).

Region	SeedFormer			Ours		
	L1_CD	L2_CD	FScore	L1_CD	L2_CD	FScore
CULS	3.61	0.43	97.39	3.75	0.47	96.19
NIBIO	3.95	0.51	95.66	4.18	0.59	94.61
RMIT	10.32	2.97	58.47	9.74	3.06	62.91
SCION	4.23	0.56	94.95	4.40	0.63	93.94
TEWIEN	6.51	1.24	84.04	6.57	1.38	83.34
Average	5.72	1.15	86.10	5.73	1.23	86.20

Experimental results show that the overall completion performance of the method in this paper is comparable to SeedFormer, with the F-Score being almost the same (86.20% vs. 86.10%), but with slightly higher L1/L2-CD (<3%). This is mainly because the method in this paper focuses more on completing the lower canopy point clouds (e.g., the tree base), which are sparse in real point clouds. As a result, when calculating the symmetric CD, the completed point cloud cannot be fully covered by the real point cloud, slightly increasing the error value.

In different forest types, the method in this paper performs well in scenes such as coniferous forests (SCION) and deciduous forests (TEWIEN), optimizing the completion of lower-layer point clouds and improving physical space coverage. However, in the low-density point cloud RMIT scene, the completion performance of both methods significantly decreases. This may be due to the sparse original scanned point clouds and insufficient training samples, making it difficult for the model to learn complex occlusion patterns. This further highlights the significant impact of point cloud quality on completion performance.

In addition, Figure 9 shows that the method in this paper can better restore the lower trunk structure, effectively alleviating the collapse issue of SeedFormer in sparse areas. However, both methods still have limitations in restoring high noise and small-scale structures (such as fine branches). Future work could explore multi-scale feature fusion or physical prior constraints to improve the detail fidelity and overall stability of point cloud completion.

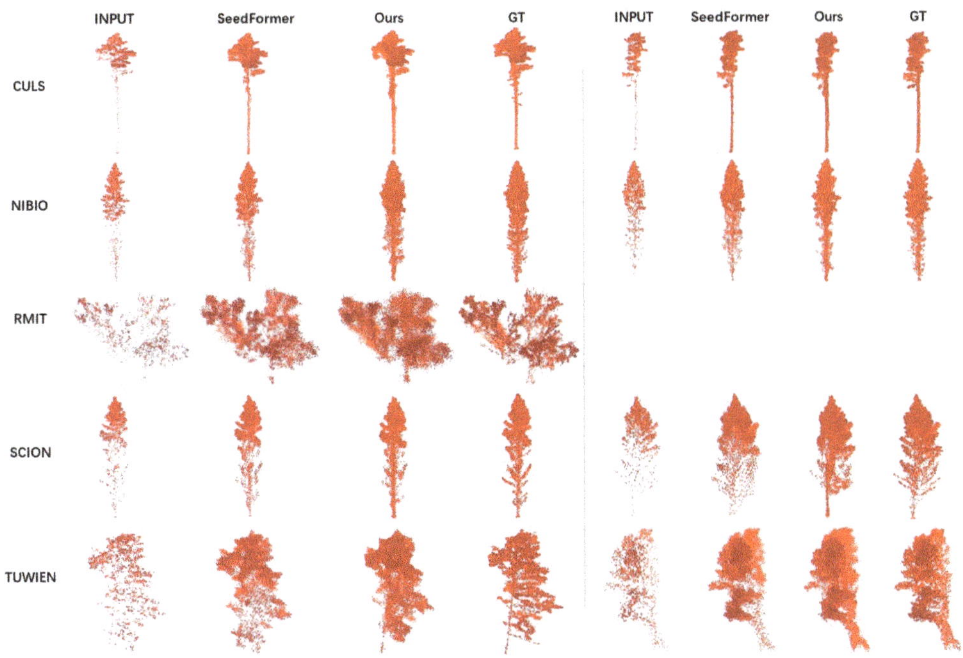

Figure 9. Model completion results based on the FOR-instance dataset. Since the RMIT region test set contains only one sample, only one result is displayed in the effect image.

4.3.2. Experiments on Xiong'an Dataset

We conducted a comparative experiment on 13 plots from the Xiong'an dataset. Using SeedFormer as the baseline model, we performed a performance comparison by incorporating the improved methods proposed in this study. Table 4 presents the quantitative evaluation results.

Table 4. Comparison of L1 CD before and after combining SeedFormer with the proposed strategy on the Xiong'an tree completion dataset. The metrics used are L1 CD $\times 1000$ (lower is better), L2 CD $\times 1000$ (lower is better) and F-Score@1% (higher is better).

PlotID	SeedFormer			Ours		
	L1_CD	L2_CD	FScore	L1_CD	L2_CD	FScore
150	5.49	0.60	93.53	5.36	0.57	94.06
151	6.93	1.32	82.75	6.90	1.39	82.28
152	16.12	5.45	30.36	14.00	4.50	49.97
153	12.42	3.86	53.71	11.45	3.21	59.31
155	11.90	2.82	70.37	11.40	2.55	73.92
157	9.99	2.02	77.86	10.06	2.28	72.94
158	7.37	1.04	84.18	7.43	0.99	85.01
159	8.37	1.52	79.29	8.13	1.46	80.32
160	8.27	1.32	82.23	8.44	1.33	82.23
161	12.08	3.30	57.05	10.45	2.29	74.77
162	9.48	1.83	78.96	9.02	1.55	79.55
163	10.38	2.09	77.33	9.04	1.51	78.02
164	11.34	2.70	71.85	11.32	2.63	72.36
Average	10.01	2.30	72.65	9.46	2.02	75.75

From the overall results, in the Xiong'an dataset, our method outperforms SeedFormer in the three evaluation metrics: L1_CD, L2_CD, and FScore-0.01. Specifically, L1_CD decreased from 10.01 to 9.46, L2_CD decreased from 2.30 to 2.02, and FScore-0.01 increased from 72.65% to 75.75%. This performance improvement may be attributed to the severe point cloud loss in the trunk region of the Xiong'an dataset, where our hierarchical random sampling strategy and loss function constraints effectively mitigated the issue of missing lower canopy point clouds, thereby enhancing the integrity and accuracy of the local geometric structure to some extent. It is worth noting that, under the same configuration, the completion performance on the FOR-instance dataset is generally superior to that on the Xiong'an dataset. This phenomenon can likely be explained by two factors: first, the more severe point cloud loss in the trunk region of the Xiong'an dataset, which limits the overall completion effect; second, the point clouds used in this experiment (drone point clouds) and the label point clouds (ground-based point clouds) are not from the same source, which may introduce some mismatching errors and affect the final completion results. The figure below shows the point cloud completion results of the two models in typical scenes. The visualization further validates the enhancing effect of the hierarchical sampling strategy on point cloud generation in sparse areas, and also demonstrates that, in the case of missing lower canopy point clouds, our method can still effectively recover key trunk structures, thus improving the overall integrity and quality of the point cloud.

4.3.3. Comparison of Sampling Methods Experiment

In the field of point cloud completion, the choice of sampling strategy significantly impacts the quality of the completion results. This experiment, based on the SeedFormer model, employs four different sampling strategies: Random Sampling (RS) [48], Furthest Point Sampling (FPS) [49], Adaptive Density Sampling (ADS) [50], and the Layered Random Sampling (LRS) proposed in this paper. Comparative experiments were conducted across multiple forest scenes in the FOR-instance dataset.

Table 5 shows that, despite using the same baseline model, different sampling strategies exhibit significant differences in point cloud detail representation. According to the data in the table, LRS (5.73) performs better than FPS (6.29) and ADS (5.92) in terms of L1 CD, second only to RS (5.71). This indicates that LRS can effectively reduce the distance error of the point cloud across the entire range. Although the F-Score of LRS (86.20%) is slightly lower than that of RS (86.54%), it is clearly superior to FPS (82.47%) and ADS (85.04%), suggesting that it has higher accuracy under the 1% threshold and is better at retaining key structural information of the point cloud.

Table 5. Comparison results of different sampling methods based on SeedFormer in various forest scenes of the FOR-instance dataset. The metrics used are L1 CD $\times 1000$ (lower is better), L2 CD $\times 1000$ (lower is better) and F-Score@1% (higher is better).

Method	CULS	NIBIO	RMIT	SCION	TUWIEN	L1_CD	L2_CD	FScore-0.01 (%)
RS	3.89	4.17	9.75	4.3	6.43	5.71	1.17	86.54
FPS	3.66	4.46	11.33	4.77	7.25	6.29	1.44	82.47
ADS	7.74	4.27	10.22	4.49	6.92	5.92	1.33	85.04
LRS (ours)	3.75	4.18	9.74	4.40	6.57	5.73	1.21	86.20

Combining Figure 10, it can be seen that the RS can better restore the tree contours. However, due to its random nature, in regions where the tree point cloud is sparse, RS may result in poor completion performance in low-density areas. This is because RS cannot effectively ensure an even distribution in sparse regions, which in turn affects the quality of point cloud completion. Through a combination of visual results and quantitative analysis, it can be concluded that the LRS proposed in this paper significantly improves the completion quality of medium- and low-density point clouds in the For-instance dataset, demonstrating strong generalization ability across different forest scenes and providing stable completion results.

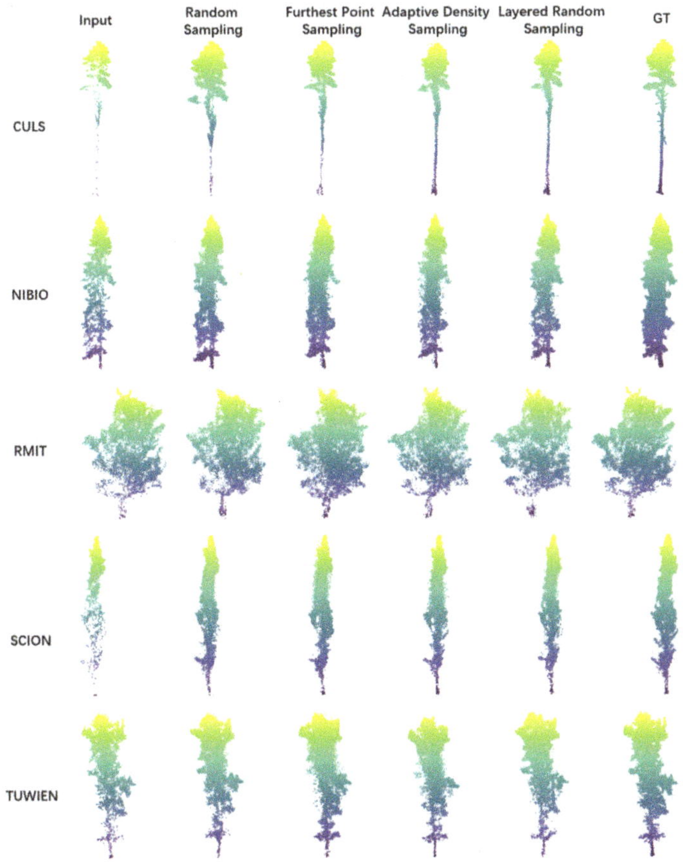

Figure 10. Completion results of different sampling methods based on SeedFormer in various forest scenes of the FOR-instance dataset. The image assigns colors to the trees based on their height.

In addition, compared to traditional RS methods, the LRS proposed in this paper does not introduce significant additional computational overhead. Although the method adds calculations of point cloud height information and layering operations, these operations only involve calculating the median and point cloud ratio, and their computational complexity is comparable to that of traditional methods. Therefore, the overall computational complexity does not increase significantly. The simple calculations in the data pre-processing stage do not affect the efficiency of the sampling process, and the random supplementation operations do not add extra computational burden, allowing the method to effectively maintain computational efficiency.

4.4. Experimental Evaluation of Point Cloud Completion in Improving the Reliability of DBH Measurement

4.4.1. Experiments on FOR-Instance Dataset

This experiment, based on five forest scenes from the FOR-instance dataset, compares the improvement effect of SeedFormer combined with our proposed method in the DBH measurement task. The evaluation metrics include root mean square error (RMSE) and the number of trees with undetected DBH (N_drop). The experimental results are shown in Table 6.

Table 6. RMSE between the predicted DBH values obtained from different point clouds and the reference values in the FOR-instance dataset (unit: cm). In the table, INPUT represents the input point cloud from the dataset, GT (ground truth) refers to the corresponding real tree point cloud, Completion denotes the completed model, and N_{drop} indicates the number of trees with undetected DBH. A smaller N_{drop} value signifies more stable DBH calculation results.

Region	Input		Completion (Seedformer)		Completion (Ours)		GT	
	RMSE	N_{drop}	RMSE	N_{drop}	RMSE	N_{drop}	RMSE	N_{drop}
CULS	3.56	35	3.84	19	3.03	15	3.12	16
NIBIO	5.55	327	5.41	251	5.31	214	5.44	216
RMIT	-	4	-	4	-	4	-	4
SCION	4.60	91	5.73	70	5.67	68	4.62	68
TEWIEN	14.59	57	14.03	45	12.70	44	11.47	43
Average	6.43	514	6.60	389	6.11	345	6.18	347

In the RMIT region, only four trees met the initial screening criteria. However, none of these trees satisfied the conditions during the DBH measurement process, making it impossible to compute a valid RMSE.

Currently, point cloud-based DBH measurement methods remain immature, making it challenging to ensure accurate DBH estimation for every tree. To evaluate the effectiveness of different methods, we introduce the N_{drop} metric, which represents the number of point clouds discarded due to unsuccessful DBH measurement. A lower N_{drop} value indicates that more trees have successfully obtained DBH measurements under the same sample size, thereby reflecting higher measurement stability.

Among all methods, the N_{drop} value for the input point cloud (Input) is the highest, likely due to the sparsity of the point cloud, which results in the lowest number of trees with a measurable DBH. After applying our method, the N_{drop} value is almost identical to that of GT, indicating that our approach effectively retains most of the useful data and performs well in reducing the number of trees with undetermined DBH.

Compared to SeedFormer, our method achieves a lower N_{drop} value, suggesting that the enhancement strategies improve the quality of the lower canopy region, making it more consistent with GT.

From the average results across all regions, the combination of SeedFormer with our enhancement method significantly reduces the N_{drop} value, while the RMSE is slightly lower than that of SeedFormer (6.11 vs. 6.43). This result indicates that our method effectively enhances the completion model's ability to recover sparse lower-layer point clouds of trees, thereby significantly increasing the number of trees with measurable DBH.

Figure 11 illustrates the scatter distribution of individual tree DBH predictions versus reference values for different point clouds.

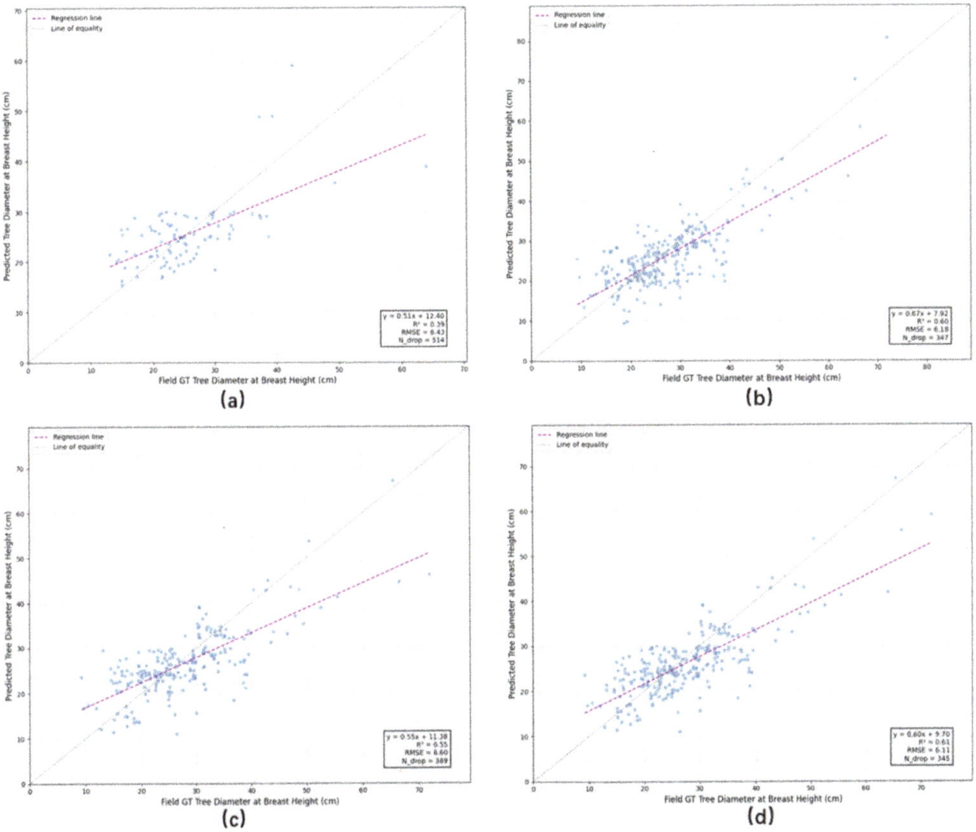

Figure 11. (**a**–**d**) Illustrate the scatter distribution of individual tree DBH predictions versus reference values across different point clouds in the FOR-instance dataset. (**a**) Shows the scatter plot of DBH predictions versus reference values obtained from sparse point clouds. (**b**) Presents the scatter plot using GT (ground truth) point clouds. (**c**) Depicts the scatter plot obtained from point clouds completed by SeedFormer. (**d**) Illustrates the scatter plot of DBH predictions based on point clouds enhanced with our proposed strategy on top of SeedFormer. N_{drop} in the figure represents the number of trees excluded due to the inability to measure DBH from the point clouds.

4.4.2. Experiments on Xiong'an Dataset

This experiment, based on 13 sample plots in the Xiong'an region, evaluates the improvement in DBH measurement performance when combining SeedFormer with our proposed method. The evaluation metrics include RMSE and the number of trees with undetected DBH (N_{drop}). The experimental results are shown in Table 7.

This experiment compares the performance of different methods in DBH measurement, evaluating them based on two metrics: RMSE and N_{drop}.

From the RMSE results, the measurement method based on the original point cloud (Input) achieves an RMSE of 3.64. However, due to missing and sparse input point cloud data, N_{drop} reaches as high as 407, indicating that a large number of trees cannot be effectively measured. In contrast, GT, serving as the reference value, has the lowest RMSE (1.22) and the smallest N_{drop} (51), demonstrating that a complete point cloud significantly improves DBH measurement accuracy and data usability.

Table 7. RMSE between predicted and reference DBH values for individual trees in the Xiong'an dataset, in cm. N_{drop} represents the number of trees with undetected DBH. A smaller N_{drop} indicates more stable DBH measurement results.

PlotID	Input		Completion (Seedformer)		Completion (Ours)		GT	
	RMSE	N_{drop}	RMSE	N_{drop}	RMSE	N_{drop}	RMSE	N_{drop}
150	3.46	39	5.53	16	5.55	11	1.63	7
151	3.62	40	5.54	7	5.14	3	1.29	1
152	-	33	5.57	12	6.2	7	1.24	4
153	-	38	4.31	3	2.84	3	0.93	0
155	-	35	3.58	24	4.12	24	1.26	2
157	-	38	3.14	12	2.99	10	1.07	4
158	5.06	23	6.38	4	4.37	1	1.05	1
159	-	28	4.39	18	4.34	12	1.35	8
160	3.17	37	4.75	8	4.77	3	1.19	2
161	1.17	27	8.49	21	5.68	17	1.75	17
162	-	19	4.62	9	3.18	6	1.69	5
163	-	29	5.35	8	4.42	0	0.67	0
164	-	21	6.15	15	6.64	8	0.68	0
Average	3.64	407	5.17	157	4.73	105	1.22	51

Among point cloud completion methods, the SeedFormer approach achieves an RMSE of 5.17, with N_{drop} reduced to 157, indicating that it enhances point cloud quality to some extent, enabling more trees to be measured. However, the relatively high RMSE suggests that structural deviations in the completed point cloud still contribute to measurement errors. When combined with our proposed enhancement method, the RMSE further decreases to 4.73, and N_{drop} drops to 105, outperforming SeedFormer on both metrics. This demonstrates that our approach not only reduces the number of undetected trees but also improves DBH prediction accuracy, bringing the measurement results closer to GT. Overall, our method exhibits better point cloud completion capability and greater stability in DBH measurement. Figure 12 illustrates the scatter distribution of individual tree DBH predictions versus reference values for different point clouds.

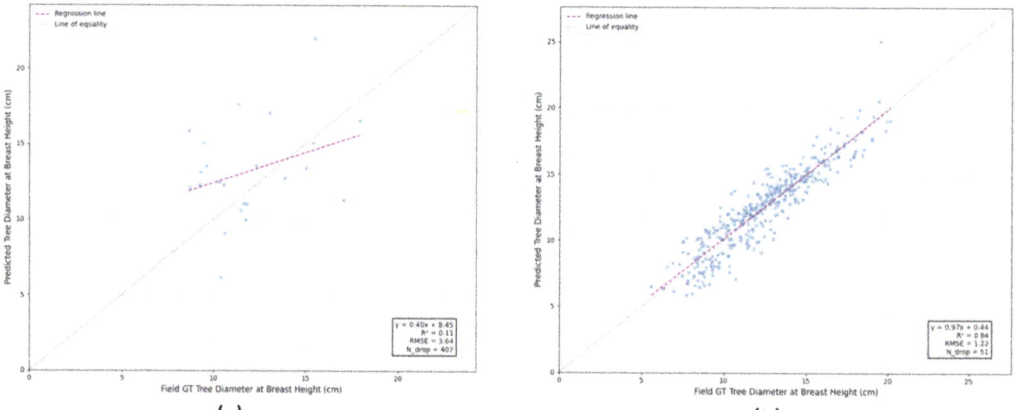

Figure 12. *Cont.*

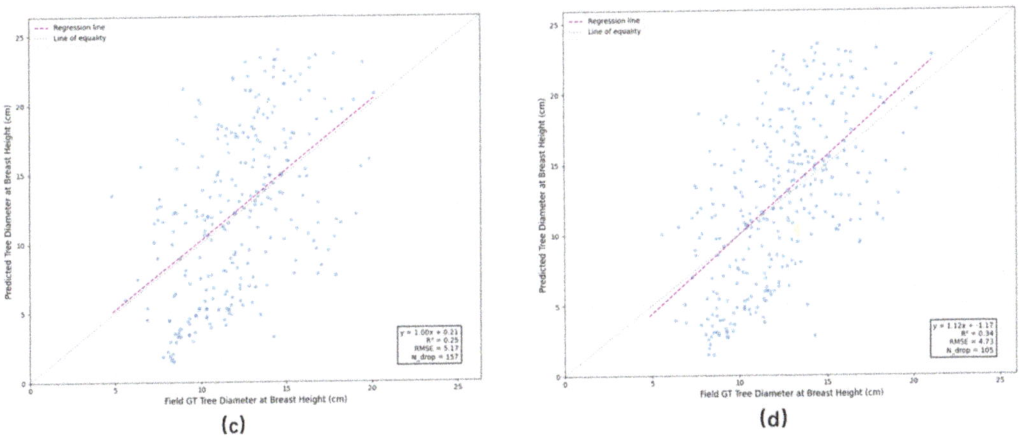

Figure 12. (**a–d**) Illustrate the scatter distribution of individual tree DBH predictions versus reference values across different point clouds in the Xiong'an dataset. (**a**) Shows the scatter plot of DBH predictions versus reference values obtained from sparse point clouds. (**b**) Presents the scatter plot using GT (Ground Truth) point clouds. (**c**) Depicts the scatter plot obtained from point clouds completed by SeedFormer. (**d**) Illustrates the scatter plot of DBH predictions based on point clouds enhanced with our proposed strategy on top of SeedFormer. N_{drop} in the figure represents the number of trees excluded due to the inability to measure DBH from the point clouds.

5. Discussion

5.1. Selection of Point Cloud Completion Models

In the selection of point cloud completion models, this study first compares and analyzes several representative point cloud completion models under optimal parameter settings, including PCN, GRNet, PoinTr, AdaPoinTr, SnowFlakeNet, and SeedFormer. Experiments were conducted on the FOR-instance dataset. The experimental results are shown in Table 8.

Table 8. Comparison of point cloud completion results using different models in various forest scenes of the FOR-instance dataset. The metrics used are L1 CD ×1000 (lower is better), L2 CD ×1000 (lower is better) and F-Score@1% (higher is better).

Method	CULS	NIBIO	RMIT	SCION	TUWIEN	L1_CD	L2_CD	FScore-0.01 (%)
PCN [29]	7.47	6.55	51.41	7.47	15.47	17.67	19.52	58.9
GRNet [18]	5.45	5.37	14.33	5.47	7.93	7.71	2.87	78.3
PoinTr [34]	5.01	4.67	13.11	5.02	8.37	7.24	2.08	76.03
AdaPoinTr [51]	5.77	5.14	23.47	5.67	9.78	9.97	6.26	72.33
SnowFlakeNet [36]	3.77	4.1	10.48	4.33	6.01	5.86	1.18	85.18
SeedFormer [37]	3.61	3.95	10.32	4.34	6.51	5.72	1.15	86.1

Based on the experimental results, PCN and GRNet are earlier point cloud completion methods. PCN uses a point-based encoder–decoder structure, while GRNet employs voxel-based global feature extraction. Both methods focus more on global structure during the completion process, but they have shortcomings in modeling local details, making it difficult to restore the complete branch and leaf structure, resulting in suboptimal overall completion performance. PoinTr and AdaPoinTr introduce the Transformer architecture, which more effectively models both global and local features of point clouds. In most scenarios, PoinTr outperforms GRNet in L1_CD and L2_CD (e.g., in the RMIT area, PoinTr's

L1_CD is 13.11, while GRNet's is 14.33), indicating better performance in handling complex point cloud structures. Notably, AdaPoinTr demonstrates significant advantages in completing complex, incomplete point clouds on public datasets (such as PCN and Completion3D) due to its adaptive query generation mechanism and denoising capabilities. The average CD error is reduced by approximately 12%, with higher training efficiency and generalization ability. However, the experimental results also show a deviation in AdaPoinTr's performance in completing medium- and low-density forestry point clouds compared to theoretical expectations. This could be due to a conflict between the denoising mechanism in AdaPoinTr and the sparse structures in the experimental data, resulting in some sparse points being mistakenly classified as noise during the denoising process. In particular, in medium- and low-density point cloud scenes collected by drones, the lower parts of the tree trunks were mistakenly filtered out due to sparse scanning, reducing the model's geometric accuracy and affecting the integrity of the tree structure. In contrast, SnowFlakeNet and SeedFormer significantly outperform other methods, especially in the restoration of tree crowns. Compared to SnowFlakeNet, SeedFormer also demonstrates better continuity in trunk restoration. Both methods, based on the Transformer structure, more accurately recover point cloud details through a layer-by-layer refinement approach and show stronger adaptability in complex forest environments, making the completed point clouds closer to the real distribution. For example, in the RMIT area, the L1_CD completion result of the PCN model is 51.41, while SeedFormer significantly reduces it to 10.32, showing its advantages in point cloud completion.

We have visually presented the completion results of the above-mentioned methods. From Figure 13, we can intuitively observe that PCN and GRNet only capture the basic shape of the trees, with the overall structure being quite vague. PoinTr and AdaPoinTr are able to generate basic tree shapes, but in some cases, the form is inaccurately captured, and their ability to recover branches and leaves is poor. SnowFlakeNet and SeedFormer clearly outperform the other methods, especially in the finer recovery of the tree canopy. SeedFormer also demonstrates better continuity in recovering the tree trunk. By comparison, we observe that the point cloud completed by SeedFormer is closest to the real point cloud, indicating its greater advantage in modeling tree topology and recovering local details. Ultimately, we selected SeedFormer as our baseline model, and when combined with the layered sampling and loss constraints proposed in this paper, it effectively improves the completeness of point cloud generation in the sparse lower areas of the trees, without significantly increasing geometric errors.

5.2. Analysis of Sampling Method Selection

In point cloud completion tasks, the sampling strategy has a significant impact on the spatial distribution characteristics of the input point cloud, which in turn affects the reconstruction performance of the completion network. In this study, we compared four different sampling strategies in the experiments: random sampling (RS), Furthest Point Sampling (FPS), Adaptive Density Sampling (ADS), and Layered Random Sampling (LRS), and analyzed the practical application of these sampling strategies.

Although the RS can uniformly cover the entire point cloud, it does not take into account the density distribution characteristics of the point cloud, which may result in insufficient sampling in low-density areas (such as tree trunks and fine branches), leading to significant information loss. When the completion network receives this uneven input, it may tend to fit the structure of high-density regions, while the completion performance in low-density areas is poorer. Therefore, in metrics such as L1 CD and F-Score, RS often shows better global matching performance, as the completed point cloud can align more accurately with the full point cloud in high-density areas, thus reducing global errors. However, in the sparsely populated parts of tree point clouds, RS struggles to provide enough structural information, resulting in incomplete preservation of the overall structure of the tree.

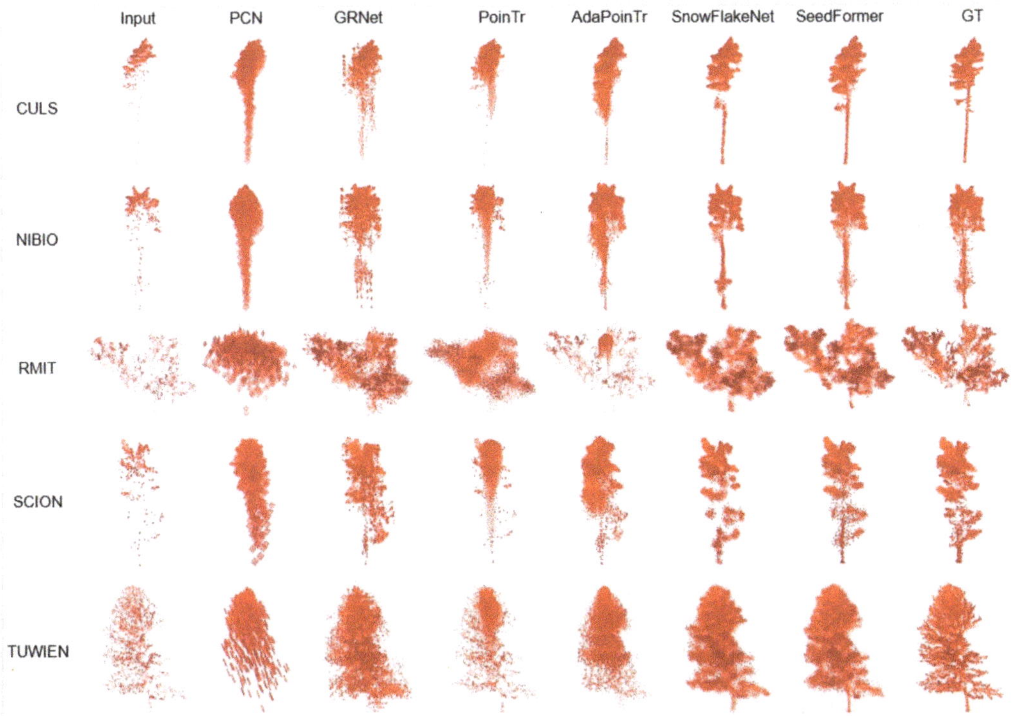

Figure 13. Completion results of different point cloud completion methods in various forest scenes of the FOR-instance dataset.

FPS is typically used for uniform point cloud sampling, achieving an even distribution in geometric space. However, for forest point clouds, especially UAV LiDAR data, the point cloud density in the canopy area is much higher than in the tree trunk area. When FPS samples in high-density regions, it tends to ignore the density characteristics of the point cloud, leading to insufficient sampling in low-density areas (such as the tree trunk), resulting in an incomplete trunk structure after completion. This is one of the reasons why FPS performs poorly in metrics like L1 Chamfer distance and F-Score.

ADS improves the completion effect by applying weighted sampling to the point cloud, giving more sampling points to low-density areas. From the visualization results, ADS achieves similar completion performance to LRS. However, compared to the LRS we propose, ADS cannot ensure the minimum proportion of low points in the point cloud, which may lead to incomplete tree trunks after completion. Additionally, since ADS primarily relies on point cloud density for adjustments, it is susceptible to local high-density outliers, causing over-sampling in certain areas, while important structures in truly low-density regions (such as the junction between the tree trunk and canopy) may still be ignored. In contrast, the LRS we propose can be seen as a special density-based sampling method that integrates the features of mid-to-low-density tree point clouds. By ensuring the proportion of low points while maintaining even coverage of the overall structure, this method is more suitable for different forest datasets. Through its layered strategy, it compensates for the shortcomings of RS and FPS, providing better stability in point cloud completion in low-density areas compared to ADS.

In UAV point clouds, the canopy part is usually denser, while the trunk part is relatively sparse. When focusing on the overall structure of the tree for completion, the key lies in whether the sparse regions can obtain enough sampling points. ADS requires calculating the local density of the point cloud during sampling, especially in the canopy area, which may introduce additional computational overhead and increase processing time. In contrast, RS does not require extra computations, resulting in lower computational costs and allowing sampling to be completed in a very short time. In the specific scenario of mid-to-low-density UAV tree point clouds, our method achieves a completion effect similar to ADS in a shorter time, balancing both efficiency and completion quality:

5.3. Limitations and Outlook

5.3.1. Improvement of Sampling Methods

The hierarchical random sampling method proposed in this paper performs excellently in the completion task of low- to medium-density UAV point clouds, improving completion quality while ensuring computational efficiency. Compared to ADS, this method avoids the additional computational overhead caused by density calculation, enabling faster sampling and providing more stable completion results in low-density areas. However, this method still relies on the rationality of the hierarchical ratio, and its completion performance may exhibit some uncertainty in scenarios with significant differences in tree species or point cloud density distribution. Therefore, future research could combine the spatial distribution characteristics of point clouds to further optimize the hierarchical strategy, making the sampling method more adaptive and improving the generalization ability of the completion model.

5.3.2. Enrichment of Evaluation Metrics

In this study, we primarily use CD and F-Score as evaluation metrics for point cloud completion. These metrics can assess the geometric similarity between the completed point cloud and the ground truth point cloud from both global and local matching perspectives, but they do not directly evaluate the integrity, morphological consistency, or biological plausibility of tree structures. Ecological plausibility is an important factor that cannot be overlooked in forest point cloud completion tasks. To further enhance the comprehensiveness of the evaluation, future research could introduce more ecologically meaningful metrics. For example, the use of trunk taper error to measure whether it aligns with the gradual narrowing feature of real trees, or assessing the overall morphological consistency through trunk axis deviation.

5.3.3. Combined Usage Strategy of ULS and TLS Data

In large-scale and complex real-world forestry datasets (such as the Xiong'an dataset), the combined use of multi-source point cloud data remains a significant challenge. Although ULS data and TLS data are registered, directly using TLS data as labels for ULS data may still introduce substantial errors, affecting the generalization ability and transferability of the completion model. Therefore, how to reasonably integrate ULS and TLS data to fully leverage the wide-area coverage advantage of ULS and the high-precision characteristics of TLS remains an area for further exploration. Future research could employ deep learning-based multi-modal data fusion, spatial alignment optimization, or adaptive completion methods to reduce the discrepancies between different data sources and improve the applicability and robustness of point cloud completion models in complex forest environments.

6. Conclusions

This paper addresses the challenges of UAV point cloud data with low to medium density, particularly the issues of self-occlusion and sparse lower canopy point clouds. Two strategies are proposed to improve point cloud completion and maintain the structural integrity of trees. First, we perform stratification based on the statistical characteristics of the input point cloud and use a stratified random sampling method to obtain the point cloud. This ensures that the proportion of lower canopy points is maintained, which better preserves the original tree structure. Second, we incorporate the CD distance between the lower and middle layers of the point clouds into the loss function, further enhancing the model's focus on the lower canopy points and improving its ability to recover structure during the completion process. We validated our method on the FOR-instance and Xiong'an datasets. The results show that the introduction of stratified random sampling and the spatial constraint loss function significantly enhances the model's ability to recover the lower trunk structure. Although the model slightly underperforms the baseline model in terms of the symmetric Chamfer distance (CD) due to systematic errors in evaluation metrics, the visual results after applying our strategies are clearly superior to those of the original model. Additionally, we made improvements to the existing DBH measurement method and calculated the DBH on two datasets. DBH fitting was performed on the point clouds before and after completion, and the fitting results were compared with the actual measurements for analysis. Experimental results show that, after point cloud completion using the proposed strategy, the measured DBH accuracy is significantly better than the baseline model SeedFormer, further validating the effectiveness and application value of our method in tree structure completion.

Author Contributions: Conceptualization, Z.C.; methodology, Y.S.; software, Y.S.; validation, Y.S. and Z.C.; formal analysis, Y.S.; investigation, Y.S.; resources, Z.C.; data curation, Y.S. and X.X.; writing—original draft preparation, Y.S.; writing—review and editing, Z.C.; visualization, Y.S.; supervision, Z.C.; project administration, Z.C.; funding acquisition, Z.C. All authors have read and agreed to the published version of the manuscript.

Funding: This research was funded by Xiong'an New Area Science and Technology Innovation. Special Project of Ministry of Science and Technology of China (2023XAGG0065).

Data Availability Statement: The data presented in this study are available on request from the corresponding author.

Conflicts of Interest: The authors declare no conflicts of interest.

References

1. Guo, Q.H.; Liu, J.; Tao, S.L.; Xue, B.; Li, L.; Xu, G.C.; Li, W.K.; Wu, F.F.; Li, Y.M.; Chen, L.H. Perspectives and prospects of Li-DAR in forest ecosystem monitoring and modeling. *Chin. Sci. Bull.* **2014**, *59*, 459–478.
2. Véga, C.; Renaud, J.; Durrieu, S.; Bouvier, M. On the interest of penetration depth, canopy area and volume metrics to improve Lidar-based models of forest parameters. *Remote Sens. Environ.* **2016**, *175*, 32–42. [CrossRef]
3. Chen, X.; Jiang, K.; Zhu, Y.; Wang, X.; Yun, T. Individual tree crown segmentation directly from UAV-borne LiDAR data using the PointNet of deep learning. *Forests* **2021**, *12*, 131. [CrossRef]

4. Quan, Y.; Li, M.; Hao, Y.; Liu, J.; Wang, B. Tree species classification in a typical natural secondary forest using UAV-borne LiDAR and hyperspectral data. *GISci. Remote Sens.* **2023**, *60*, 2171706. [CrossRef]
5. Guo, Q.; Su, Y.; Hu, T.; Zhao, X.; Wu, F.; Li, Y.; Liu, J.; Chen, L.; Xu, G.; Lin, G. An integrated UAV-borne lidar system for 3D habitat mapping in three forest ecosystems across China. *Int. J. Remote Sens.* **2017**, *38*, 2954–2972. [CrossRef]
6. Xu, D.; Chen, G.; Jing, W. A single-tree point cloud completion approach of feature fusion for agricultural robots. *Electronics* **2023**, *12*, 1296. [CrossRef]
7. Ge, B.; Chen, S.; He, W.; Qiang, X.; Li, J.; Teng, G.; Huang, F. Tree Completion Net: A Novel Vegetation Point Clouds Completion Model Based on Deep Learning. *Remote Sens.* **2024**, *16*, 3763. [CrossRef]
8. Xu, H.; Huai, Y.; Zhao, X.; Meng, Q.; Nie, X.; Li, B.; Lu, H. SK-TreePCN: Skeleton-Embedded Transformer Model for Point Cloud Completion of Individual Trees from Simulated to Real Data. *Remote Sens.* **2025**, *17*, 656. [CrossRef]
9. Yurtseven, H.; Çoban, S.; Akgül, M.; Akay, A.O. Individual tree measurements in a planted woodland with terrestrial laser scanner. *Turk. J. Agric. For.* **2019**, *43*, 192–208. [CrossRef]
10. Xu, H.; Chen, W.; Liu, H. Single-wood DBH and tree height extraction using terrestrial laser scanning. *J. Forest Environ.* **2019**, *39*, 524–529.
11. Ko, B.; Park, S.; Park, H.; Lee, S. Measurement of tree height and diameter using terrestrial laser scanner in coniferous forests. *J. Environ. Sci. Int.* **2022**, *31*, 479–490. [CrossRef]
12. Liu, Y.; Wu, X.-Q. 3D shape completion via deep learning: A method survey. *J. Graph.* **2023**, *44*, 201–215.
13. Li, Y.; Wu, X.; Chrysathou, Y.; Sharf, A.; Cohen-Or, D.; Mitra, N.J. Globfit: Consistently fitting primitives by discovering global relations. In Proceedings of the ACM SIGGRAPH 2011, SIGGRAPH 2011, Vancouver, BC, Canada, 7–11 August 2011; pp. 1–12.
14. Kazhdan, M.; Hoppe, H. Screened poisson surface reconstruction. *ACM Trans. Graph. (TOG)* **2013**, *32*, 29. [CrossRef]
15. Nan, L.; Xie, K.; Sharf, A. A search-classify approach for cluttered indoor scene understanding. *ACM Trans. Graph. (TOG)* **2012**, *31*, 137. [CrossRef]
16. Kim, J.; Kwon, H.; Yang, Y.; Yoon, K. Learning Point Cloud Completion without Complete Point Clouds: A Pose-Aware Approach. In Proceedings of the 2023 International Conference on Computer Vision, Paris, France, 2–3 October 2023; pp. 14157–14167.
17. Li, Y.; Dai, A.; Guibas, L.; Nießner, M. Database-assisted object retrieval for real-time 3d reconstruction. In *Computer Graphics Forum*; Wiley Online Library: Hoboken, NJ, USA, 2015; pp. 435–446.
18. Xie, H.; Yao, H.; Zhou, S.; Mao, J.; Zhang, S.; Sun, W. Grnet: Gridding residual network for dense point cloud completion. In Proceedings of the European Conference on Computer Vision, Glasgow, UK, 23–28 August 2020; Springer: Berlin/Heidelberg, Germany, 2020; pp. 365–381.
19. Wang, X.; Ang, M.H.; Lee, G.H. Voxel-based network for shape completion by leveraging edge generation. In Proceedings of the IEEE/CVF International Conference on Computer Vision, Montreal, BC, Canada, 11–17 October 2021; pp. 13189–13198.
20. Yi, L.; Gong, B.; Funkhouser, T. Complete & label: A domain adaptation approach to semantic segmentation of lidar point clouds. In Proceedings of the IEEE/CVF Conference on Computer Vision and Pattern Recognition, Nashville, TN, USA, 19–25 June 2021; pp. 15363–15373.
21. Hu, T.; Han, Z.; Shrivastava, A.; Zwicker, M. Render4Completion: Synthesizing multi-view depth maps for 3D shape completion. In Proceedings of the IEEE/CVF International Conference on Computer Vision Workshops, Seoul, Republic of Korea, 27–28 October 2019.
22. Hu, T.; Han, Z.; Zwicker, M. 3D shape completion with multi-view consistent inference. In Proceedings of the AAAI Conference on Artificial Intelligence, New York, NY, USA, 7–12 February 2020; pp. 10997–11004.
23. Tang, J.; Han, X.; Pan, J.; Jia, K.; Tong, X. A skeleton-bridged deep learning approach for generating meshes of complex topologies from single rgb images. In Proceedings of the IEEE/CVF Conference on Computer Vision and Pattern Recognition, Long Beach, CA, USA, 15–20 June 2019; pp. 4541–4550.
24. Li, Y.; Yu, Z.; Choy, C.; Xiao, C.; Alvarez, J.M.; Fidler, S.; Feng, C.; Anandkumar, A. Voxformer: Sparse voxel transformer for camera-based 3d semantic scene completion. In Proceedings of the IEEE/CVF Conference on Computer Vision and Pattern Recognition, Vancouver, BC, Canada, 17–24 June 2023; pp. 9087–9098.
25. Wen, X.; Xiang, P.; Han, Z.; Cao, Y.; Wan, P.; Zheng, W.; Liu, Y. Pmp-net: Point cloud completion by learning multi-step point moving paths. In Proceedings of the IEEE/CVF Conference on Computer Vision and Pattern Recognition, Nashville, TN, USA, 19–25 June 2021; pp. 7443–7452.
26. Li, S.; Gao, P.; Tan, X.; Wei, M. Proxyformer: Proxy alignment assisted point cloud completion with missing part sensitive transformer. In Proceedings of the IEEE/CVF Conference on Computer Vision and Pattern Recognition, Vancouver, BC, Canada, 17–24 June 2023; pp. 9466–9475.

27. Rong, Y.; Zhou, H.; Yuan, L.; Mei, C.; Wang, J.; Lu, T. Cra-pcn: Point cloud completion with intra-and inter-level cross-resolution transformers. In Proceedings of the AAAI Conference on Artificial Intelligence, Vancouver, BC, Canada, 26–27 February 2024; pp. 4676–4685.
28. Qi, C.R.; Su, H.; Mo, K.; Guibas, L.J. Pointnet: Deep learning on point sets for 3d classification and segmentation. In Proceedings of the IEEE Conference on Computer Vision and Pattern Recognition, Honolulu, HI, USA, 21–26 July 2017; pp. 652–660.
29. Yuan, W.; Khot, T.; Held, D.; Mertz, C.; Hebert, M. Pcn: Point completion network. In Proceedings of the 2018 International Conference on 3D Vision (3DV), Verona, Italy, 5–8 September 2018; IEEE: Piscataway, NJ, USA, 2018; pp. 728–737.
30. Yang, Y.; Feng, C.; Shen, Y.; Tian, D. Foldingnet: Point cloud auto-encoder via deep grid deformation. In Proceedings of the IEEE Conference on Computer Vision and Pattern Recognition, Salt Lake City, UT, USA, 18–23 June 2018; pp. 206–215.
31. Wen, X.; Li, T.; Han, Z.; Liu, Y. Point cloud completion by skip-attention network with hierarchical folding. In Proceedings of the IEEE/CVF Conference on Computer Vision and Pattern Recognition, Seattle, WA, USA, 13–19 June 2020; pp. 1939–1948.
32. Huang, Z.; Yu, Y.; Xu, J.; Ni, F.; Le, X. Pf-net: Point fractal network for 3d point cloud completion. In Proceedings of the IEEE/CVF Conference on Computer Vision and Pattern Recognition, Seattle, WA, USA, 13–19 June 2020; pp. 7662–7670.
33. Vaswani, A.; Shazeer, N.; Parmar, N.; Uszkoreit, J.; Jones, L.; Gomez, A.N.; Kaiser, A.; Polosukhin, I. Attention is all you need. In Proceedings of the 31st Conference on Neural Information Processing Systems (NIPS 2017), Long Beach, CA, USA, 4–9 December 2017.
34. Yu, X.; Rao, Y.; Wang, Z.; Liu, Z.; Lu, J.; Zhou, J. Pointr: Diverse point cloud completion with geometry-aware trans-formers. In Proceedings of the IEEE/CVF International Conference on Computer Vision, Montreal, BC, Canada, 11–17 October 2021; pp. 12498–12507.
35. Zhao, H.; Jiang, L.; Jia, J.; Torr, P.H.; Koltun, V. Point transformer. In Proceedings of the IEEE/CVF International Conference on Computer Vision, Montreal, BC, Canada, 11–17 October 2021; pp. 16259–16268.
36. Xiang, P.; Wen, X.; Liu, Y.; Cao, Y.; Wan, P.; Zheng, W.; Han, Z. Snowflakenet: Point cloud completion by snowflake point deconvolution with skip-transformer. In Proceedings of the IEEE/CVF International Conference on Computer Vision, Montreal, BC, Canada, 11–17 October 2021; pp. 5499–5509.
37. Zhou, H.; Cao, Y.; Chu, W.; Zhu, J.; Lu, T.; Tai, Y.; Wang, C. Seedformer: Patch seeds based point cloud completion with upsample transformer. In Proceedings of the European Conference on Computer Vision, Tel Aviv, Israel, 23–27 October 2022; Springer: Berlin/Heidelberg, Germany, 2022; pp. 416–432.
38. Fan, H.; Su, H.; Guibas, L.J. A point set generation network for 3d object reconstruction from a single image. In Proceedings of the IEEE Conference on Computer Vision and Pattern Recognition, Honolulu, HI, USA, 21–26 July 2017; pp. 605–613.
39. Jucker, T.; Caspersen, J.; Chave, J.; Antin, C.; Barbier, N.; Bongers, F.; Dalponte, M.; van Ewijk, K.Y.; Forrester, D.I.; Haeni, M. Allometric equations for integrating remote sensing imagery into forest monitoring programmes. *Glob. Change Biol.* **2017**, *23*, 177–190. [CrossRef]
40. Xu, J.; Su, M.; Sun, Y.; Pan, W.; Cui, H.; Jin, S.; Zhang, L.; Wang, P. Tree Crown Segmentation and Diameter at Breast Height Prediction Based on BlendMask in Unmanned Aerial Vehicle Imagery. *Remote Sens.* **2024**, *16*, 368. [CrossRef]
41. Fakhri, A.; Latifi, H.; Samani, K.M.; Fassnacht, F.E. CaR3DMIC: A novel method for evaluating UAV-derived 3D forest models by tree features. *ISPRS J. Photogramm. Remote Sens.* **2024**, *208*, 279–295. [CrossRef]
42. Bogdanovich, E.; Perez-Priego, O.; El-Madany, T.S.; Guderle, M.; Pacheco-Labrador, J.; Levick, S.R.; Moreno, G.; Carrara, A.; Martín, M.P.; Migliavacca, M. Using terrestrial laser scanning for characterizing tree structural parameters and their changes under different management in a Mediterranean open woodland. *For. Ecol. Manag.* **2021**, *486*, 118945. [CrossRef]
43. Puliti, S.; Pearse, G.; Surový, P.; Wallace, L.; Hollaus, M.; Wielgosz, M.; Astrup, R. For-instance: A uav laser scanning benchmark dataset for semantic and instance segmentation of individual trees. *arXiv* **2023**, arXiv:2309.01279.
44. Puliti, S.; McLean, J.P.; Cattaneo, N.; Fischer, C.; Astrup, R. Tree height-growth trajectory estimation using uni-temporal UAV laser scanning data and deep learning. *Forestry* **2023**, *96*, 37–48. [CrossRef]
45. Kuželka, K.; Slavík, M.; Surový, P. Very high density point clouds from UAV laser scanning for automatic tree stem detection and direct diameter measurement. *Remote Sens.* **2020**, *12*, 1236. [CrossRef]
46. Wieser, M.; Mandlburger, G.; Hollaus, M.; Otepka, J.; Glira, P.; Pfeifer, N. A case study of UAS borne laser scanning for measurement of tree stem diameter. *Remote Sens.* **2017**, *9*, 1154. [CrossRef]
47. Girardeau-Montaut, D. *CloudCompare*; EDF R&D Telecom ParisTech: Paris, France, 2016.
48. Noor, S.; Tajik, O.; Golzar, J. Simple random sampling. *Int. J. Educ. Lang. Stud.* **2022**, *1*, 78–82.
49. Moenning, C.; Dodgson, N.A. *Fast Marching Farthest Point Sampling*; University of Cambridge, Computer Laboratory: Cambridge, UK, 2003.

50. Lin, Y.; Habib, A. An adaptive down-sampling strategy for efficient point cloud segmentation. In Proceedings of the ASPRS 2015 Annual Conference, Tampa, FL, USA, 4–8 May 2015.
51. Yu, X.; Rao, Y.; Wang, Z.; Lu, J.; Zhou, J. AdaPoinTr: Diverse point cloud completion with adaptive geometry-aware transformers. *IEEE Trans. Pattern Anal. Mach. Intell.* **2023**, *45*, 14114–14130. [CrossRef]

forests

MDPI

Article

Monitoring Forest Dynamics and Conducting Restoration Assessment Using Multi-Source Earth Observation Data in Northern Andes, Colombia

Carlos Pedraza [1,2], Nicola Clerici [2,*], Marcelo Villa [1], Milton Romero [3], Adriana Sarmiento Dueñas [3], Dallan Beltrán Rojas [3], Paola Quintero [4], Mauricio Martínez [4] and Josef Kellndorfer [5]

1 DreamGIS, Bogotá 110231, Colombia; carlos@dreamgis.com (C.P.); marcelo@dreamgis.com (M.V.)
2 Department of Biology, Faculty of Natural Sciences, Universidad del Rosario, Bogotá 111221, Colombia
3 4D Elements Consultores, Bogotá 111321, Colombia; 4delements.consultores@gmail.com (M.R.); adrianasarmi@gmail.com (A.S.D.); sdallan.db@gmail.com (D.B.R.)
4 Urra S.A E.S.P., Environmental Unit, Monteria 230002, Colombia; ambiental3@urra.com.co (P.Q.); mauriciojavier.ml@gmail.com (M.M.)
5 Earth Big Data LLC, Woods Hole, MA 02543, USA; josef@earthbigdata.com
* Correspondence: nicola.clerici@urosario.edu.co; Tel.: +57-313-348-57-26

Abstract: Examining the efficacy of current assessment methodologies for forest conservation and restoration initiatives to align with global and national agendas to combat deforestation and facilitate restoration efforts is necessary to identify efficient and robust approaches. The objective of this study is to understand forest dynamics (1996–2021) and assess restoration implementations at the Urra's supplying basin hydroelectric reservoir in Colombia. The processing approach integrates optical and radar Earth Observation (EO) data from Sentinel-2 and Landsat for forest mapping and multi-temporal forest change assessment (1996–2021), and a Sentinel-1 backscatter time-series analysis is conducted to assess the state of forest restoration implementations. The processing chain was scaled in a cloud-based environment using the Nebari and SEPPO software and the Python language. The results demonstrate an overall substantial decrease in forested areas in the 1996–2000 period (37,763 ha). An accuracy assessment of multi-temporal forest change maps showed a high precision in detecting deforestation events, while improvements are necessary for accurately representing non-forested areas. The forest restoration assessment suggests that the majority of the 270 evaluated plots are in the intermediate growth state (82.96%) compared to the reference data. This study underscores the need for robust and continuous monitoring systems that integrate ground truth data with EO techniques for enhanced accuracy and effectiveness in forest restoration and conservation endeavors.

Keywords: forest management; forest restoration; Sentinel-1; Sentinel-2; Landsat; time series; cloud computing

1. Introduction

In the context of the scientific and technical progresses in remote sensing, Colombia is taking advantage of the agreements made by the United Nations Framework Convention on Climate Change (UNFCCC) during the Conferences of the Parties in 2009 and 2010 (COP 15 and 16, respectively), and recently in Warsaw (COP 19), which require developing countries to establish national forest monitoring systems that allow for the quantification of greenhouse gas emissions and absorptions and changes in the extension of forests and forest carbon stocks. For this purpose, forest cover in Colombia was defined as "Land occupied mainly by trees that may contain shrubs, palms, *guaduas*, herbs and lianas, in which tree cover predominates with a minimum canopy density of 30%, a minimum canopy height (in situ) of 5 m at the time of identification, and a minimum area of 1.0 ha" [1]. The tree cover of commercial forest plantations, palm plantations, and trees planted for agricultural production are excluded. Deforestation is defined as "the direct and/or induced conversion of forest cover to another type of land cover in a given period of time", and restoration is defined as "the recovery of forest cover in areas where it was not present in the past" [1]. These definitions are consistent with the criteria defined by the UNFCCC in its 11/COP 7 decision (FAO, Terms and definitions FRA 2020. 2020) and with the definition adopted by Colombia under the Kyoto Protocol.

Thus, in 2011, the governmental Colombian Forest and Carbon Monitoring System (SMByC) evaluated different remote sensing data processing techniques for the detection of forest cover and its respective changes over time, generating, as a result, the "Protocol for digital image processing for the quantification of deforestation in Colombia at the national level—coarse and fine scale" [1] which has been the methodological approach for the historical quantification of deforestation at the national level, e.g., for the 1990–2000, 2000–2005, and 2005–2010 periods and annually in the past decade from 2011 to 2022. These guidelines have become the Colombian standard for the evaluation of spatiotemporal forested and deforested dynamics, which is applied in the case of national regulations for companies that must report their effectiveness in protecting forests.

Since the first missions that integrated sensors on satellites to generate Earth Observation data, e.g., Landsat in 1972, a large number of orbiting missions have been launched for environmental applications with different characteristics, providing information that can be integrated into multiple applications to interpret land surface patterns and monitor changes in land use and land cover [2]. Satellite-based remote sensing has been proposed as an especially cost-effective way to provide reliable data on forest change dynamics [3–5]. Multiple methods have been developed based on satellite Earth Observation; nevertheless, there are limitations associated with each approach, including how to balance the integration of multiple sensors and the available resources for their implementation [6]. Some of these challenges are related to complex patterns and rapid changes associated with land cover trajectories [7–10], inconsistencies in Earth Observation data series among different missions [11], and, specifically in tropical areas, the continuous presence of clouds, which generate uncertainties in mapping tasks. The first launch of a global pathfinder optical mission with advanced observation strategies (i.e., Landsat) has been providing consistent EO data; however, it has the limitation of cloud interference. Recently, Synthetic Aperture Radar (SAR) sensors (e.g., ALOS-1 PALSAR-1 and Sentinel-1) have provided persistent data with the advantage of cloud-free EO data, which allow multiple challenges to be resolved in time series analyses to detect and quantify forest trajectories. In fact, the use of SAR sensors to generate EO imagery has increased in recent years in tropical regions, where radars provide day/night imagery without interference from atmospheric conditions [12–14]. This impulse has started in the past decade, mainly with the staging of ALOS-1 PALSAR-1, ALOS-2 PALSAR-2, and Sentinel-1 sensors, which can obtain EO data with a high acquisition frequency in the same area of interest when multiple missions are integrated in the analysis.

In Colombia, with respect to the increasing use of EO data from SAR sensors, most of the research has been focused on wetlands and flooding mapping [1,15–17], land use/land cover mapping (LULC) [18–20], and forest mapping and deforestation monitoring [21–24]. The growing application of EO data to monitor restoration has proven that this is a cost-effective means to select restoration sites and observe their progress over time [25], evaluate success and understand restoration recovery trajectories [26], generate specific restoration metrics [26] while monitoring vegetation structure trajectory to guide adaptive management [27], and determine specific spatial and temporal changes associated with degradation and recovery [28]. Nevertheless, with the increasing application of standardized protocols and methodologies based on remote EO data for restoration and forest monitoring, it is necessary to assess the level of uncertainty and their applicability in diverse regions facing multiple challenges, such as the Colombian Andes, which is characterized by almost permanent cloud cover and complex topographies, which may limit the application of EO data. The implementation of forest restoration initiatives is considered as one of the most promising climate actions to rapidly remove CO_2 from the atmosphere, especially in tropical regions [29]; thus, analytical approaches based on remote sensing to monitor and assess forest restoration initiatives need to be improved.

The objectives of this work are as follows: (i) to detect forest change dynamics in a 25-year timeframe (the 1996–2000–2005–2010–2015–2021 period) based on optical sensors in area of influence of the hydroelectric dam (Colombia) based on national standards for forest and deforestation monitoring and (ii) to design a methodological approach to assess the status of forest restoration plots implemented as environmental offsets to compensate for impacts generated by the construction of the hydroelectric dam using a 5-year time series analysis of Sentinel-1 C-SAR imagery. Thus, this study proposes an operational monitoring and assessment approach for initiatives aimed to preserve forests, reduce deforestation, and restore forests by applying scalable and cost-effective technologies and methods.

2. Materials and Methods

2.1. Study Area

The study area, covering almost 89.000 ha, is located in the northwestern sector of Colombia, between 8°8′ N 7°47′ N and 76°27′ to 76°05′ W. It is characterized by an altitude ranging between 0 and 1250 m above sea level. Annual rainfall varies between 1.500 and 4.000 mm, with a dry season running from December to the end of March. The mean annual temperature fluctuates between 21.1 °C and 28 °C, with a warm semi-humid regime climate [30]. Within this region, three major biomes are identified: (i) The tropical wet zone biome covers 88% of the area and is the largest one, corresponding to areas where a humid rainforest predominates at elevations below 800 m above sea level (a.s.l.). (ii) The pedobiome of the tropical wet zone biome corresponds to areas where vegetation and flora types are determined by soil and humidity conditions [26]. This region is mainly represented by the Urra reservoir area, covering 9.3% of the region. (iii) The sub-Andean orobiome corresponds to an altitude range between 800 and 1800 m a.s.l., where a temperate climatic zone predominates along with the following humidity provinces: humid, semi-humid, and super humid [30].

The study area includes 90 villages within the municipalities of Tierralta and Valencia in the department of Córdoba and within Turbo, Apartadó, Carepa, and San Pedro de Urabá in the department of Antioquia. In 1996, the construction of the Urra hydropower facility, which collects the water of the Sinú river, began in the region, and the operations began in 2000 (Figure 1).

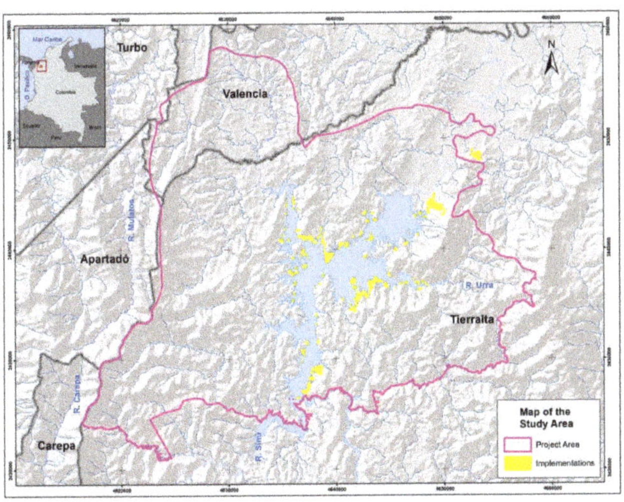

Figure 1. The study area around the influence region of the Urra hydropower project, located in Cordoba Department, Colombia.

2.2. EO from Passive (Landsat and Sentinel-2) and Active (Sentinel-1) Sensors

A 25-year (1996–2000–2005–2010–2015–2021) forest change analysis was performed based on the long-term availability of EO data from optical sensors (Landsat and Sentinel-2). The Landsat program has been providing continuous observations with 8- or 9-day repeat intervals for about 50 years [31], and it provides key advantages for long-term monitoring tasks, such as temporal and radiometric consistency throughout multiple Landsat programs [32]. To ensure a complete assessment from 1996 to 2021 could be conducted, multiple Landsat mission products were used, e.g., Landsat 5 TM, Landsat 7 ETM+, and Landsat 8 OLI, with a 30 m spatial resolution for visible and infrared bands. ESA Sentinel-2 MSI Level2A imagery was also included for the year 2021; imagery from this sensor was included to increase data availability by densifying the temporal coverage. Thus, revisit frequency can be improved from 8 or 9 days (based on Landsat) to 2.9 days (integrating Sentinel-2) [33], which contributes to this study by reducing spatial gaps caused by cloud cover, a major problem in Colombia. Also, Sentinel-2 imagery was integrated into the analysis to increase and improve the spatial resolution of thematic products through the generation of the 2021 forested/non-forested baseline, achieving a minimum mapping area between 0.25 and 0.5 ha, which was required to be decreased to a 1:50,000 scale for this study.

Landsat (Collection 2 Level-2) and Sentinel-2 (Level 2A) imagery was accessed through the Microsoft Planetary Computer platform (https://planetarycomputer.microsoft.com, accessed on 23 August 2023). All available imagery for each year with less than 50% cloud cover was transferred to the project's cloud-based storage infrastructure (AWS S3 bucket in the us-east-1 region). Based on all available imagery for each year, each band associated with a single acquisition was stacked to generate a multispectral multi-band image (Figure 2A). Cloud and shadow masking procedures were performed for each single acquisition scene from every dry season to generate a cloud-free mosaic for each year, avoiding atmospheric interference, especially during the rainy season [34]. Every pixel from the last acquisition of the corresponding year with the presence of clouds or mountain shadows was discarded and replaced by the closest cloud and shadow-free pixel to generate a yearly last pixel mosaic. The "last pixel" corresponds to the latest pixel acquired without cloud interference near the end of the year (31 December) for the corresponding year. Orthorectification and co-registration procedures were performed for the stacked images to obtain accurate georeferenced information (Figure 2A). To detect changes in the time series, it is key that the images are accurately co-registered and orthorectified so that the images acquired from different sensors and dates can be directly compared and processed; for the construction of the time series, it is essential to have an accurate pixel-level co-registration among all acquired images for each scene. This adjustment was performed by measuring the difference between pixels and re-projecting the coordinate of the displaced image ends in this same magnitude (considering that the downloaded images are in a projection system with metric units). The UTM projected reference system was maintained during the whole process until the final products to prevent the loss of co-registration between pixels when applying the geometric adjustments of the cartographic re-projection models.

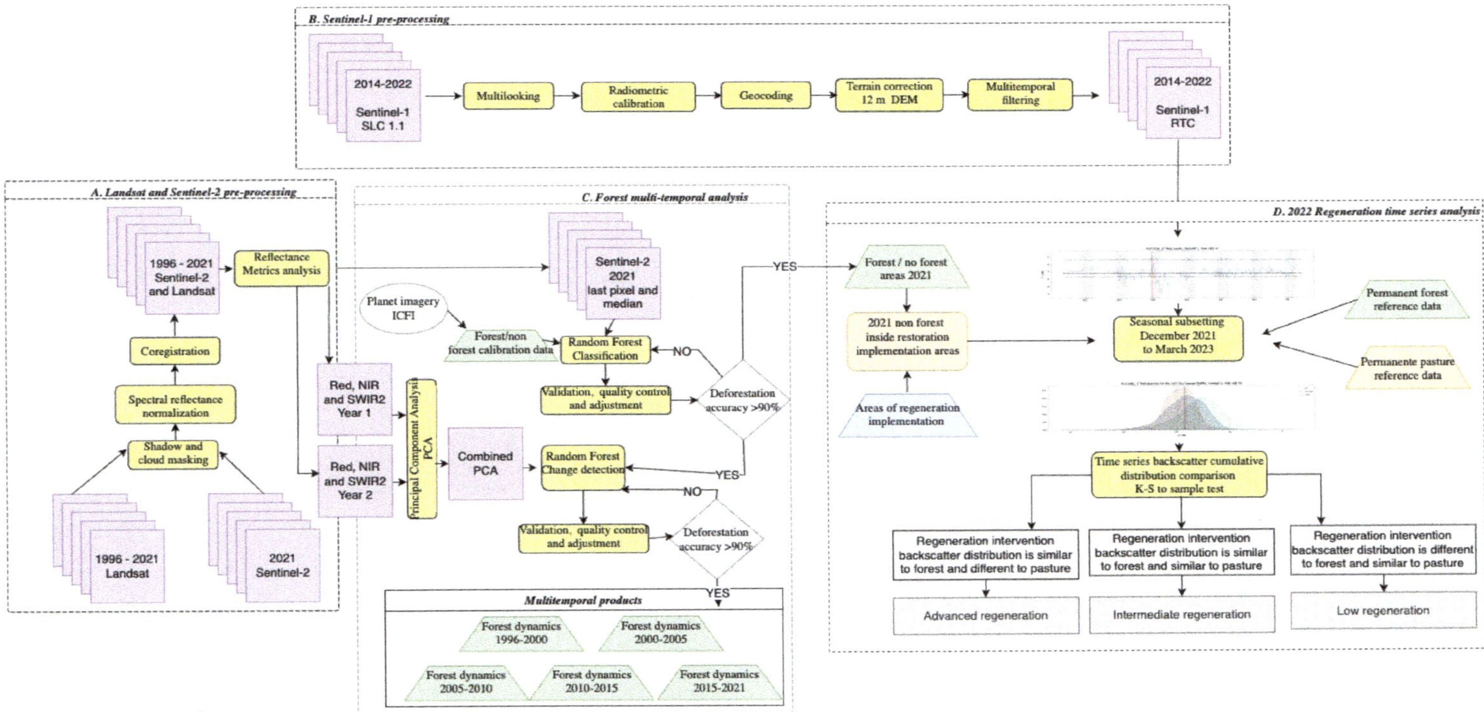

Figure 2. Processing workflow: (**A**) (optical pre-processing) and (**B**) (SAR pre-processing) describe all imagery pre-processing generated as inputs to be integrated in (**C**) to generate multitemporal analysis of forest dynamics, and (**D**) is used to assess restoration implementations around the Urra hydropower facility.

Precise reflectance and geometric image rectifications are necessary to generate quality data and to ensure the delivery of consistent products [35], and it is necessary to avoid measurement differences and false detections, i.e., due to geometric distortions or sensor calibrations. An adjustment to the radiometric signals of the images was thus performed to ensure consistency with each other (Figure 2A), i.e., through radiometric calibration, where digital numbers are transformed to surface reflectance, which, in the case of a tropical forest, is driven by species composition and vegetation structure characteristics [36]. Relative normalization was applied under the assumption that the relationship between the radiances recorded by the sensor on two different dates is spatially homogeneous (Figure 2A). In order to detect deforestation changes in multi-temporal forest dynamics analysis, a multivariate alteration detection algorithm, based on canonical correlation [37], was implemented, where the 2021 image was taken as baseline, and the rest of the images (1996, 2000, 2005, 2010, and 2015) were normalized with respect to this one. Quality control was carried out to ensure that the scenes were correctly co-registered and that they would allow effective multi-temporal analysis. This process guaranteed the use of different sensors and spatial resolutions so that they could be used in a homogeneous analysis.

Pan-sharpening techniques were employed on multi-year Landsat imagery (level 1) with spatial resolutions ranging from 15 to 30 m to facilitate fusion with Sentinel-2 data, which offer a higher spatial resolution at 10 m. Pan sharpening is a crucial preprocessing step aimed at enhancing the spatial details of lower-resolution multispectral imagery by integrating them with higher-resolution panchromatic imagery [38]. Through this process, we ensured compatibility and consistency between the Landsat and Sentinel-2 datasets, enabling seamless integration for subsequent analysis and interpretation. The application of pan-sharpening techniques not only preserves the spectral characteristics of Landsat imagery but also enhances its spatial resolution to match that of Sentinel-2 data, thereby facilitating comprehensive and accurate assessments of landscape dynamics over time. Pan sharpening was implemented with GDAL Pansharpen library (https://gdal.org/programs/gdal_pansharpen.html, accessed 20 August 2023) using the nearest neighbor resampling algorithm without weights. Procedures were applied to the Landsat-7 imagery due to the failure of the scan line corrector (SLC) [39], resulting in gap strips on the imagery and leading to 22% of the image being missing. A Phase II approach [39] was implemented, integrating multiple SLC-off scenes to fill the scan gaps by incorporating cloud-free pixels from the closest scenes.

The Copernicus Sentinel-1 mission launched by the European Space Agency (ESA) is based on a constellation of the two satellites, Sentinel-1A (launched in 2014) and Sentinel-1B (launched in 2016), with a revisit time of 12 days each, carrying a SAR C-band sensor on board. Sentinel-1 SAR imagery was available from the Alaska Satellite Facility (ASF) platform, a mirror of the ESA Scientific Sentinel-1 hub. All available data from May 2015 to June 2022 were accessed through the ASF-DACC data repository, which resides at an AWS S3 bucket in the us-west-2 region. Figure 2B summarizes the five methodological routines for preprocessing Sentinel-1 imagery, which allow the Single Look Complex (SLC) images to be converted into 12 m resolution multi-look geocoded images, radiometrically and terrain-corrected to generate gamma-naught backscatter intensity ($\gamma 0$) images, obtained after pre-processing is performed with the Gamma® software V2.6 [40]. To reduce speckle effects associated with SAR imagery, an enhanced Lee multitemporal filter was applied to the VH polarization by means of least squares of the signal intensity in a kernel area of 3×3 pixels [41].

2.3. Forest Multi-Temporal Analysis

All surface reflectance imagery processed based on Sentinel-2 and Landsat data were used to generate annual median and last pixel metrics for the Red, NIR, and SWIR2 spectral bands, which are the most relevant in vegetation change analysis [42] (Figure 2C). A forested/non-forested map was generated for 2021 as baseline using a pixel-based supervised Random Forest machine learning algorithm (MLA) included on the sklearn.ensemble

Python module (V 1.4.1) [43] (Figure 2C). The Random Forest (RF) algorithm has been widely used in classifying remote sensing data [44–47], which generally handles noise and overfitting of the data well and has high data dimensionality [47]. The RF classifier was initially parameterized using training samples (250 forested and 275 non-forested point samples) generated based on the visual inspection of 2021 high-resolution Planet imagery of Red, NIR, and SWIR surface reflectance metrics (median and last pixel). Once the first version of the 2021 forested/non-forested map was generated by the RF classifiers, a visual assessment of the classified map was carried out using Planet imagery as quality control, validation, and adjustment procedure (Figure 2C). If deforestation accuracy of at least 90% was not obtained, new samples were generated in regions where misclassifications were observed. Then, the model was recalculated, and this process was iterated until a 90% accuracy was obtained (Figure 2C). Sample data and quality control procedures followed the forest definition from the Colombian Forest and Carbon Monitoring System (SMByC).

To derive multitemporal forest change products and estimate deforestation extent between for the 1996–2000, 2000–2005, 2005–2010, and 2015–2021 periods, a semi-automated RF approach was used (Figure 2C). For the change class, training points between 2015 and 2021 were generated based on a visual interpretation of deforested areas from Sentinel-2, Landsat, and Planet imagery. Red, NIR, and SWIR bands from 2021 and 2015 were combined into principal component bands within a single image (Figure 2C) and integrated as predictive variables in the RF model to identify highly and lowly correlated areas. Principal component analysis has been demonstrated to capture maximum variances in a finite number of orthogonal components based on eigenvector analysis from the correlation matrix, providing a robust and simple approach to generate input variables for change detection analysis [48]. The principal component bands generated based on two different years are highly correlated between unchanged areas (forested and non-forested stable areas), while a low correlation between significantly changed areas is expected (e.g., deforestation). The RF classifier was configured with the following parameters: n_estimators set to 100, representing the number of trees in the forest; criterion set to "gini" to measure the quality of a split based on Gini impurity; max_depth set to none to expand nodes until all leaves are pure or contain less than the specified minimum samples for splitting; and min_samples_split set to 2, indicating the minimum number of samples required for node splitting, with the option to set it as a fraction of the total number of samples. No weight was assigned to trees or predictable bands, and iterations continued until an output error of 5% was achieved. Once the first version of the 2015–2021 deforestation map was derived using the RF classifiers, a visual verification was performed by iterating validation, generating new training points and a new model until accuracies higher or equal than 90% for the deforestation class were obtained (Figure 2C). An accuracy threshold of 90% for the deforestation category was defined for the producer and user accuracy estimator to reduce the under and overestimation of deforested area. Once the expected accuracy was achieved, the forested/non-forested map of the following year (2015) and the change map (2015–2021) were updated. This process was repeated for each change period analyzed.

Sample-Based Estimation of Area and Accuracy

Due the importance of estimating forest, non-forest, and deforestation extent based on the thematic maps generated and the need for reproducible protocols for forest and deforestation mapping, an accuracy quantification is needed for reporting areas associated with each class of thematic maps [49]. A quantitative accuracy assessment for each class of thematic products was carried out to estimate and understand uncertainties on area estimation-based bias attributable to omission and commission classification errors. Olofsson et al.'s approach [50] was implemented for assessing and estimating land change product accuracy, since it is a well-accepted standard for land change accuracy assessment besides the national standard in Colombia implemented by the SMByC [42]. The initial validation sample size (n) was determined by setting the expected standard error to Si = 0.01, the area proportions were mapped for each class, and theoretical validation user accuracy was defined by the SMByC standards for each class, i.e., 99 validation points were estimated for forested areas, 6 for deforested areas, and 129 for non-forested areas. Nevertheless, the final validation size per category was balanced to a total of 100 points for each class due to the low numbers obtained for deforestation class explained by the proportion area of this category. The spatial distribution of validation points was determined by an automated stratified random seed to ensure no bias due to the spatial distribution of validation points. Target user accuracy (U_i) was set to 0.9 for both stable forest and non-stable forest covers, since these classes are usually considered to have a high accuracy, while for the deforestation class, Ui = 0.8 was assigned.

2.4. Forest Restoration Status Assessment

Time series analysis using Sentinel-1 imagery was carried out to study vegetation seasonal variations and to assess the status of areas where restoration initiatives were implemented since 2014. The boundaries of the implementation areas of restoration by the Urra company were integrated (Figure 2D) and overlapped to the forested/non-forested areas of the thematic map for 2021 (Figure 2C). The masking process filters the non-forested areas within the intervention plots to perform a restoration status evaluation analysis (Figure 2D). For many biomes, seasonal stratification of the time series will improve the detection of change events, for example, when dry/wet season conditions introduce significant changes in backscatter due the presence of surface water or phenology variations [51]. In this sense, to minimize the influence of seasonal variation in the assessment of structural trajectories of restoration areas, a seasonal trend analysis was introduced in the assessment. Therefore, the first step was to subset the time series data by identifying the dry season to minimize the precipitation effects on backscatter levels. Dry season months were identified within the time series backscatter from Sentinel-1 from the years prior to 2021 (2017–2020); dry season (December to March) corresponds to yearly consistent low backscatter values that were validated with precipitation data [52]. Imagery from December to March was selected to be included in the analysis as the dry season period (Figure 3).

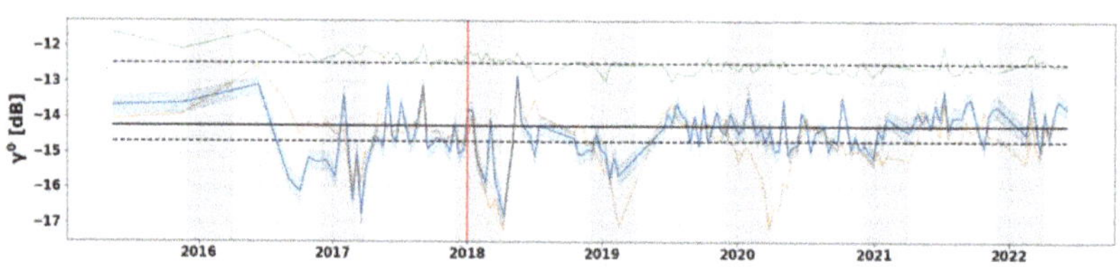

Figure 3. Sentinel-1 VH cross-polarization backscatter time series profile for assessed plot (blue line) with implementation year 2018 (vertical red line) contrasted with forest reference time series (green line) and pasture reference time series (yellow line). Dry season was defined as December to March (grey areas).

To assess the current state of the restoration process and determine if the current state of the vegetation structure is more similar to the reference coverages of well-preserved natural forests or, on the contrary, has a greater affinity with the backscatter response of the pastures, a comparative analysis of the dynamics of the vegetation structure from plots undergoing restoration was performed based on available Sentinel-1 time series data (Figure 2B). The assessment was based on a Kolmogorov–Smirnov test, which allows the distributions of the backscatter of the implementations to be compared with forest and pasture reference data, where the following hypotheses were proposed:

H0. *The sample of the radar signal backscatter of the regeneration plot and the sample from the reference data belong to the same distribution.*

H1. *The sample of the radar signal backscatter of the regeneration plot and the sample from the reference data DO NOT belong to the same distribution.*

A total of 270 polygons of intervened plots, which have different years of implementation of reforested and/or revegetated measures (2004–2010 and 2012–2018), were analyzed through a comparative analysis of the dynamics in the vegetation structure in plots undergoing restoration with respect to two types of reference areas: pasture and forest (Figure 4). Reference data were generated based on the visual inspection of multi-year optical imagery products from Landsat, Sentinel-2, and Planet (1996, 2000, 2005, 2010, 2015, and 2021). Forest reference data were defined as consistent forested areas identified in all years, and pasture reference data were defined as consistent pasture areas from 2005 to 2021.

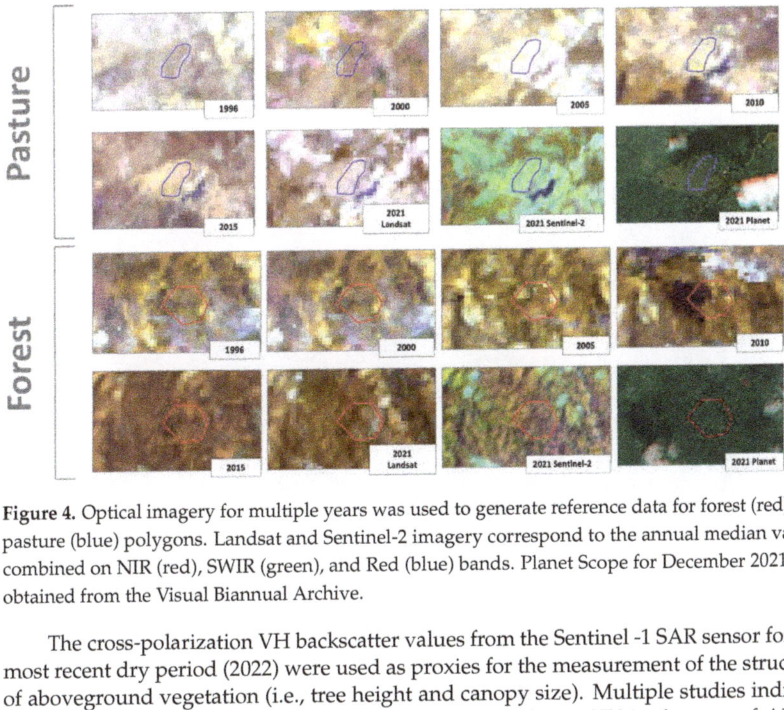

Figure 4. Optical imagery for multiple years was used to generate reference data for forest (red) and pasture (blue) polygons. Landsat and Sentinel-2 imagery correspond to the annual median values combined on NIR (red), SWIR (green), and Red (blue) bands. Planet Scope for December 2021 was obtained from the Visual Biannual Archive.

The cross-polarization VH backscatter values from the Sentinel -1 SAR sensor for the most recent dry period (2022) were used as proxies for the measurement of the structure of aboveground vegetation (i.e., tree height and canopy size). Multiple studies indicate that SAR cross-polarization (VH in the case of Sentinel-1 or HV in the case of ALOS-PALSAR) shows a higher correlation with biomass compared with co-polarization (VV in the case of Sentinel-1 or HH in the case of ALOS-PALSAR) to detect vegetation structure changes [53–57]. Backscatter from co-polarization is usually sensible to surface scattering components [57], and it is frequently applied to detect surface water [56].

A 95% confidence interval was chosen; therefore, if the p-value is less than 0.05 (or 5%), it provides sufficient evidence to reject the null hypothesis in favor of the alternative. This determines the status of the forest restoration implementation based on the following rules: (i) if it is in an incipient state, it has a greater similarity to pasture; (ii) if it is in an intermediate state, it has no similarity to pasture nor forest; and (iii) if it is in an advanced state, it has a greater similarity to forest (Figure 2D).

2.5. Computing Infrastructure

The Nebari open-source data science platform (https://www.nebari.dev/, v2023.5.1, accessed on 12 August 2023) was used to develop and implement Python routines for restoration assessment, integrating the Sentinel-1 time series backscatter generated using the proprietary software for Earth Big Data Processing, Prediction, Modeling, and Organization (SEPPO v5.0.0) to access the cloud Sentinel-1 imagery and perform all pre-processing procedures necessary to generate useful time series EO data for future analysis. The Nebari platform was configured by Earth Science Information Partners (ESIP; https://www.esipfed.org/ accessed on 12 August 2023), through the Earth Big Data LLC partnership, to run XLarge Instances (8CPU/32 GB), implementing a parallel computing auto-scaling approach with a maximum of 20 workers (CPU).

3. Results

3.1. *Optical and Radar Imagery Processing*

The images correspond to those available in the second half of the year corresponding to the analysis date. For example, for the analysis of the year 2010, imagery from 1 July to 31 December 2010 was selected to obtain a composite for this year from the historical series of images, metrics, and statistics. A total of 63 images were obtained for the 1996–2021 period to generate the yearly median and last pixel partially cloud-free mosaics for the Red, NIR, and SWIR bands (Figure 5); some pixels for specific years consistently presented the presence of clouds.

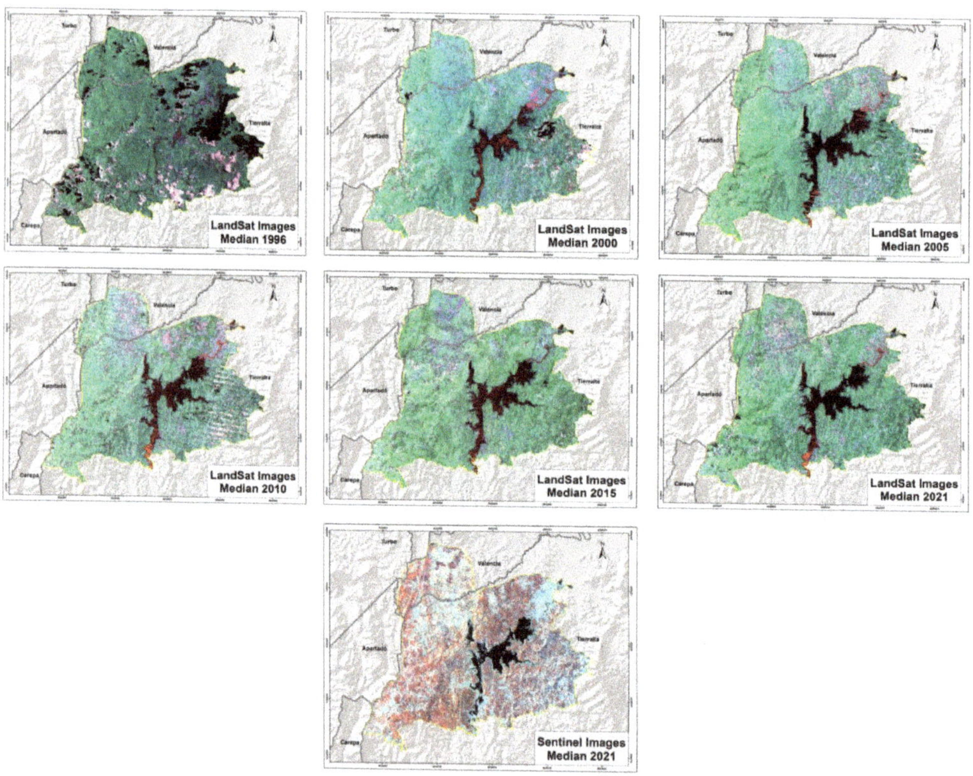

Figure 5. Optical pre-processing results for Sentinel-2 (2021) and Landsat (1996, 2000, 2005, 2010, 2015, and 2021) imagery. The mosaics generated correspond to annual median values combined on NIR (red), SWIR (green), and Red (blue) bands.

A total of 285 Sentinel-1 radiometric and terrain-corrected images were used for the 2014–2022 period, of which 136 correspond to Sentinel-1A and 149 correspond Sentinel-1B, providing Earth's surface observations every six days (Figure 6). All images, with a final spatial resolution of 12 m, were stacked in Zarr files, a Python library for chunked, compressed, and parallelized N-dimensional arrays (Figure 6), providing a flexible and scalable approach to store large datasets in a compressed format, allowing for the efficient storage and retrieval of data in a cloud-based computing environment.

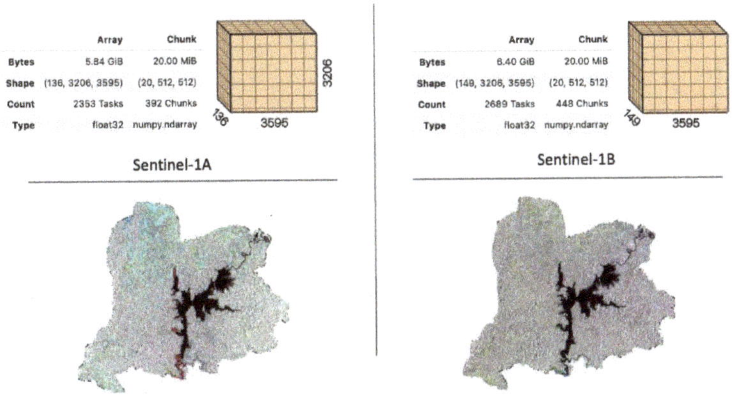

Figure 6. Sentinel-1 radiometric and terrain products obtained from pre-processed imagery from both satellites (S-1A and S-1B). Single images were stacked in multidimensional time series array of earth observations.

3.2. *Forested and Deforested Spatiotemporal Dynamics Using Optical Sensors*

Based on the analysis of the forested/non-forested maps for the year of 1996, it was found that the initial baseline for forested areas covered a significant portion of the study area, corresponding to 45,186.9 hectares, which represents 50.7% of the total area. The remaining 49.3% of the area is represented by non-forested areas. However, when examining the dynamics of forest coverage change from 1996 to 2021 (Figure 7), a noticeable decreasing trend in forested areas was found.

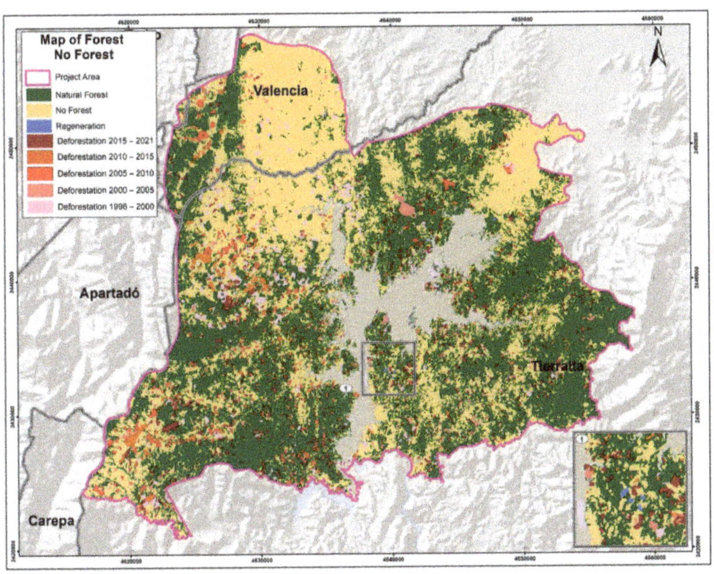

Figure 7. Map of forested, non-forested, deforested, and restored forest areas between 1996 and 2021 in the study area. Different red areas represent deforestation events during different periods.

Through the analyzed period, the reported forested area experienced a decline, decreasing from 43,350 ha in the 1996–2000 period to 37,763 ha in the 2015–2021 period. Concurrently, the extent of deforestation increased from 1837 ha in the 1996–2000 period to 2178 ha in the 2015–2021 period. In contrast, non-forested areas expanded, growing from 44,004 ha in the 1996–2000 period to 49,223 ha in the 2015–2021 period (Figure 8).

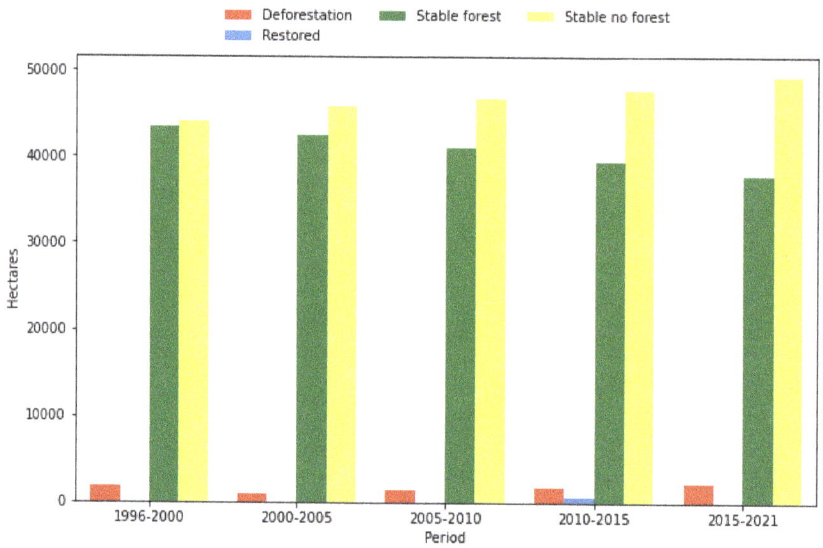

Figure 8. Estimated extent of forest change dynamics (hectares) for each of the periods analyzed (1996–2021).

Notably, the analysis also identified reforested and/or revegetated areas, which were first reported in the last two analysis periods. A total of 583 hectares were recorded by the year 2015, and an additional 27 hectares were identified by the year 2021, representing restored forested areas. By the year 2021, forests constituted 42.4% of the study area, covering a total of 37,789 hectares. The remaining 57.63% of the area consisted of non-forest coverages, accounting for 51,401 hectares (Figure 8).

Validation of Forest Change Dynamics for 1991–2021 Products

Table 1 shows the locations of the 300 points that served as samples for the validation of the forest surface change maps for the analyzed study periods. These points were spatially randomly generated for each of the classes: forest, non-forest, and forest loss. No validation points were integrated in the accuracy assessment approach for the reforestation category, as these areas represented a minimal percentage of the study area, and the areas were concentrated in specific sites that were completely visually inspected.

The classification accuracy assessment revealed variations in the performance of the maps across different time periods (Table 1). Table 1 shows that during the 1996–2000 period, the overall accuracy of the forest change map was 89%, the user's accuracy for the forest class was 78%, and the producer's accuracy for the forest class was 96%. Conversely, the non-forest class achieved a user's accuracy of 100% and a producer's accuracy of 80%. The deforestation class exhibited a user's accuracy of 90% and a producer's accuracy of 95%. In the subsequent period, spanning 2000–2005, the overall accuracy of the map improved to 96% compared with the change map for the 1996–2000 period. The user's accuracy for the forest class was 93%, indicating a high percentage of correctly classified forest pixels. Additionally, the producer's accuracy for the forest class was 98%, suggesting that the map accurately represented the distribution of forested areas. The deforestation class presented a user's accuracy of 95% and a producer's accuracy of 99%, indicating the map's effectiveness in detecting and representing deforestation events. The non-forest class user's accuracy was 98%, and the producer's accuracy was 90%, indicating a slight overestimation of non-forested areas in the map. These results demonstrate the improved accuracy and reliability of the map in capturing land cover changes during this specific period compared to the 1996–2000 period.

Table 1. An accuracy assessment sample for the 1996–2000, 2005–2010, 2010–2015, and 2015–2021 periods' forested/deforested/non-forested maps for the area of influence of Urra's hydropower facility. The map categories are rows, while the reference categories are columns. The accuracy measures are presented with a 95% confidence interval.

Class	Forested	Deforested	Non-Forested	Total	Wi (%)	User's	Producer's	Overall
1996–2000								
Forested	78	4	18	100	0.38	0.78 ± 0.4	0.96 ± 0.3	0.89
Deforested	3	90	7	100	0.2	0.90 ± 0.02	0.95 ± 0.04	
Non-Forested	0	0	100	100	0.42	1 ± 0.4	0.8 ± 0.5	
Total	81	94	125	300	1			
2000–2005								
Forested	93	1	6	100	0.48	0.93 ± 0.4	0.98 ± 0.4	0.96
Deforested	0	95	5	100	0.01	0.95 ± 0.01	0.99 ± 0.01	
Non-Forested	2	0	98	100	0.51	0.98 ± 0.5	0.9 ± 0.5	
Total	95	96	109	300	1			
2005–2010								
Forested	92	0	8	100	0.46	0.92 ± 0.4	0.99 ± 0.4	0.95
Deforested	0	99	1	100	0.02	0.99 ± 0.01	0.98 ± 0.02	
Non-Forested	1	0	97	100	0.52	0.97 ±0.5	0.92 ± 0.5	
Total	93	101	106	300				
2010–2015								
Forested	92	1	7	100	0.45	0.92 ± 0.4	0.97 ± 0.4	0.95
Deforested	0	95	5	100	0.2	0.95 ± 0.1	0.99 ± 0.02	
Non-Forested	3	0	97	100	0.53	0.97 ± 0.5	0.89 ± 0.5	
Total	95	96	109	300				
2015–2021								
Forested	92	1	7	100	0.43	0.94 ± 0.4	0.97 ± 0.4	0.96
Deforested	0	95	5	100	0.2	0.94 ± 0.02	1 ± 0.02	
Non-Forested	3	0	97	100	0.55	0.97 ± 0.5	0.89 ± 0.5	
Total	95	96	109	300				

For the 2005–2010 period, the overall accuracy of the map remained consistently high at 95%. The user's accuracy for the forest class was 92%, indicating a high level of agreement between the map classification and the actual forested areas. The producer's accuracy for the forest class was 99%, suggesting that the map accurately represented the distribution of forest pixels. The deforestation class's user's accuracy was 99% with a producer's accuracy of 98%, indicating the map's effectiveness in detecting and representing deforestation events with a high level of precision. The non-forest class showed a user's accuracy of 97% and a producer's accuracy of 92%, suggesting a slight overestimation of non-forested areas in the map. These results highlight the accuracy and reliability of the map in capturing land cover changes during this specific time period.

From 2010 to 2015, the overall accuracy of the map remained consistently high at 95%. The user's accuracy for the forest class was 92%, indicating a high percentage of correctly classified forested pixels. The producer's accuracy for the forest class was 97%, indicating that the map accurately represented the distribution of forested areas. The deforestation class achieved a user's accuracy of 95% and a producer's accuracy of 99%, demonstrating the map's effectiveness in identifying and representing deforestation events. The non-forest class demonstrated a user's accuracy of 97% and a producer's accuracy of 89%, suggesting a slight underestimation of non-forested areas in the map.

In the most recent time period analyzed, spanning 2015–2021, the overall accuracy of the map remained consistently high at 94.7%. The user's accuracy for the forest class was 92%, indicating a high percentage of correctly classified forested pixels. The producer's accuracy for the forest class was 97%, suggesting that the map accurately represented the distribution of forested areas. The deforestation class achieved a user's accuracy of 95% and a producer's accuracy of 100%, demonstrating the map's effectiveness in detecting and representing deforestation events. The non-forest class showed a user's accuracy of 97% and a producer's accuracy of 89%, indicating a slight underestimation of non-forested areas in the map.

3.3. Restoration Implementation Assessment Using Sentinel-1

The previous analysis of the forested/non-forested dynamics showed that three of the implementation plots were found to reach a completed forest succession state, indicating that, as of the analysis cutoff date in March 2022, these specific plots exhibited forest coverage in accordance with the definition provided by the SMByC.

Based on the time series analysis performed using VH polarization from Sentinel-1, cumulative distributions derived for each of the 270 implementation areas were contrasted to forest and pasture reference data (Figure 9). The two-sample Kolgomorov–Smirnov test revealed that 16 plots (5.93%) demonstrated an incipient restoration state, as evidenced by the vegetation structure estimates that displayed a greater resemblance to grassland areas compared to forests (Figure 10 provides an incipient example). Among the intervened plots subjected to evaluation (totaling 224), a significant majority of 82.96% (Table 2) was found to be in an intermediate state of the restoration process. These plots did not exhibit significant differences when compared to the two reference coverages: grasslands and forests (Figure 10 exemplifies the intermediate state). Furthermore, 27 of the evaluated plots, accounting for 10% of the total (as outlined in Table 2), displayed significant differences compared to the grassland reference data, but they did not exhibit significant differences in comparison to forests. Consequently, these plots were classified as being in an advanced state of reforestation (Figure 10).

Table 2. Results of restoration state assessment for 2022 dry season for 270 implementation plots evaluated based on Sentinel-1 VH backscatter.

Regeneration State	Number of Implementations
Incipient	16
Intermediate	224
Advanced	27
Completed	3

When the number and state of restoration were inspected in relation to the implementation years, some years showed a higher number of implementation plots in an advanced state of restoration. The years 2018, 2016, 2013, and 2014 demonstrated a substantial presence of seven, five, five, and four plots in the advanced state, respectively. Additionally, the year 2008 showed two plots in the advanced implementation state, while for the years 2009, 2012, 2015, and 2017, each one had a single plot in this state (Supplementary Figure S1 and Table S1).

No implementation plots from the years 2004, 2005, 2006, 2007, and 2010 were observed to achieve an advanced state in the restoration process. Instead, most plots from these years exhibited intermediate states within the reforestation process, indicating ongoing progress towards full restoration. To explore the relationship between the restoration state and the implementation year, a Chi-squared test was performed. The results of the test indicated that there is no statistically significant association between the restoration state and the implementation year (Chi-squared statistic: 44.2823; degrees of freedom: 39; and *p*-value: 0.2586), suggesting that the restoration state is independent of the specific year of implementation.

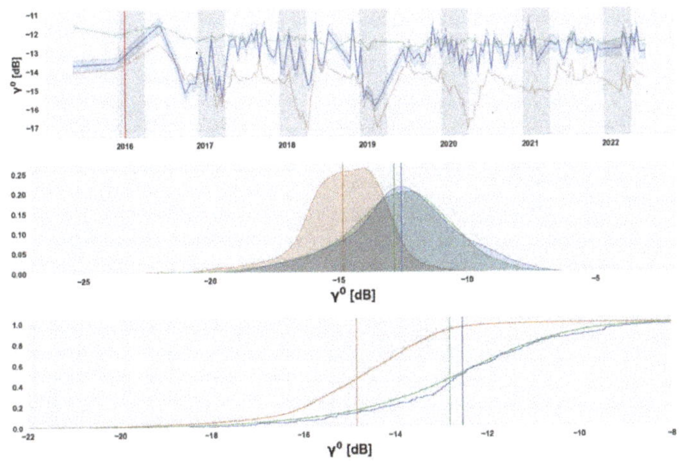

Figure 9. The Synthetic Aperture Radar (SAR) time series mean cross-polarized (VH) backscatter values (**top**) and cumulative distributions (**middle** and **bottom**) associated with reference areas, forest (green) and pasture (orange) areas, and assessed implementation plots (blue). This implementation was classified as being in an advanced state of restoration where the cumulative distribution is closer to the forest distribution. The implementation of the restoration date (red) and dry season period (December, January, February, and March) (grey) are included.

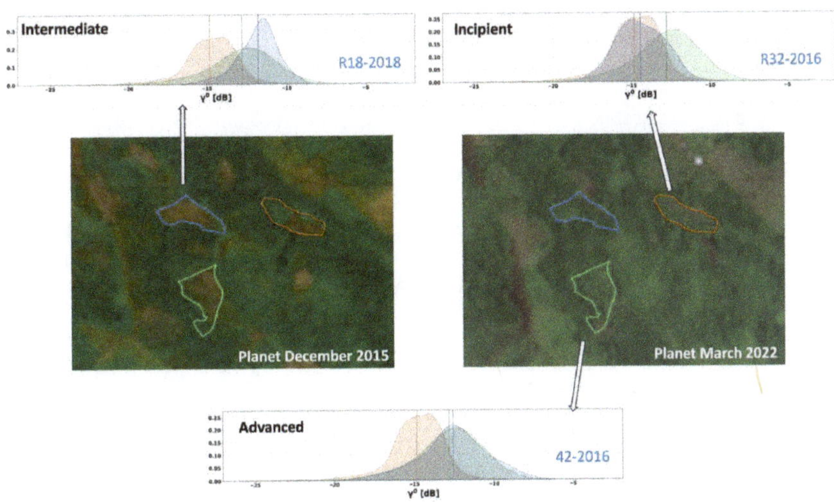

Figure 10. An assessment of three implementation plots based on the backscatter cumulative distributions and reference data for the 2022 dry season. The 42-2016 plot (green polygon) exhibited similarity to the forest reference (green backscatter distribution). The R18-2018 plot (blue polygon) indicated an intermediate restoration state due to differences in the backscatter distribution (blue) compared to forest (green backscatter distribution) and pasture (orange backscatter distribution). The R32-2016 plot (orange polygon) exhibited similarity to the pasture reference (orange backscatter distribution).

4. Discussion

4.1. Forested and Deforested Spatiotemporal Dynamics Using Optical Sensors

The results demonstrate that the proposed method allows for an efficient identification of forest loss within the study area. The developed method enables the detection of changes occurring within forest coverage, providing a quantitative measure and the spatial locations of the observed change events. This is achieved through the creation of composite images that reduce areas without information and extract pixels to generate information-rich images, removing the gaps generated by clouds and sensor malfunctions [30,49]. In accordance with a national [1] deforestation spatiotemporal trend analysis, this research reveals the highest deforestation trend during the period from 1996 to 2000. The initial baseline in 1996 showed that forests covered a significant portion of the study area, accounting for 50.7% of the total area. However, over the analyzed period, there was a noticeable decrease in forested areas, with a corresponding increase in non-forested areas. Deforestation played a significant role in this trend, as forested areas declined from 43,350 hectares in the 1996–2000 period to 37,763 hectares in the 2015–2021 period, while non-forested areas expanded from 44,004 hectares to 49,223 hectares during the same period. By 2021, forests constituted 42.4% of the study area, resulting in a loss of approximately 7397 hectares, which corresponds to 8.3% of the total area. Deforestation has primarily concentrated in the western part of the study area. The effects of deforestation are particularly evident near rural areas located far from the water reservoir. This indicates that there are actors and actions (especially related to land use) that influence the study area besides the hydropower reservoir. Consequently, the ten closest most affected rural areas in terms of forest loss area are Nain, Murmullo Alto, Maria de Jesus, Casajales, Kilometro 14, Zumbona, Chispas, La Osa, Alto Quimary, and Angostura (Figure 7). Evidence of forest recovery is observed in areas that are not associated with or adjacent to the intervention plots carried out by the company. This finding suggests the presence of ecological processes, e.g., natural succession, that are promoting restoration. It is important to note that, although the largest areas of forest recovery were detected in the year 2015, this does not necessarily imply that the recovery occurred specifically between 2010 and 2015. The visual inspection, which identified pastures in 1996, indicates that the recovery could have started before the previous period.

An accuracy assessment of the forest change map revealed variations in its performance across different time periods. The overall accuracy of the map consistently remained high for most of the periods, except in the 1996–2000 period, which had the lowest overall user's and producers' accuracies. This could be attributed to the Landsat 7 mission having a less frequent revisit rate compared to subsequent missions [5] and SLC-off malfunction [38,39]. As a result, it becomes challenging to produce cloud-free mosaics in tropical regions based on optical sensors, demonstrating the limitations of using optical sensors in tropical areas with high cloud cover [30] and the need to integrate active sensors for cloud-free EO data generation [5,10,14,15]. With the arrival of future missions, such as Landsat-8 and Landsat-9, the frequency of revisits increases, enhancing the probability of generating cloud-free mosaics. However, along with the information gaps caused by the SLC-off effect [38], these gaps in the data can be observed in mosaic generation with cloud and banding for the years 1996, 2000, 2005, and 2010; this can represent a limitation in the application of optic sensors to generate EO data for forest and deforestation monitoring in tropical and mountainous regions with a high frequency of clouds, especially during the rainy season. This was evident in the 1996 and 2000 mosaics, where certain areas lacked cloud-free pixels. Furthermore, since the first Landsat missions in 1972, several enhancements were made to improve the spatial, radiometric, and spectral resolutions [5]. These improvements can have varying effects on the final products, such as mosaics, which, in turn, can influence the accuracy assessment. The user's accuracy for the forest class also remained consistently high, ranging from 78% to 92%, indicating a high percentage of correctly classified forested pixels. The producer's accuracy for the forest class ranged from 96% to 100%, suggesting that the map accurately represented the distribution of forested areas. Notably, the map showed a high level of effectiveness in detecting and representing deforestation events, with user's accuracies ranging from 90% to 99% and producer's accuracies ranging from 95% to 100% for the deforestation class. However, there were slight discrepancies in the representation of non-forested areas, with user's accuracies ranging from 97% to 100% and producer's accuracies ranging from 89% to 92%. These results demonstrate the overall high accuracy and reliability of the map in capturing land cover changes, particularly in detecting deforestation events. Nevertheless, there is room for improvement in accurately representing non-forested areas. Overall, the findings highlight the importance of regularly updating and refining classification methods to enhance the accuracy and precision of forest change maps.

The analysis of forested/non-forested dynamics and the time series analysis of the implementation plots revealed that three implementation plots reached a completed state, indicating successful forest restoration according to the provided definition. Furthermore, a small percentage (5.93%) of plots demonstrated an incipient restoration state, resembling grassland areas more than forests. Most evaluated plots (82.96%) were in an intermediate state, showing no significant differences compared to both grassland and forest reference data. Additionally, 10% of the plots displayed significant differences compared to grasslands but not forests, indicating an advanced state of reforestation.

4.2. *Restoration Implementation Assessment Using Sentinel-1*

When examining the implementation years, certain years showed a higher number of plots in the advanced state, such as 2018, 2016, 2013, and 2014. However, there was no statistically significant association between the restoration state and the implementation year, suggesting that the restoration status is not associated with the implementation year. The integration of additional variables associated with the type of implementations (active or passive) and species composition, among other factors, in conjunction with continuous vegetation structure time series provided by SAR data, could generate more insights for the evaluation of restoration process.

To effectively evaluate the success and impact of restoration strategies, it is crucial to combine on-the-ground data with remote sensing observations. Integrating these two sources of information allows for a comprehensive assessment of the restoration process, including the identification of factors that contribute to successful restoration and the identification of areas that may require further intervention. Implementing monitoring systems that integrate ground truth data and EO data enables a more robust and accurate evaluation of restoration efforts. By continuously assessing the progress and outcomes of restoration strategies, decision makers and stakeholders can make informed decisions regarding resource allocation, adaptive management, and the refinement of restoration approaches.

5. Conclusions

The proposed method for detecting and quantifying deforestation events using cloud-based and open-source science platforms proved to be effective in monitoring forest dynamics in the study area. The method allowed for the creation of composite images that facilitated the extraction of information-rich images, enabling the detection of spatially and quantitatively significant changes in forest coverage. The impacts of deforestation were particularly evident near rural areas, emphasizing the influence of actors and actions beyond the scope of the hydropower reservoir. However, this study also highlighted areas where forest recovery was observed, suggesting that the presence of ecological processes promote restoration.

The accuracy assessment of the forest change map demonstrated its high performance in detecting and representing deforestation events, with consistently high user's and producer's accuracies. While the accuracy for the forest class remained high, there were slight discrepancies in the representation of non-forested areas, indicating room for improvement. Regular updates and refinements to classification methods are crucial to enhance the accuracy and precision of forest change maps.

The analysis of implementation plots revealed that, currently, most of the implementation plots are in an intermediate state. Certain years showed a higher number of plots in an advanced state of reforestation, but no statistically significant association was found between the restoration state and the implementation year. The further integration of additional variables, such as the implementation type and species composition, along with continuous time series data from SAR, could provide deeper insights into the evaluation of the restoration process.

This research emphasizes that the application of these methods can be extended to tropical regions, and certain techniques can mitigate spatiotemporal gaps to produce annual mosaics. It is anticipated that integrating Landsat-7 may result in decreased accuracy due to its infrequent revisits and SLC-off malfunction. However, the reported lower accuracy estimates do not pose a significant issue when generating yearly mosaics inclusive of dry seasons. Yet, this limitation becomes pertinent when only imagery from the rainy season is integrated, necessitating the consideration of seasonality in future analyses.

To effectively evaluate the success and impact of restoration strategies, combining on-the-ground data with remote sensing observations is crucial. This integrated approach allows for a comprehensive assessment of the restoration process, aiding in the identification of factors contributing to successful restoration and areas requiring further intervention. By continuously monitoring and evaluating restoration efforts, decision makers and stakeholders can make informed decisions regarding resource allocation, adaptive management, and the refinement of restoration approaches.

Supplementary Materials: The following supporting information can be downloaded at: https://www.mdpi.com/article/10.3390/f15050754/s1, Figure S1: Number of restoration plots assessed grouped by implementation state and year of implementation; Table S1: Number of restoration plots assessed, grouped by implementation state and year if implementation.

Author Contributions: Conceptualization, C.P. and N.C.; methodology, C.P., N.C. and M.V.; software, M.V. and J.K.; validation, M.R. and D.B.R.; formal analysis, C.P., M.V. and M.R.; investigation, C.P. and N.C.; data curation, P.Q. and M.M.; resources, P.Q. and M.M.; data curation, resources, P.Q. and M.M.; writing—original draft preparation, C.P., M.R. and A.S.D.; Writing—review & editing, C.P. and N.C.; visualization, M.V.; supervision, C.P. and N.C.; project administration M.R. All authors have read and agreed to the published version of the manuscript.

Funding: We thank the Colciencias scholarship program for the 727-2016 doctoral studies, which allowed the main author to integrate this work as part of his academic process.

Data Availability Statement: Data are contained within the article.

Acknowledgments: We would like to extend our heartfelt gratitude to the Urra company for generously providing the implementation data, which played a pivotal role in our analysis. We thank the Colciencias scholarship program for the 727-2016 doctoral studies, which allowed the main author to integrate this work as part of his academic process. Additionally, we appreciate all of the support staff across the institutions involved.

Conflicts of Interest: Carlos Pedraza and Marcelo Villa were employed by DreamGIS. Milton Romero, Adriana Sarmiento Dueñas and Dallan Beltrán Rojas were employed by 4D Elements Consultores. Paola Quintero and Harold Mauricio Martínez were employed by Urra S.A E.S.P. Josef Kellndrofer was employed by Earth Big Data LLC. The remaining author declares that the research was conducted in the absence of any commercial or financial relationships that could be construed as a potential conflict of interest.

References

1. Cabrera, E.; Vargas, D.M.; Galindo, G.; García, M.C.; Ordoñez, M.F. *Memoria Técnica de la Cuantificación de la Deforestación Histórica Nacional–Escalas Gruesa y Fina*; Instituto de Hidrología, Meteorología, y Estudios Ambientales-IDEAM: Bogotá, Colombia, 2011; p. 91.
2. Ustin, S.L.; Middleton, E.M. Current and Near-Term Advances in Earth Observation for Ecological Applications. *Ecol. Process.* **2021**, *10*, 1. [CrossRef] [PubMed]
3. Onoda, M.; Young, O.R. *Satellite Earth Observations and Their Impact on Society and Policy*; Springer: Singapore, 2017. [CrossRef]

4. Hansen, M.C.; Stehman, S.V.; Potapov, P.V. Quantification of Global Gross Forest Cover Loss. *Proc. Natl. Acad. Sci. USA* **2010**, *107*, 8650–8655. [CrossRef] [PubMed]
5. Reiche, J.; Lucas, R.; Mitchell, A.L.; Verbesselt, J.; Hoekman, D.H.; Haarpaintner, J.; Kellndorfer, J.M.; Rosenqvist, A.; Lehmann, E.A.; Woodcock, C.E.; et al. Combining Satellite Data for Better Tropical Forest Monitoring. *Nat. Clim. Chang.* **2016**, *6*, 120–122. [CrossRef]
6. Mitchell, A.L.; Rosenqvist, A.; Mora, B. Current Remote Sensing Approaches to Monitoring Forest Degradation in Support of Countries Measurement, Reporting and Verification (MRV) Systems for REDD+. *Carbon Balance Manag.* **2017**, *12*, 9. [CrossRef] [PubMed]
7. Lewis, S.L.; Edwards, D.P.; Galbraith, D. Increasing Human Dominance of Tropical Forests. *Science* **2015**, *349*, 827–832. [CrossRef]
8. Murillo-Sandoval, P.J.; Kilbride, J.; Tellman, E.; Wrathall, D.; Van Den Hoek, J.; Kennedy, R.E. The Post-Conflict Expansion of Coca Farming and Illicit Cattle Ranching in Colombia. *Sci. Rep.* **2023**, *13*, 1965. [CrossRef]
9. Joshi, N.; Mitchard, E.T.A.; Woo, N.; Torres, J.; Moll-Rocek, J.; Ehammer, A.; Collins, M.; Jepsen, M.R.; Fensholt, R. Mapping Dynamics of Deforestation and Forest Degradation in Tropical Forests Using Radar Satellite Data. *Environ. Res. Lett.* **2015**, *10*, 34014. [CrossRef]
10. Joshi, N.; Baumann, M.; Ehammer, A.; Fensholt, R.; Grogan, K.; Hostert, P.; Jepsen, M.R.; Kuemmerle, T.; Meyfroidt, P.; Mitchard, E.T.A.; et al. A Review of the Application of Optical and Radar Remote Sensing Data Fusion to Land Use Mapping and Monitoring. *Remote Sens.* **2016**, *8*, 70. [CrossRef]
11. van Oostende, M.; Hieronymi, M.; Krasemann, H.; Baschek, B.; Röttgers, R. Correction of Inter-Mission Inconsistencies in Merged Ocean Colour Satellite Data. *Front. Remote Sens.* **2022**, *3*, 882418. [CrossRef]
12. Shimada, M.; Itoh, T.; Motooka, T.; Watanabe, M.; Shiraishi, T.; Thapa, R.; Lucas, R. New Global Forest/Non-Forest Maps from ALOS PALSAR Data (2007–2010). *Remote Sens. Environ.* **2014**, *155*, 13–31. [CrossRef]
13. Lucas, R.; Rebelo, L.M.; Fatoyinbo, L.; Rosenqvist, A.; Itoh, T.; Shimada, M.; Simard, M.; Souza-Filho, P.W.; Thomas, N.; Trettin, C.; et al. Contribution of L-Band SAR to Systematic Global Mangrove Monitoring. *Mar. Freshw. Res.* **2014**, *65*, 589–603. [CrossRef]
14. Persaud, H.; Cabrera, I. Eficiencia de Las Imágenes de Radar Para El Monitoreo a Tiempo Casi Real de Bosques Tropicales En Guyana. *Arnaldoa* **2021**, *28*, 577–592. Available online: http://www.scielo.org.pe/scielo.php?pid=S2413-32992021000300577&script=sci_arttext (accessed on 11 December 2023).
15. Palomino-Ángel, S.; Anaya-Acevedo, J.A.; Simard, M.; Liao, T.-H.; Jaramillo, F. Analysis of Floodplain Dynamics in the Atrato River Colombia Using SAR Interferometry. *Water* **2019**, *11*, 875. [CrossRef]
16. Estupinan-Suarez, L.M.; Florez-Ayala, C.; Quinones, M.J.; Pacheco, A.M.; Santos, A.C. Detection and Characterizacion of Colombian Wetlands Using Alos Palsar and MODIS Imagery. *Int. Arch. Photogramm. Remote Sens. Spatial Inf. Sci.* **2015**, *XL-7/W3*, 375–382. [CrossRef]
17. Ayala, C.F.; Suárez, L.M.E.; Rojas, S.; Aponte, C. Identification and Mapping of Colombian Inland Wetlands. *Biota Colomb.* **2016**, *17*, 1–22. Available online: http://repository.humboldt.org.co/handle/20.500.11761/9352 (accessed on 9 September 2023).
18. Hoekman, D.H.; Quinones, M.J. Land Cover Type and Forest Biomass Assessment in the Colombian Amazon. In Proceedings of the IGARSS'97, 1997 IEEE International Geoscience and Remote Sensing Symposium Proceedings. Remote Sensing-A Scientific Vision for Sustainable Development, Singapore, 3–8 August 1997.
19. Hoekman, D.H.; Quinones, M.J. Land Cover Type and Biomass Classification Using AirSAR Data for Evaluation of Monitoring Scenarios in the Colombian Amazon. *IEEE Trans. Geosci. Remote Sens.* **2000**, *38*, 685–696. [CrossRef]
20. Anaya, J.A.; Rodríguez-Buriticá, S.; Londoño, M.C. Clasificación de cobertura vegetal con resolución espacial de 10 metros en bosques del Caribe colombiano basado en misiones Sentinel 1 y 2. *Rev. Teledetec.* **2023**, 29–41. [CrossRef]
21. Hoekman, D.; Quiñones, M.J. Biophysical Forest Type Characterization in the Colombian Amazon by Airborne Polarimetric SAR. *IEEE Trans. Geosci. Remote Sens.* **2002**, *40*, 1288–1300. [CrossRef]
22. Quiñones, M.J.; Vissers, M.; Pacheco-Pascaza, A.M.; Flórez, C.; Estupiñán-Suárez, L.M.; César, A.; Úrsula, J.; Claudia, H.; Dirk, H. Un enfoque ecosistémico para el análisis de una serie densa de tiempo de imágenes de radar Alos PALSAR, para el mapeo de zonas inundadas en el territorio continental colombiano. *Biota Colomb.* **2016**, *16*, 63–84. [CrossRef]
23. Pedraza, C.; Clerici, N.; Forero, C.F.; Melo, A.; Navarrete, D.; Lizcano, D.; Zuluaga, A.F.; Delgado, J.; Galindo, G. Zero Deforestation Agreement Assessment at Farm Level in Colombia Using ALOS PALSAR. *Remote Sens.* **2018**, *10*, 1464. [CrossRef]
24. Anaya, J.A.; Gutiérrez-Vélez, V.H.; Pacheco-Pascagaza, A.M.; Palomino-Ángel, S.; Han, N.; Balzter, H. Drivers of Forest Loss in a Megadiverse Hotspot on the Pacific Coast of Colombia. *Remote Sens.* **2020**, *12*, 1235. [CrossRef]
25. Klemas, V. Using Remote Sensing to Select and Monitor Wetland Restoration Sites: An Overview. *J. Coast. Res.* **2013**, *289*, 958–970. [CrossRef]
26. Reif, M.K.; Theel, H.J. Remote Sensing for Restoration Ecology: Application for Restoring Degraded, Damaged, Transformed, or Destroyed Ecosystems. *Integr. Environ. Assess. Manag.* **2017**, *13*, 614–630. [CrossRef] [PubMed]
27. Camarretta, N.; Harrison, P.A.; Bailey, T.; Potts, B.; Lucieer, A.; Davidson, N.; Hunt, M. Monitoring Forest Structure to Guide Adaptive Management of Forest Restoration: A Review of Remote Sensing Approaches. *New For.* **2020**, *51*, 573–596. [CrossRef]
28. Xie, L.; Wu, W.; Huang, X.; Ou, P.; Lin, Z.; Zhiling, W.; Song, Y.; Lang, T.; Huangfu, W.; Zhang, Y.; et al. Mining and Restoration Monitoring of Rare Earth Element (REE) Exploitation by New Remote Sensing Indicators in Southern Jiangxi, China. *Remote Sens.* **2020**, *12*, 3558. [CrossRef]
29. Koch, A.; Kaplan, J.O. Tropical Forest Restoration under Future Climate Change. *Nat. Clim. Chang.* **2022**, *12*, 279–283. [CrossRef]

30. Rodriguez, N.; Armenteras, D.; Morales, M.; Romero, M. *Ecosistemas de Los Andes Colombianos*; Instituto de Investigación de Recursos Biológicos Alexander von Humboldt: Bogotá, Colombia, 2006.
31. Masek, J.G.; Wulder, M.A.; Markham, B.; McCorkel, J.; Crawford, C.J.; Storey, J.; Jenstrom, D.T. Landsat 9: Empowering Open Science and Applications through Continuity. *Remote Sens. Environ.* **2020**, *248*, 111968. [CrossRef]
32. Helder, D.L.; Karki, S.; Bhatt, R.; Micijevic, E.; Aaron, D.; Jasinski, B. Radiometric Calibration of the Jandsat MSS Sensor Series. *IEEE Trans. Geosci. Remote Sens.* **2012**, *50*, 2380–2399. [CrossRef]
33. Li, J.; Roy, D.P. A Global Analysis of Sentinel-2a, Sentinel-2b and Landsat-8 Data Revisit Intervals and Implications for Terrestrial Monitoring. *Remote Sens.* **2017**, *9*, 902. [CrossRef]
34. Prudente, V.H.R.; Martins, V.S.; Vieira, D.C.; Silva, N.R.d.F.e.; Adami, M.; Sanches, I.D.A. Limitations of Cloud Cover for Optical Remote Sensing of Agricultural Areas across South America. *Remote Sens. Appl. Soc. Environ.* **2020**, *20*, 100414. [CrossRef]
35. Jafarbiglu, H.; Pourreza, A. Impact of Sun-View Geometry on Canopy Spectral Reflectance Variability. *ISPRS J. Photogramm. Remote Sens.* **2023**, *196*, 270–286. [CrossRef]
36. Asner, G.P. Automated Mapping of Tropical Deforestation and Forest Degradation: CLASlite. *J. Appl. Remote Sens.* **2009**, *3*, 033543. [CrossRef]
37. Nielsen, A.; Conradsen, K.; Simpson, J.J. Multivariate Alteration Detection (MAD) in Multispectral, Bi-Temporal Image Data: A New Approach to Change Detection Studies. *Remote Sens. Environ.* **1998**, *1*, 1–28. [CrossRef]
38. Rahaman, K.; Hassan, Q.; Ahmed, M. Pan-Sharpening of Landsat-8 Images and Its Application in Calculating Vegetation Greenness and Canopy Water Contents. *ISPRS Int. J. Geo-Inf.* **2017**, *6*, 168. [CrossRef]
39. Storey, J.; Scaramuzza, P.; Schmidt, G.; Barsi, J. Landsat 7 scan line corrector-off gap-filled product development. *Proceeding Pecora* **2005**, *16*, 23–27.
40. Werner, C.; Wegmüller, U.; Strozzi, T.; Wiesmann, A. *GAMMA SAR and Interferometric Processing Software*; European Space Agency (Special Publication) ESA SP.: São Paulo, Brazil, 2000; No. 461; pp. 211–219.
41. Lopes, A.; Touzi, R.; Nezry, E. Adaptive Speckle Filters and Scene Heterogeneity. *IEEE Trans. Geosci. Remote Sens.* **1990**, *28*, 992–1000. [CrossRef]
42. Galindo, G.; Espejo, O.J.; Rubiano, J.C.; Vergara, L.K.; Cabrera, E. *Protocolo de Procesamiento Digital de Imágenes Para La Cuantificación de La Deforestacion En Colombia V.2*; IDEAM, Instituto de Hidrología, Metereología y Estudios Ambientales: Bogotá, Colombia, 2014; pp. 1–225. [CrossRef]
43. Baranwal, A.; Bagwe, B.R.; Vanitha, M. Machine Learning in Python. *J. Mach. Learn. Res.* **2019**, *12*, 128–154. [CrossRef]
44. Tatsumi, K.; Yamashiki, Y.; Canales Torres, M.A.; Taipe, C.L.R. Crop Classification of Upland Fields Using Random Forest of Time-Series Landsat 7 ETM+ Data. *Comput. Electron. Agric.* **2015**, *115*, 171–179. [CrossRef]
45. Wang, J.; Zhao, Y.; Li, C.; Yu, L.; Liu, D.; Gong, P. Mapping Global Land Cover in 2001 and 2010 with Spatial-Temporal Consistency at 250m Resolution. *ISPRS J. Photogramm. Remote Sens.* **2015**, *103*, 38–47. [CrossRef]
46. Pal, M. Random forests for land cover classification. Mahesh Pal Department of civil engineering National Institute of technology, Kurukshetra. *Symp. A Q. J. Mod. Foreign Lit.* **1996**, *26*, 3510–3512. [CrossRef]
47. Teluguntla, P.; Thenkabail, P.S.; Oliphant, A.; Xiong, J.; Gumma, M.K.; Congalton, R.G.; Yadav, K.; Huete, A. A 30-m Landsat-Derived Cropland Extent Product of Australia and China Using Random Forest Machine Learning Algorithm on Google Earth Engine Cloud Computing Platform. *ISPRS J. Photogramm. Remote Sens.* **2018**, *144*, 325–340. [CrossRef]
48. Deng, J.S.; Wang, K.; Deng, Y.H.; Qi, G.J. PCA-based Land-use Change Detection and Analysis Using Multitemporal and Multisensor Satellite Data. *Int. J. Remote Sens.* **2008**, *29*, 4823–4838. [CrossRef]
49. DeFries, R.; Achard, F.; Brown, S.; Herold, M.; Murdiyarso, D.; Schlamadinger, B.; De Souza, C. Earth Observations for Estimating Greenhouse Gas Emissions from Deforestation in Developing Countries. *Environ. Sci. Policy* **2007**, *10*, 385–394. [CrossRef]
50. Olofsson, P.; Foody, G.M.; Stehman, S.V.; Woodcock, C.E. Making Better Use of Accuracy Data in Land Change Studies: Estimating Accuracy and Area and Quantifying Uncertainty Using Stratified Estimation. *Remote Sens. Environ.* **2013**, *129*, 122–131. [CrossRef]
51. Ruiz-Ramos, J.; Marino, A.; Boardman, C.; Suarez, J. Continuous Forest Monitoring Using Cumulative Sums of Sentinel-1 Timeseries. *Remote Sens.* **2020**, *12*, 3061. [CrossRef]
52. Urrea, V.; Ochoa, A.; Mesa, O. Seasonality of Rainfall in Colombia. *Water Resour. Res.* **2019**, *55*, 4149–4162. [CrossRef]
53. Susan Moran, M. Principles and Applications of Imaging Radar, Manual of Remote Sensing, 3rd Edition, Volume 2. *Eos Trans. Am. Geophys. Union* **1999**, *80*, 67.
54. Rüetschi, M.; Small, D.; Waser, L. Rapid Detection of Windthrows Using Sentinel-1 C-Band SAR Data. *Remote Sens.* **2019**, *11*, 115. [CrossRef]
55. Eriksson, L.E.B.; Fransson, J.E.S.; Soja, M.J.; Santoro, M. Backscatter Signatures of Wind-Thrown Forest in Satellite SAR Images. In Proceedings of the 2012 IEEE International Geoscience and Remote Sensing Symposium, Munich, Germany, 22–27 July 2012; IEEE: Munich, Germany, 2012; pp. 6435–6438. [CrossRef]

56. Kasischke, E.S.; Melack, J.M.; Dobson, M.C. The Use of Imaging Radars for Ecological Applications—A Review. *Remote Sens. Environ.* **1997**, *59*, 141–156. [CrossRef]
57. Flores, A.; Herndon, K.; Thapa, R.; Cherrington, E. *Synthetic Aperture Radar (SAR) Handbook: Comprehensive Methodologies for Forest Monitoring and Biomass Estimation*; Alabama University Huntsville: Huntsville, AL, USA, 2019. [CrossRef]

forests

MDPI

Article

Detection and Analysis of Forest Clear-Cutting Activities Using Sentinel-2 and Random Forest Classification: A Case Study on Chungcheongnam-do, Republic of Korea

Sol-E Choi [1], Sunjeoung Lee [1,*], Jeongmook Park [1], Suyeon Lee [2], Jongsu Yim [1] and Jintaek Kang [1]

1 Forest ICT Research Center, National Institute of Forest Science, 57, Hoegi-ro, Dongdaemun-gu, Seoul 02455, Republic of Korea; pine0630@gmail.com (S.-E.C.); pjm7@korea.kr (J.P.); yimjs@korea.kr (J.Y.); beg8bune@korea.kr (J.K.)

2 Department of Forestry, Environment and Systems, Kookmin University, 77, Jeongneung-ro, Seongbuk-gu, Seoul 02707, Republic of Korea; slee240201@gmail.com

* Correspondence: sunjleei@korea.kr; Tel.: +82-961-2955

Abstract: This study provides the methodology for the development of sustainable forest management activities and systematic strategies using national spatial data, satellite imagery, and a random forest machine learning classifier. This study conducts a regional province-scale approach that can be used to analyze forest clear-cutting in South Korea; we focused on the Chungcheongnam-do region. Based on spatial information from digital forestry data, Sentinel-2 satellite imagery, random forest (RF) classifier, and digital forest-type maps (DFTMs), we detected and analyzed the characteristics of clear-cut areas. We identified forest clear-cut areas (accounting for 2.48% of the total forest area). The methodology integrates various vegetation indices and the RF classifier to ensure the effective detection of clear-cut areas at the provincial level with an accuracy of 92.8%. Specific leaf area vegetation index (SLAVI) was determined as the most important factor for accurately detecting clear-cut areas. Moreover, using a DFTM, we analyzed clear-cutting characteristics in different forest types (including private, national, natural, and planted forests), along with age class and diameter-at-breast-height class. Our method can serve as a basis for forest management and monitoring by analyzing tree-cutting trends in countries with forest areas, such as Republic of Korea.

Keywords: forest clear-cutting; random forest classifier; vegetation indices; forest management; digital forestry data; integrating spatial data and satellite imagery

1. Introduction

Forests provide important basic provisional, ecosystem, and socioeconomic services that are essential for life at the global scale [1–3]. Forests also act as significant sinks for atmospheric carbon dioxide and play an important role in mitigating climate change. Hence, there is growing global recognition of the importance of forests, with respect to adapting to climate change and the effective regulation of atmospheric carbon dioxide concentrations. With respect to adapting to climate change, the importance of forest management has been increasingly emphasized. In response, various strategies have been deployed to monitor and assess forest management while exploring its role in environmental sustainability [4]. Globally, forests cover 4.06 billion ha, accounting for 31% of the total land area, and the area with forest management plans significantly increased from 233 million ha in 2000 to 2.05 billion ha by 2020 [5]. Generally, forest management activities follow a cyclical process of afforestation, forest tending, tree-cutting (harvesting), and reforestation (to produce forest products) to maintain the health of the forest ecosystem. Among forest management activities, tree-cutting is closely related to carbon emissions on a global scale, and regionally, it is deeply related to timber use and income. Therefore, tree-cutting should be performed comprehensively, while considering the forest's carbon sequestration and its economic and commercial situation. Furthermore, strategically implemented tree-cutting can ensure the sustained effectiveness of forests as carbon sinks. However, there is a global trend of increasing illegal tree-cutting. Therefore, quantitative tracking of tree-cutting at the national and regional levels is required to manage carbon emissions and conserve forest ecosystems.

Previous studies analyzed the impact of tree-cutting on carbon dioxide emissions and assessed the effectiveness of reforestation efforts at the regional and national scales. Such studies were conducted globally using remote sensing (RS) because RS is useful for large-scale detection and monitoring. These studies used multispectral sensors from MODIS, Landsat, and Sentinel-2 satellite imagery at moderate and finer spatial resolutions (100 m, 30 m, 10 m) to detect tree-cutting areas [6–12], used very-high-spatial-resolution imagery (e.g., WorldView-4 ~0.3 m, Pleiades' (1A/1B) ~0.5 m, and Quick bird ~1 m) to detect small-scale or selective tree-cutting, and applied methods to detect forest cover changes by classifying land use changes using multispectral bands or vegetation indices (VIs). The application of remote sensing to derive VIs, including the normalized difference vegetation index (NDVI) and enhanced vegetation index (EVI), is the most commonly adopted method for monitoring forest change. These studies confirmed that tree-cutting patterns can be accurately detected with high accuracy using multispectral bands; this is mainly because in a temporal image dataset, this type of cover change in forested land is generally expressed as a large spectral contrast [6,7]. In addition, clear-cutting studies that cover large geographical areas and are based on relatively dense and freely available multispectral imagery [13]. Recent studies focus on integrating VIs with machine-learning techniques to achieve accurate land use and forest change detection [14–18]. Machine learning techniques have attracted considerable interest from multidisciplinary fields, especially the field of remote sensing, because these techniques provide a high classification accuracy and are robust to noise [19–24]. Maximum likelihood, random forest (RF), deep learning, artificial neural networks (ANNs), conventional neural networks, and fuzzy classification are used to analyze forest and land cover changes. Among these classifiers, RF has gained attention and evolved due to its improved accuracy using training data, efficiency, and comparatively low computational process [16]. Therefore, the RF classifier has frequently been applied in forest distribution attribution algorithms; in general, it includes a suite of predictor variables to optimize the classification results. Notably, RF classification has proven to be a successful and powerful method for separating classes [25]. Moreover, the RF classifier is a popular ensemble method used for various remote sensing applications, e.g., vegetation mapping. The basic concept of an ensemble classifier is that a combination of multi-base classifiers can provide better prediction results than a single classifier; RF classifiers have been successfully applied for various classifications, using remote sensing data [19]. To achieve successful classification using RF, it is important to have sufficient and accurate data for training. Previous studies have only used ground truth data from areas where tree-cutting has occurred, identified through Google Earth, visual interpretation, and small size of field data [11,16,17,19,26,27]. The collection of ground truth data can help to calibrate other sources of data, such as remote sensing data, using geographic information systems (GISs), and also support the interpretation and analysis of satellite imagery and aerial photography, among other methods of remote data collection [28]. In the Republic of Korea, ground base national spatial data that can be used for training data to detect tree clear-cutting are collected. Therefore, using such spatial data would enable the accurate detection of clear-cutting.

The Republic of Korea shares the Yellow Sea with the People's Republic of China to the west and the East Sea and Korean Straits with Japan to the east and the south (125°04′ and 131°52′ E longitude, 38°27′ N latitude). The Republic of Korea's total forest area is 6.29 million ha, accounting for 62.2% of the country's land area; thus, the country requires robust forestry and effective activities to manage forest resources. Recently, in the Republic of Korea, with the increasing emphasis on the importance of the role of forests in climate change and the need for sustainable management, the country has initiated a policy shift toward environmentally sustainable harvesting methods and is currently discussing sustainable forest management to achieve carbon neutrality and the nationally determined contribution (NDC) targets. In the Republic of Korea, 67% of forests are privately owned [29], making the active participation of private forest owners essential for the effective implementation of related policies. However, most owners prefer clear-cutting to maximize profits. Therefore, a systematic approach is necessary to manage the cyclical system of clear-cutting, reforestation, and utilization for sustainable forest management. The Korea Forest Service implements a system to collect and manage spatiotemporal management information from local governments. As a national business, they have performed projects such as construction of Digital Forestry (DF) data, which spatializes information on forest management activities conducted in private forests. However, it has yet to be able to be used as a national data source due to large variations in collection rates by region. Local governments are responsible for permitting and managing tree-cutting activities in private forests; the collection and management of data on this level are insufficient, resulting in a lack of comprehensive data on tree-cutting in national spatial data. Therefore, collecting systematic data is necessary to manage carbon sink and establish sustainable forest management policies at the national level. Furthermore, the government should attempt to mediate and change forest management activities by arranging consultations with the aim of developing and promoting effective and sustainable forest management activities and policies.

The purpose of this study was to provide a methodology to detect and analyze clear-cut areas used by national spatial data (DF data) and VIs from Sentinel-2 imagery and an RF classifier. Furthermore, we aimed to understand forest management activities and support policymaking by identifying the characteristics of clear-cut areas through integration with DFTMs.

2. Materials and Methods

2.1. Study Area

In this study, the detection of tree-cutting was conducted in Chungcheongnam-do, and the training area for the RF was Buyeo-gun, a regional unit within Chungcheongnam-do. Chungcheongnam-do is one of the 17 provinces in the Republic of Korea, located in the west of the country's center (latitude: 35°59′02″–37°04′19″ N; longitude: 126°06′39″–127°37′23″ E), with a total forest area of 404,927 ha, accounting for 6.4% of the Republic of Korea's total forest area. In Chungcheongnam-do's forests, deciduous forests represent 37.0%, coniferous forests 46.0%, and mixed forests 12.3%. Medium-sized trees represent 66.0% of the tree size class, small-sized trees 24.0%, saplings 9.0%, and large-sized trees 1.0%. In addition, 87.0% of the forests in Chungcheongnam-do are privately owned, leading to various activities being performed to gain income from forests. Chungcheongnam-do was selected as the study area because it has a relatively large amount of Digital Forestry (DF) data (data constructed from spatial data between 2015 and 2022 that contains all aspects of forestry information at the local and national levels in the Republic of Korea and provides forest management activity information) containing information on forest management activities in the Republic of Korea compared to other provinces. Therefore, sufficient training and validation data can be used for RF.

The training area, Buyeo-gun (latitude: 36°16′35″ N; longitude: 126°54′40″ E), covers a land area of 62,450 ha, which amounts to 8% of the total land area of Chungcheongnam-do; 31,964 ha (47%) of land in Buyeo-gun is forested. Buyeo-gun was selected for its efficiency in running the RF classifier, as it had the highest amount of DF data per forest area within Chungcheongnam-do (Figure 1).

By using this scale-up method, it will be possible to detect tree-cutting in larger areas with only small amounts of data from field surveys or small-scale regions, thereby providing local governments and others with valuable data on clear-cut areas.

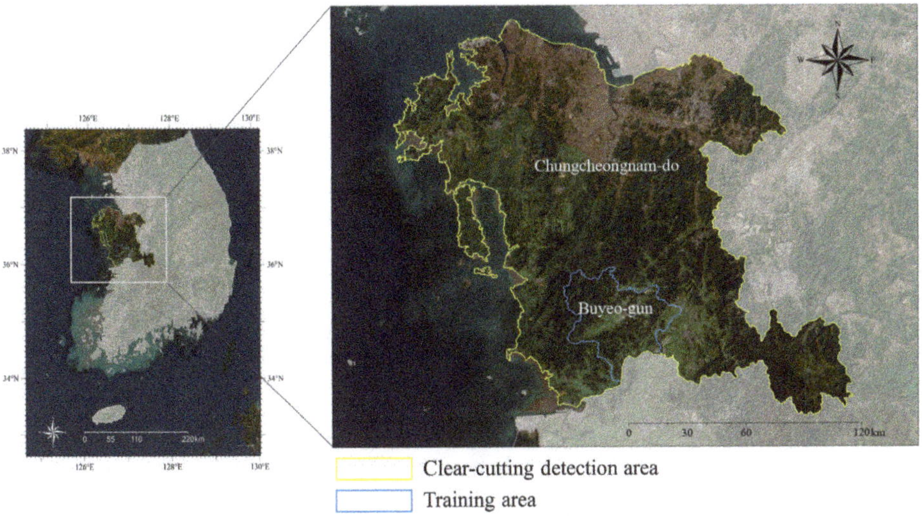

Figure 1. Map portraying clear-cut areas in the Republic of Korea, with the location and forest cover.

We performed the detection process in areas where the forest area in digital forest-type maps (DFTMs) overlapped with the forest land in cadastral maps (CDMs). In the Republic of Korea, forest areas have different definitions based on the land cover and land use concept, resulting in the construction of different spatial datasets. In DFTMs, using the land cover concept, forests are defined as covered forests if they cover areas of at least 0.5 ha, or plantation forests if they cover an area of more than 0.1 ha of trees [29]. However, CDMs, using the concept of land use and management, define forests as areas that contain trees, including rocky areas and small-scale residential areas within forest boundaries [30]. Forest management activities in the Republic of Korea are conducted based on CDMs. However, within CDMs, there are areas where the actual land cover is not forested, which can lead to the false detection of tree-cutting in these areas. Therefore, in this study, based on these two spatial datasets, we selected the areas that were uniquely classified as "forests". This approach ensures an accurate analysis of clear-cutting and avoids potential misclassifications of urban areas or other non-forested areas as clear-cut areas, thereby increasing the accuracy and relevance of the results. Table 1 presents the data considered for setting the spatial range for this study.

Table 1. Data considered for setting the spatial range for our analysis.

Data	Description	Source	Data Type	Year
Digital Forest-type Maps (scale: 1:5000)	Republic of Korea's forest information constructed from digitizing orthoimages	Forest Geospatial Information System (http://fgis.forest.go.kr, accessed on 3 December 2023)	Vector	2018
Forest land category in Cadastral Maps	Map dividing the national land into 28 categories, according to land use and status	National Spatial Infrastructure Portal (https://www.vworld.kr/dtna/dtna_guide_s001.do, accessed on 3 December 2023)	Vector	2018

The details of the forest areas considered in this study and their characteristics within the spatial data range in Chungcheongnam-do are shown in Table 2. The total forest area covers 329,042 ha, with national forests accounting for 40,275 ha and private forests accounting for 288,789 ha, with the latter accounting for 88% of the total forest area.

Table 2. Characteristics of forest area in Chungcheongnam-do Province *.

(Unit: ha)						
Division		**Conifer**	**Deciduous**	**Mixed**	**Total**
National Forests	Planted	5487	1903	87	7476
	Natural	8789	18,356	5622	32,767
	Subtotal	14,276	20,259	5708	40,243
Private Forests	Planted	51,333	33,572	1610	86,515
	Natural	70,341	98,128	33,813	202,284
	Subtotal	121,674	131,700	35,423	288,798
Total		135,950	151,960	41,132	329,042

* The forest areas were analyzed for overlapping areas in digital forest-type (DFTMs) maps and cadastral maps (CDMs).

2.2. *Workflow Overview*

The workflow consisted of the following steps: (1) obtaining vegetation indices, (2) acquiring the training and validation datasets from the vegetation index values that were calculated based on DF data for clear-cut areas (for RF classification), (3) detecting the clear-cut areas with RF and validating the detection results through accuracy assessment, and (4) analyzing the characteristics of the clear-cut areas using DFTMs (Figure 2).

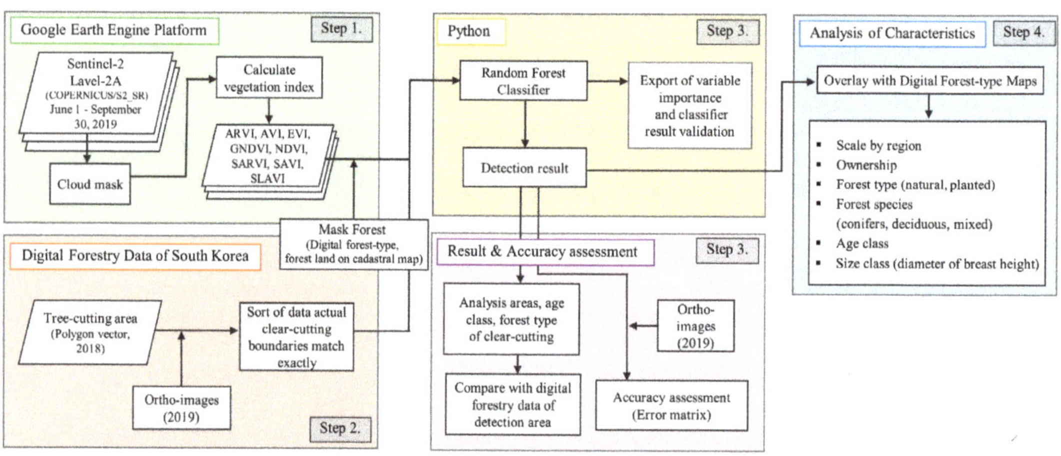

Figure 2. Flowchart portraying the methodology of this study. Abbreviations: atmospherically resistant vegetation index (ARVI), advanced vegetation index (AVI), soil and atmospherically resistant vegetation index (SARVI), soil-adjusted vegetation index (SAVI), enhanced vegetation index (EVI), green normalized difference vegetation index (GNDVI), normalized difference vegetation index (NDVI), specific leaf area vegetation index (SLAVI).

2.3. *Data Collection*

2.3.1. Vegetation Indices (VIs)

VIs were used to train the RF classifier to ensure the accurate detection of clear-cut areas. In this study, VIs were based on different wavelength reflectance in the electromagnetic spectrum for each target, achieved using both active and passive sensors [26]. These indices were designed to highlight specific properties of vegetation [27]. We used the VIs that minimized the effects of the atmosphere and soil, i.e., atmospherically resistant vegetation index (ARVI), advanced vegetation index (AVI), soil and atmospherically resistant vegetation index (SARVI), and soil-adjusted vegetation index (SAVI). Several previous studies developed VIs to minimize the effects of atmospheric aerosols and soil reflectance [28–32]. In addition, we used the green normalized difference vegetation index (GNDVI), normalized difference vegetation index (NDVI), enhanced vegetation index (EVI), and specific leaf area vegetation index (SLAVI), which are widely used to assess biomass and vegetation density [33,34]. We used these eight indices to determine the ones that were the most suitable for analyzing the areas affected by clear-cutting. The VIs were computed using the spectral bands of Sentinel-2 on the Google Earth Engine (GEE) platform (Table 2). Sentinel-2 data, which have a high spatial resolution of 10 m for bands in the visible and near-infrared spectra, allowed for the detection of changes in smaller areas and the identification of individual clear-cutting events; GEE provided the fundamental calculation functions for the raster and vector data, along with the application programming interfaces (APIs) for JavaScript and Python [35]. We calculated the VIs using the spectral bands of Sentinel-2 Level-2A (COPERNICUS/S2_SR) in the GEE. Level-2A processing included corrections for atmospheric gases, aerosols, and Rayleigh scattering effects, thus significantly improving the data quality. This enhanced quality was beneficial to accurately computing the VIs. The VIs were computed using the Level-2A images from 1 June 2019 to 30 September 2019, when the cloud cover was less than 10%. We used the median values of these indices from the images collected for the analysis. Table 3 presents the details of the VIs calculated for this study using Sentinel-2 data.

Table 3. Vegetation indices (VIs) calculated based on Sentinel-2 data.

Vegetation Indices	Remarks	Formula	References
Atmospherically Resistant Vegetation Index (ARVI)	Minimizes the impact of atmospheric scattering from aerosols	$\text{ARVI} = \frac{\text{NIR} - (2 \times \text{Red}) + \text{Blue}}{\text{NIR} + (2 \times \text{Red}) + \text{Blue}}$	[36]
Advanced Vegetation Index (AVI)	Provides vegetation cover by using the infrared spectral band to sensitively assess vegetation density	$\text{AVI} = (\text{NIR} \times (1 - \text{Red}) \times (\text{NIR} - \text{Red}))^{\frac{1}{3}}$	[36]
Enhanced Vegetation Index (EVI)	Corrects for atmospheric effects and soil background	$\text{EVI} = 2.5 \times \frac{\text{NIR} - \text{Red}}{\text{NIR} + 6 \times \text{Red} - 7.5 \times \text{Blue} + 1}$	[37]
Green Normalized Difference Vegetation Index (GNDVI)	More sensitive to changes in chlorophyll content than NDVI	$\text{GNDVI} = \frac{\text{NIR} - \text{Green}}{\text{NIR} + \text{Green}}$	[38]
Normalized Difference Vegetation Index (NDVI)	Presents vegetation cover ratio and vegetation characteristics such as biomass and chlorophyll content	$\text{NDVI} = \frac{\text{NIR} - \text{Red}}{\text{NIR} + \text{Red}}$	[39]
Soil and Atmospherically Resistant Vegetation Index (SARVI)	Minimizes the effect of atmospheric aerosols and reduces the influence of soil	$\text{SARVI} = \frac{\text{NIR} - ((\text{Red} - 1) \times (\text{Blue} - \text{Red}))}{\text{NIR} + ((\text{Red} - 1) \times (\text{Blue} - \text{Red}))} \times (1 + 0.5)$	[34]
Soil-Adjusted Vegetation Index (SAVI)	Minimizes spectral reflection variations based on soil types	$\text{SAVI} = 1.5 \times (\frac{\text{NIR} - \text{Red}}{\text{NIR} + \text{Red} + 0.5})$	[35]
Specific Leaf Area Vegetation Index (SLAVI)	Estimates the leaf area ratio within a forest	$\text{SLAVI} = \frac{\text{NIR}}{\text{NIR} + \text{SWIR}}$	[40]

Abbreviations: For the Sentinel-2 satellite, near-infrared (NIR) corresponds to band 8 (842 nm), 'Red' to band 4 (665 nm), 'Blue' to band 2 (490 nm), 'Green' to band 3 (560 nm), and short-wave infrared (SWIR) to band 11 (1610 nm) and band 12 (2190 nm).

2.3.2. Training and Validation Dataset

The training and validation datasets were obtained from the Republic of Korea's DF data for 2018. This dataset consisted of polygonal spatial data containing information on forest management, e.g., afforestation, tending, and harvest-cutting in national, public, and private forests. The purpose of the data collection was to systematically analyze and manage the forest management activities in the Republic of Korea, including the identification of target areas and prevention of duplication of forest-management activities. To detect clear-cutting in the Republic of Korea, we used the available Sentinel-2 satellite imagery. However, due to the spatial resolution limitations of the satellite images (10 m), detecting selective logging and forest thinning activities proved challenging. Therefore, our study focused on detecting clear-cutting. However, the spatial information for clear-cutting in this dataset is not well-detailed, because the data are based on forestland registers and the cadastral numbers of areas where forest management activities were conducted. To address this issue, we compared the DF data with orthoimages acquired from National Geospatial-Intelligence Platform (https://map.ngii.go.kr/, accessed on 24 December 2023) to identify the areas for which clear-cutting and spatial information matched exactly. In the Republic of Korea, clear-cutting generally begins after September and lasts for less than six months. The analysis used DF data from 2018, which is earlier than the acquisition time of the Sentinel-2 satellite imagery in 2019. From the 533 original data polygons collected for Buyeo-gun (the target area for training), we used 127 polygons as the training data.

2.4. Detection of Clear-Cut Areas Using Random Forest Classifier

We used an RF classifier that has been successfully applied for various classifications using remote sensing data [19]. The RF algorithm generates and combines multiple decision tree classifiers for classification and prediction and is known for its immunity to overfitting (relative to other models), making it a robust choice for such analyses. In addition, RF provides a simple method for evaluating the importance of each variable in the prediction process. This feature was used to identify the relevant spectral bands or derive the indices that were important for detecting the areas of change. The RF algorithm was used with VIs from the training datasets (based on DF data) to classify the clear-cut areas. The DF data were divided into an 80% training set and 20% validation set for processing. A five-fold cross-validation method was used within the training set to ensure robustness. The performance of the RF classifier was evaluated using various metrics, such as accuracy, precision, recall, and F1 score. The variable importance of VIs were evaluated using the mean decrease in impurity (MDI) importance method. This method measures the importance of each feature in a decision tree. The process is used to determine the importance of different features based on the decrease in data impurities. MDI importance is often used in tree-based machine learning models because it provides an intuitive understanding and straightforward interpretation of feature importance. Therefore, in this study, MDI was used to analyze the importance of VIs and determine the indices that should be prioritized when analyzing clear-cut areas. In this study, clear-cut areas were identified by the overlapping spatial extent of forest land in CDMs and DFTMs, thereby setting the detection area. Note that we excluded results for areas smaller than 0.5 ha. This exclusion was based on the regulatory context in the Republic of Korea, wherein clear-cutting activities within small areas (<0.5 ha) do not require national or local government permission. This exclusion was essential because data for smaller areas were generally not included in the dataset, leading to an increase in the error and inaccuracy of the study results.

2.5. Accuracy Assessment

To assess the accuracy of the RF classifier in detecting clear-cut areas, a confusion matrix was constructed using validation samples. This matrix was used to calculate the overall accuracy and kappa coefficients of the classifier. The process of constructing the confusion matrix included a comparative analysis of the 2019 orthoimages, focusing on sampled points within both clear-cut and non-clear-cut areas.

Visual interpretation methods were used for comparison. Clear-cut areas were primarily identified by specific land cover, e.g., grassland and the presence of harvested trees in rows (which are characteristic indicators of clear-cutting activities). The sample size for accuracy assessment was determined by random selection from each class (clear-cut and non-clear-cut areas). This sample size was determined while achieving a 95% confidence level and 3% sampling error for each pixel count, resulting in 1100 sample points for each class of more than 1068 clear-cut area and 1063 non-clear-cut area samples, respectively.

2.6. Characteristics Analysis of the Clear-Cut Areas

In this study, we used DFTMs to analyze the detection results. The maps included the spatial distribution and detailed attributes of forests in the Republic of Korea, e.g., tree species, types, age class (age classes are classified in 10-year intervals), and diameter at breast height (DBH) class. These data provide an understanding of the characteristics of the clear-cut areas identified by the RF classifier, e.g., ownership type (national or private forest), forest composition (natural or planted forest), age class, and DBH class. Thus, our study can serve as a foundation for the development of systematic management strategies for future clear-cutting activities in the Republic of Korea.

3. Results

3.1. Detection of Clear-Cut Areas

The analysis of clear-cut areas in Chungcheongnam-do (conducted using the RF classifier) detected that the total clear-cut area was 8164.0 ha, accounting for 2.5% of the total forested area (Figure 3). Compared to national and DF data statistics, our result indicates a more extensive coverage of clear-cut areas in Chungcheongnam-do. The area of clear-cutting detected by the RF classifier exceeded the estimations of 7300.6 ha reported in the DF data and those of 6456.0 ha reported in national statistical data [41].

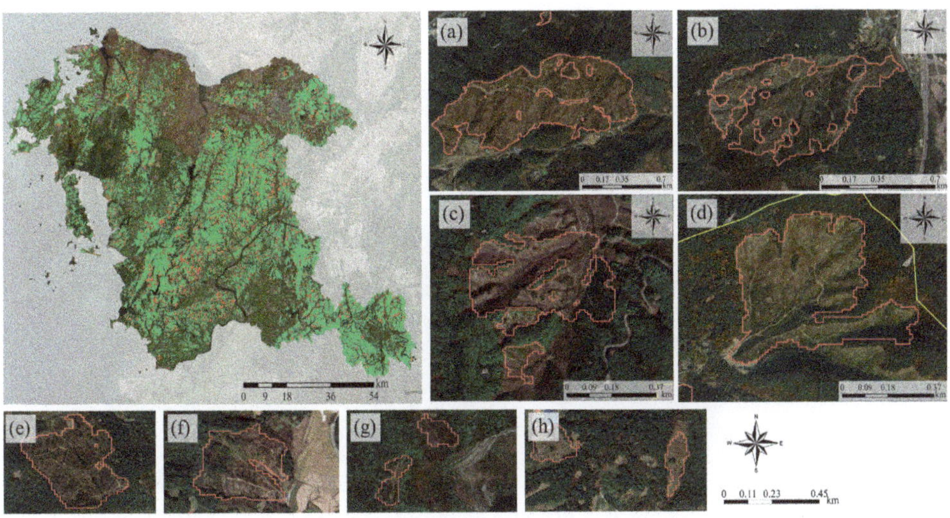

Figure 3. Detection result of clear-cut areas in Chungcheongnam-do province using a random-forest (RF) classifier. Scale of clear-cut areas: (**a**,**b**) 30 ha; (**c**,**d**) 10 to 29 ha; (**e**,**f**) 5 to 9 ha; (**g**,**h**) 1 to 4 ha.

For Chungcheongnam-do, the analysis of clear-cutting detection results at the regional level indicated that the total clear-cut area composed of areas of 1 to 4 ha in scale was 3806.1 ha, accounting for 46.6% of the total clear-cut area in Chungcheongnam-do, the highest ratio. In the Republic of Korea, forest management tasks are directly managed by local governments. As a result, local governments are responsible for authorizing clear-cutting activities and providing related data to the government. However, they do not provide sufficient data on areas where clear-cutting is performed. Therefore, this study analyzed the characteristics of clear-cutting in various regions of Chungcheongnam-do. The results showed that large-scale clear-cut areas, ranging from 30 ha to 60 ha, were observed in Cheongyang-gun (119.0 ha), Gongju-si (85.2 ha), and Seosan-gun (32.4 ha). In particular, Gongju-si contained the highest clear-cut area (1470.0 ha) and Gyeryong-si had the smallest total clear-cut area (45.8 ha) (Table 4). This result has confirmed that there are differences in the scale of clear-cutting among the various regions of Chungcheongnam-do. Areas where large-scale clear-cutting has been conducted, as well as regions with relatively large clear-cut areas, have been identified. Furthermore, this result enables targeted management and supervision within the province for those areas that require it.

Table 4. Scale and extent of clear-cut areas in different regions of Chungcheongnam-do Province.

(Unit: ha)						
Region	**<1ha**	**1~4 ha**	**5~9 ha**	**10~29 ha**	**30~60 ha**	**Total**
Asan-si	64	188	103	11	-	366
Boryeong-si	102	400	173	67	-	742
Buyeo-gun	224	580	173	314	-	1291
Cheongyang-gun	130	412	234	89	119	984
Dangjin-si	94	239	47	50	-	431
Cheonan-si	97	341	123	132	-	692
Geumsan-gun	56	185	120	84	-	444
Gongju-si	208	578	315	283	85	1470
Gyeryong-si	18	14	14	-	-	46
Hongseong-gun	76	206	39	22	-	344
Nonsan-si	62	154	77	45	-	339
Seocheon-gun	55	155	7	25	-	241
Seosan-si	53	148	72	19	32	325
Taean-gun	50	64	30	-	-	145
Yesan-gun	63	142	50	48	-	304
Total	1354	3806	1577	1190	237	8164

3.2. Accuracy Assessment, Variable Value, and Importance

With respect to the performance of the RF classifier for Buyeo-gun, the accuracy metrics indicated high precision and effectiveness. The classifier achieved precision, recall, and F1 scores of 1.0, indicating perfect classification accuracy. In the case of clear-cut area class detection accuracy, the precision, recall, and F1 score were 0.99, 0.76, 0.86, respectively. The overall average for both clear-cut area and forest area classes was impressive, with a precision of 0.99, a recall of 0.88, and an F1 score of 0.93. Furthermore, the accuracy assessment for detecting clear-cut areas in Chungcheongnam-do using the error matrix showed an overall accuracy of 92.8% and a kappa coefficient of 85.6%. These results indicated an "almost perfect" level of accuracy in the classifier's performance. In detail, the producer's accuracy for clear-cut areas was 91.2% and the user's accuracy was 90.9%. For non-clear-cut areas, the producer and user accuracies were 94.5 and 94.7%, respectively. These results demonstrate the robustness and reliability of the RF classifier in detecting clear-cut and non-clear-cut forest areas.

The differences in VIs between clear-cut and non-clear-cut areas in Buyeo-gun were statistically verified. In non-clear-cut forest areas, the mean values of ARVI, AVI, EVI, GNDVI, and NDVI were 0.5–0.8, with standard deviations of 0.08–0.1. For SARVI and SAVI, the mean was 1.1, with a standard deviation of 0.2. The mean value of SLAVI was 2.8, with a standard deviation of 0.87 (Figure 4a). In clear-cut areas, the mean values of ARVI, AVI, and EVI were 0.3–0.4, with standard deviations of 0.1–0.2. The GNDVI and NDVI had mean values of 0.5, with standard deviations of 0.1 and 0.2, respectively. The SARVI and SAVI had mean values of 0.6 and 0.7, respectively, with standard deviations of 0.3, respectively. The mean SLAVI value was 1.2, with a standard deviation of 0.7 (Figure 4b).

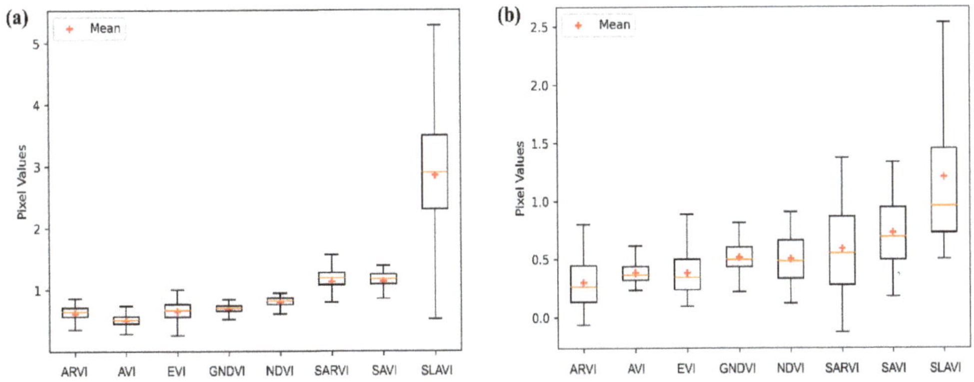

Figure 4. Vegetation index (VI) statistics for non-clear-cut (**a**) and clear-cut areas (**b**) in the training dataset. Abbreviations: atmospherically resistant vegetation index (ARVI), advanced vegetation index (AVI), soil and atmospherically resistant vegetation index (SARVI), soil-adjusted vegetation index (SAVI), enhanced vegetation index (EVI), green normalized difference vegetation index (GNDVI), normalized difference vegetation index (NDVI), specific leaf area vegetation index (SLAVI).

When assessing the importance of the variables within the model, SLAVI was ranked as the most important index, whereas EVI and AVI were ranked as the least important indices. This difference may be attributed to SLAVI's wide range of values compared to other VIs and its clear differentiation between values, specifically for the clear-cut areas. In contrast, EVI and AVI received lower importance scores because of the minor differences between values for clear-cut and non-clear-cut areas. The results indicate the effectiveness of SLAVI in identifying clear-cut areas; we conclude that it is an effective vegetation index for forest management monitoring. Table 5 presents the variable importance of the eight variables considered in this study based on the mean decrease in impurity (MDI) method.

Table 5. Variable importance as determined by the mean decrease in impurity (MDI) method.

Variables	Importance
SLAVI	0.261940
SARVI	0.193277
NDVI	0.171838
ARVI	0.136113
SAVI	0.079224
GNDVI	0.064720
EVI	0.049265
AVI	0.043622

Abbreviations: Atmospherically resistant vegetation index (ARVI); advanced vegetation index (AVI); soil and atmospherically resistant vegetation index (SARVI); soil-adjusted vegetation index (SAVI); enhanced vegetation index (EVI); green normalized difference vegetation index (GNDVI); normalized difference vegetation index (NDVI); specific leaf area vegetation index (SLAVI).

3.3. Characteristics of Clear-Cut Areas

The results obtained using the RF classifier and analyzed using the DFTMs demonstrated the effectiveness of the proposed method for characterizing clear-cut areas. In Chungcheongnam-do, 2.4% of privately owned forests and 0.1% of national forests were clear-cut areas relative to the total forest area. The analysis showed that within national forests, clear-cut areas were more prominent in natural forests than in planted forests, with the most extensive clear-cut areas observed in age class V (Figure 5a). In private forests, larger clear-cut areas were observed in age classes IV, V, and I, compared to those observed in other age classes. Furthermore, within privately owned forests, clear-cut areas were more extensive in natural forests than in planted forests (Figure 5b). However, the results of clear-cut areas divided by natural and planted forests in Chungcheongnam-do show that, in private forests, the proportion of clear-cut areas in planted forests was 3.4% of the total planted forest area, while the proportion of clear-cut areas in natural forests was 1.9% of the total natural forest area (Table 6). Moreover, the total clear-cut area in planted forests was higher, at 3.5%, than the area in natural forests, at 2.1%. The reason for this is that although natural forests represent 71.4% of the total forest area in Chungcheongnam-do, the proportion of the actual clear-cut area is lower than that of planted forests.

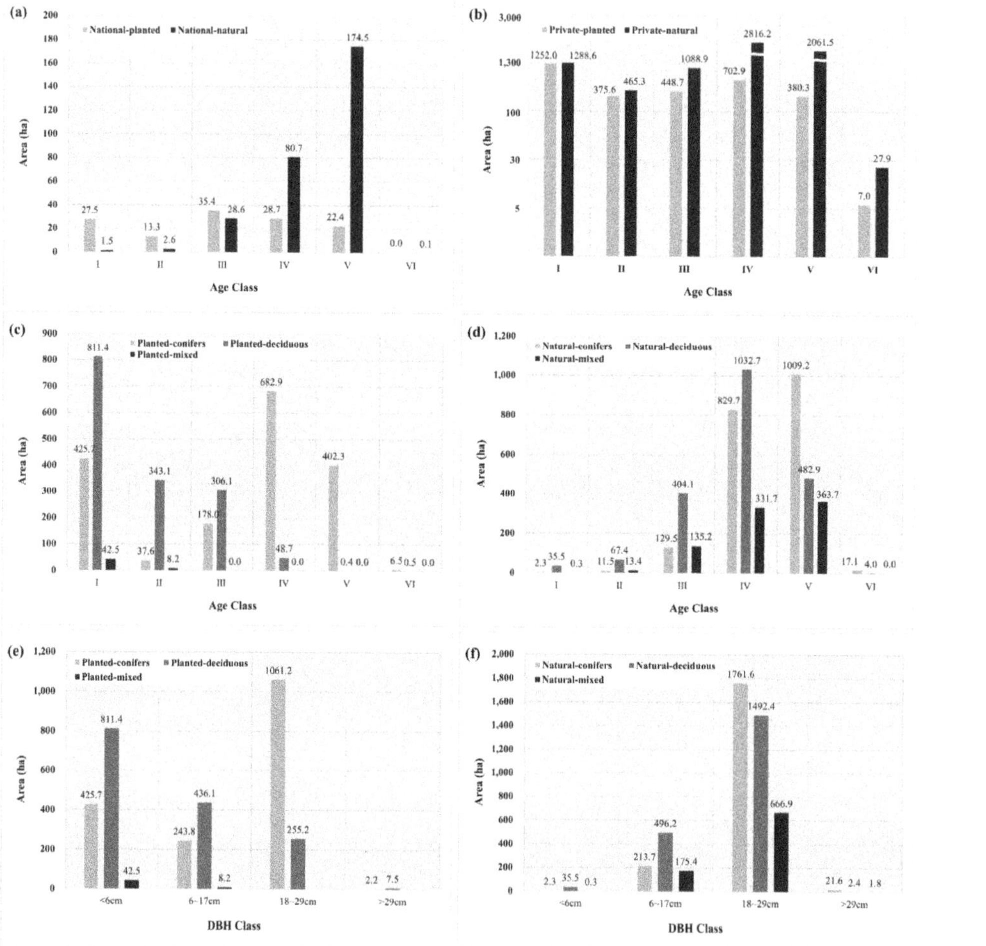

Figure 5. Characteristics of clear-cut areas in the study region obtained using digital forest-type maps. Characteristics of natural and planted forests by (**a**) national and (**b**) privately owned forests. Tree species and age class in (**c**) planted and (**d**) natural forests. Tree species and DBH class in (**e**) planted and (**f**) natural forests.

Table 6. Ratio of clear-cut area to total area of planted and natural forests by age class in Chungcheongnam-do.

(Unit: %)

Division		I	II	III	IV	V	VI	Ratio to Total Area
Planted	National forest	0.1	0.1	0.2	0.1	0.2	0.0	0.1
	Private forest	4.2	3.3	2.8	2.8	3.3	4.4	3.4
	Total	4.3	3.5	3.0	2.9	3.5	4.4	3.5
Natural	National forest	0.1	0.0	0.1	0.1	0.2	0.0	0.1
	Private forest	2.0	1.7	1.9	1.8	2.3	0.7	1.9
	Total	2.1	1.7	2.0	1.9	2.5	0.8	2.1

In planted forests, clear-cutting mainly occurred in age class I, with deciduous species being cut more frequently than coniferous species. The largest clear-cut area in planted forests was observed in age class I deciduous species; we noted a decreasing trend in clear-cut areas in older age classes. In coniferous forests, age class IV experienced the most extensive clear-cutting activities, followed by classes I and V (Figure 5c). In natural forests, the highest clear-cutting occurred in age classes IV and V, with deciduous and coniferous species undergoing clear-cutting mainly in age classes IV and class V, respectively (Figure 5d). However, the ratio of clear-cut trees by species was found to be similar, at 1.8% for conifers and 1.6% for deciduous trees, and the ratio of harvested area by species was found to be the highest, at 4.0%, for age class VI conifers in planted forests. In natural forests, conifers and deciduous trees underwent clear-cutting at a similar rate of 0.9%, with the highest rate of clear-cutting at 1.9% for age class I deciduous trees (Table 7).

Table 7. Ratio of clear-cut area to total area by tree species and age class in Chungcheongnam-do.

(Unit: %)

Division		I	II	III	IV	V	VI	Ratio to Total Area
Planted	Coniferous	1.4	0.3	1.1	2.7	3.5	4.0	1.8
	Deciduous	2.7	3.1	1.9	0.2	0.0	0.3	1.6
	Mixed	0.1	0.1	0.0	0.0	0.0	0.0	0.1
	Total	4.3	3.5	3.0	2.9	3.5	4.4	3.5
Natural	Coniferous	0.1	0.2	0.4	0.7	1.4	0.6	0.9
	Deciduous	1.9	1.2	1.2	0.9	0.7	0.1	0.9
	Mixed	0.0	0.2	0.4	0.3	0.5	0.0	0.4
	Total	2.1	1.7	2.0	1.9	2.5	0.8	2.1

With respect to DBH class in clear-cut areas in planted and natural forests, a higher proportion of clear-cut areas was found in classes with medium-diameter trees (18–29 cm), accounting for 40 and 80.5% of planted and natural forests, respectively (Figure 5e,f).

According to DBH class, the highest proportion of clear-cut areas in planted forests was found for small-diameter trees, which accounted for 4.1%, followed by large-diameter trees, at 3.2%. In natural forests, the highest proportion of young growth reproduction was found to be 2.4%, followed by 2.1% in the medium-diameter tree class (Table 8). The result of clear-cutting trends obtained through overlay analysis with DFTMs shows that while there are differences between natural and planted forests in terms of area and tree species when considering the composition of forests within Chungcheongnam-do, a higher rate of clear-cutting is observed in planted forests compared to natural forests. The rates of clear-cutting are highest in age classes V and VI, and tree species are clear-cut relatively similarly. Additionally, in terms of DBH class results, clear-cutting occurred prominently in young growth reproduction, followed by higher rates in larger diameter classes.

Table 8. Ratio of clear-cut area to total area by tree species and DBH class in Chungcheongnam-do.

(Unit: %)						
Division		**<6 cm**	**6~17 cm**	**18~29 cm**	**>29 cm**	**Ratio to Total Area**
Planted	Coniferous	1.3	0.9	2.5	0.7	1.7
	Deciduous	2.6	1.6	0.6	2.4	1.5
	Mixed	0.1	0.0	0.0	0.0	0.1
	Total	4.1	2.5	3.1	3.2	3.3
Natural	Coniferous	0.1	0.4	0.9	0.9	0.8
	Deciduous	2.3	1.0	0.8	0.1	0.8
	Mixed	0.0	0.3	0.3	0.1	0.3
	Total	2.4	1.8	2.0	1.1	2.0

4. Discussion

4.1. Discussion of Result

This study used spatial data on forest management from the Republic of Korea to detect and analyze clear-cut areas at the provincial level. VIs from Sentinel-2 imagery and an RF classifier were applied to achieve this. Overlay analysis with DFTMs further identified clear-cutting characteristics, enhancing our understanding of forest management's spatial dynamics.

Although the RF classifier in this study was highly accurate, it detected discrepancies of 863.4 ha and 1708.0 ha in clear-cut areas in Chungcheongnam-do when compared to national statistics and DF data, respectively. As indicated by Ham et al. [42], these discrepancies may be due to inconsistencies between spatial and management registry data. National statistical data on forest management in the Republic of Korea are based on cadastral registry data, which may not be consistent with spatially focused data. In a scenario wherein accurate registry data on forest management activities are unavailable, DF data are generally based on the area of cadastral parcels, resulting in discrepancies from the actual areas of activities. Registry data and spatial discrepancies indicate potential variances in actual forest management activities. The significance of a well-defined training dataset is critical in machine learning models, affecting model accuracy and reliability. Therefore, for accuracy and consistency, obtaining precise spatial and location information through comparative analysis with remote sensing results [42,43], as shown in this study, is important. Enhancing the accuracy of national statistics and detection outcomes necessitates boosting DF data collection rates and ensuring precise field and spatial data gathering. The RF classifier's accuracy is significantly influenced by training data characteristics, such as sample size, class proportions, and spatial autocorrelation. Utilizing a larger training sample has been shown to reduce both out-of-bag and independent assessment errors. Therefore, larger training sample sizes are recommended to enhance RF classifier accuracy and stability [44]. This underscores the importance of extensive field data collection and the continuous enhancement of digital forest management resources. Such data collection and processing advancements facilitate more precise and reliable assessments and can support effective forest management and policymaking.

This study detected clear-cut areas effectively using VIs. The importance of the variables in the RF classifier was demonstrated. Compared to previous studies, which primarily relied on NDVI and EVI, we included additional indices to reduce soil and atmospheric effects, improving accuracy [45–50]. These VIs enhanced land cover classification for forests and other land types. Including diverse datasets broadened the spectral signatures for RF classifier training, enhancing classification accuracy with training data variability [51,52].

Among the eight VIs evaluated, SLAVI, which reflects changes in plant leaf surface and photosynthetic activity, was the most important. Morcillo-Pallarés et al. [53] confirmed the effectiveness of SLAVI in identifying vegetation conditions, as it has a positive correlation with NIR and SWIR bands and a negative correlation with the red band. SLAVI's design also makes it unaffected by cloud shadows, providing reliable detection capabilities [54]. Bolyn et al. [55] and Boonprong et al. [56] noted the critical role of SWIR in mapping forest species and monitoring forest regeneration, with SLAVI being the only index to include the SWIR band. Although EVI and AVI were useful for evaluating vegetation productivity and coverage, SLAVI was found to be superior in detecting vegetation presence and reducing soil reflectance effects, making it ideal for identifying sparse vegetation areas [40]. Future research should focus on exploring SLAVI's applicability in detecting clear-cutting.

In the analyzing of characteristics, the majority of clear-cutting was noted in age classes III–V, which corresponded to the legal harvest age. However, we noted a considerable amount of clear-cutting of deciduous species within age class I in planted forests. Related to these age class results, it was also found that the rate of clear-cut areas with a DBH class of less than 6 cm is high. These results are considered to be errors due to limitations in detecting clear-cut areas due to the resolution and time-series of satellite imagery, where areas already afforested as age class I do not appear to have a dense tree cover. These errors may also be related to the update cycle of DFTMs. In the Republic of Korea, attribute information for areas with forest management activities is updated annually. For other areas, attribute information is updated for one region per year, and it takes up to five years for the entire dataset to be updated. Therefore, future research should consider the regional update cycles of DFTMs to detect and analyze the characteristics of clear-cut areas more accurately. Using additional data from afforestation areas can help identify clear-cutting characteristics more accurately.

As a result, the detection, verification, and characteristics analysis of clear-cut areas require the use of data with more accurate information. To improve data precision and provide more meaningful and officially usable results at the national level, it is essential to construct data on forest management activities in a complementary approach, as was performed in this study, through the analysis of field or administrative records (documents or spatial data) and remote sensing data.

4.2. Limitation

This study has limitations. First, detection was limited to a single period, which may have caused discrepancies with national statistics. Previous research has shown the effectiveness of the RF classifier in detecting land use change and forest dynamics [16,57,58]; however, these studies have been limited to single-period analysis. This approach may include areas where clear-cutting occurred just before the base year of 2018, such as in 2017 or 2016, without reforestation. Therefore, time-series analysis is needed to accurately identify clear-cut areas over time [6,7,26,59–61]. Our study encountered limitations in conducting a time-series analysis due to the beginning of clear-cutting data construction in 2018 DF data, with no previous data. Future studies should focus on time-series analysis starting in 2018 to detect clear-cut areas.

Second, in this study, while high-accuracy detection was achieved through the RF classifier, this study identified false positives in areas with lower forest density and exposed ground, such as rocky areas and forest roads. Therefore, it is possible to identify a more suitable machine learning approach for detecting clear-cut areas through comparative analysis with other methods (e.g., CNNs, U-net, AdaBoost, K-NN, ANN). These machine learning methods can effectively learn high-dimensional features, improve generalization abilities, and iteratively refine training results for enhanced detection accuracy. Therefore, future research should compare these machine learning approaches to evaluate their utility and accuracy. Additionally, our result indicates the necessity of incorporating ancillary data, such as land cover and forest road network maps, in future research to enhance detection accuracy.

Third, the spatial resolution limitations of remote sensing data can impact the accuracy of the detected boundaries and areas. The detectable size of clear-cut areas varies with the resolution of remote sensing data. Junaid et al. [16] detected large tree-cutting areas at a 50 m resolution and identified smaller areas at a 20 m resolution. Sentinel-2 imagery offers meaningful information at the national and regional scales, as supported by various studies [17,62–64]. In this study, we detected small clear-cut areas, but spatial resolution may have led to inaccuracies in boundary detection, especially at the edges. These discrepancies can be rectified with higher-resolution aerial imagery. If high-resolution remote sensing imagery is used, it is useful to combine a RF classifier with texture analysis [65]. Clear-cut areas reveal the ground surface, which differs in characteristics from areas with grassland and shrubbery or regions with rocks. This integration can help reduce the false positives associated with resolution and provide a more accurate analysis by including additional factors related to the object's characteristics.

4.3. Implications

Integrating remote sensing imagery and RF classifier detection results with spatial datasets is a powerful approach for collecting comprehensive data for analyzing and managing clear-cut areas. Forests are characterized by diverse activities and varying ownership, covering extensive areas. Collecting data on forest management activities can be challenging due to regional diversity. Integrating remote sensing clear-cutting detection with spatial data containing forest characteristics information (DFTMs) provides valuable insights into forest land use and cover. Additionally, analyzing ancillary spatial data such as forest roads, elevation, and slope can provide insights into clear-cutting practices and help identify illegal clear-cut areas. This methodology can provide deeper insights into forest conditions and management practices and enhance our understanding of the management status of clear-cutting at the regional and national levels for forest management decision-making. This approach improves our understanding of forest dynamics and aids in developing sustainable management strategies that balance ecological, economic, and social needs, ensuring the preservation of forest ecosystems. It is a methodology that can be adapted to different regions and countries to enhance forest management and detect areas targeted by REDD+ (reducing emissions from deforestation and forest degradation plus) in various countries.

5. Conclusions

In this study, we analyzed clear-cut areas in the Chungcheongnam-do region in the Republic of Korea. We accurately detected clear-cut areas by integrating spatial forest management data (DF data) with high-resolution Sentinel-2 satellite imagery using an RF classifier. The proposed approach, which combined various VIs and machine learning techniques, proved effective in detecting and analyzing clear-cut areas at the regional and provincial levels. We identified clear-cutting patterns in different forest types and age classes using DFTMs. Furthermore, we verified the importance of specific VIs, particularly SLAVI, for accurately detecting clear-cut areas. However, the results portrayed a greater extent of clear-cutting than that reported in national statistics. This discrepancy may be related to limitations and inaccuracies in national and spatial data. The discrepancies between spatial and registry data and the limitations associated with single-period analyses may be the causes of this issue. However, these limitations provide opportunities for future studies. Notably, our study confirms the importance of conducting time-series analyses, including additional ancillary data, and considering spatial resolution when detecting clear-cut areas in forests.

Our study offers an effective methodology for monitoring and analyzing clear-cutting activities, thus contributing to forest management. This study presents an effective methodology for continuous data collection for forest clear-cut areas. The collected data can be used for the establishment of forest policies at the regional and national levels. Monitoring clear-cut areas is important for improving the efficiency of national forest restoration projects and establishing sustainable forest management strategies. The insights from our study can support policymaking and the development and implementation of sustainable and systematic forest management strategies at the national and regional levels. Notably, our method can serve as a basis for forest management and monitoring by analyzing tree-cutting trends in countries with complex forest conditions, such as Republic of Korea. Furthermore, this methodology could usefully be applied in identifying target areas for carbon reduction, such as REDD+.

Author Contributions: Conceptualization and methodology, S.-E.C., J.P. and S.L. (Sunjeoung Lee); investigation and analysis, S.-E.C. and S.L. (Suyeon Lee); writing, S.-E.C. and J.P.; review and editing, S.L. (Sunjeoung Lee) and J.Y.; funding acquisition, S.L. (Sunjeoung Lee); and supervision, J.K. All authors have read and agreed to the published version of the manuscript.

Funding: This study was funded by the R&D Program for Forest Science Technology (Project No. FM0200-2023-01-2023), provided by the National Institute of Forest Science.

Data Availability Statement: Data are contained within the article.

Conflicts of Interest: The authors declare no conflicts of interest.

References

1. Potapov, P.; Yaroshenko, A.; Turubanova, S.; Dubinin, M.; Laestadius, L.; Thies, C.; Aksenov, D.; Egorov, A.; Yesipova, Y.; Glushkov, I.; et al. Mapping the World's intact forest landscapes by remote sensing. *Ecol. Soc.* **2008**, *13*, 51. [CrossRef]
2. Food and Agriculture Organization of the United Nations (FAO). *Global Forest Resources Assessment 2010*; Food and Agriculture Organization of the United Nations (FAO): Rome, Italy, 2010.
3. Hansen, M.C.; Potapov, P.V.; Moore, R.; Hancher, M.; Turubanova, S.A.; Tyukavina, A.; Thau, D.; Stehman, S.V.; Goetz, S.J.; Loveland, T.R.; et al. High-Resolution Global Maps of 21st-Century Forest Cover Change. *Science* **2013**, *342*, 850–853. [CrossRef]
4. Agrawal, A.; Hajjar, R.; Liao, C.; Rasmussen, L.V.; Watkins, C. Editorial overview: Forest governance interventions for sustainability through information, incentives, and institutions. *Curr. Opin. Environ. Sustain.* **2018**, *32*, A1–A7. [CrossRef]
5. Food and Agriculture Organization of the United Nations (FAO). *Global Forest Resources Assessment 2020*; Food and Agriculture Organization of the United Nations (FAO): Rome, Italy, 2020.
6. Olsson, H. Changes in satellite-measured reflectances caused by thinning cuttings in boreal forest. *Remote Sens. Environ.* **1994**, *50*, 221–230. [CrossRef]
7. Nilson, T.; Olsson, H.; Anniste, J.; Lükk, T.; Praks, J.; Change, T.-C. Thinning-caused change in Reflectance of ground vegetation in boreal forest. *Int. J. Remote Sens.* **2001**, *22*, 2763–2776. [CrossRef]
8. Bucha, T.; Stibig, H.J. Analysis of MODIS imagery for detection of clear cuts in the boreal forest in North-West Russia. *Remote Sens. Environ.* **2008**, *112*, 2416–2429. [CrossRef]
9. Santoro, M.; Pantze, A.; Fransson, J.E.S.; Dahlgren, J.; Persson, A. Nation-wide clear-cut mapping in Sweden using ALOS PALSAR strip images. *Remote Sens.* **2012**, *4*, 1693–1715. [CrossRef]
10. Cohen, W.B.; Healey, S.P.; Yang, Z.; Stehman, S.V.; Brewer, C.K.; Brooks, E.B.; Gorelick, N.; Huang, C.; Hughes, M.J.; Kennedy, R.E.; et al. How similar are forest disturbance maps derived from different landsat time series algorithms? *Forests* **2017**, *8*, 98. [CrossRef]
11. Palaiologou, P.; Essen, M.; Hogland, J.; Kalabokidis, K. Locating forest management units using remote sensing and geostatistical tools in north-central Washington, USA. *Sensors* **2020**, *20*, 2454. [CrossRef] [PubMed]
12. Chen, X.; Zhao, W.; Chen, J.; Qu, Y.; Wu, D.; Chen, X. Mapping large-scale forest disturbance types with multi-temporal CNN framework. *Remote Sens.* **2021**, *13*, 5177. [CrossRef]
13. Xie, Y.; Sha, Z.; Yu, M. Remote sensing imagery in vegetation mapping: A review. *J. Plant Ecol.* **2008**, *1*, 9–23. [CrossRef]
14. Wang, S.; Yang, D.; Li, Z.; Liu, L.; Huang, C.; Zhang, L. A global sensitivity analysis of commonly used satellite-derived vegetation indices for homogeneous canopies based on model simulation and random forest learning. *Remote Sens.* **2019**, *11*, 2547. [CrossRef]
15. Ferchichi, A.; Ben Abbes, A.B.; Barra, V.; Farah, I.R. Forecasting vegetation indices from spatio-temporal remotely sensed data using deep learning-based approaches: A systematic literature review. *Ecol. Inform.* **2022**, *68*, 101552. [CrossRef]
16. Junaid, M.; Sun, J.; Iqbal, A.; Sohail, M.; Zafar, S.; Khan, A. Mapping, LULC Dynamics and its potential implication on forest cover in malam Jabba region with landsat time series imagery and random forest classification. *Sustainability* **2023**, *15*, 1858. [CrossRef]
17. Ma, T.; Hu, Y.; Wang, J.; Beckline, M.; Pang, D.; Chen, L.; Ni, X.; Li, X. A novel vegetation index approach using Sentinel-2 data and random forest algorithm for estimating forest stock volume in the Helan Mountains, Ningxia, China. *Remote Sens.* **2023**, *15*, 1853. [CrossRef]
18. Potić, I.; Srdić, Z.; Vakanjac, B.; Bakrač, S.; Đorđević, D.; Banković, R.; Jovanović, J.M. Improving forest detection using machine learning and remote sensing: A Case study in Southeastern Serbia. *Appl. Sci.* **2023**, *13*, 8289. [CrossRef]
19. Saini, R. Integrating vegetation indices and spectral features for vegetation mapping from multispectral satellite imagery using AdaBoost and random forest machine learning classifiers. *Geomat. Environ. Eng.* **2023**, *17*, 57–74. [CrossRef]

20. Liu, J.; Feng, Q.; Gong, J.; Zhou, J.; Liang, J.; Li, Y. Winter wheat mapping using a random forest classifier combined with multi-temporal and multi-sensor data. *Int. J. Digit. Earth* **2018**, *11*, 783–802. [CrossRef]
21. Adam, E.; Mutanga, O.; Odindi, J.; Abdel-Rahman, E.M. Land-use/cover classification in a heterogeneous coastal landscape using RapidEye imagery: Evaluating the performance of random forest and support vector machines classifiers. *Int. J. Remote Sens.* **2014**, *35*, 3440–3458. [CrossRef]
22. Tigges, J.; Lakes, T.; Hostert, P. Urban vegetation classification: Benefits of multitemporal RapidEye satellite data. *Remote Sens. Environ.* **2013**, *136*, 66–75. [CrossRef]
23. Micheletti, N.; Foresti, L.; Robert, S.; Leuenberger, M.; Pedrazzini, A.; Jaboyedoff, M.; Kanevski, M. Machine learning feature selection methods for landslide susceptibility mapping. *Math. Geosci.* **2014**, *46*, 33–57. [CrossRef]
24. Saini, R.; Ghosh, S.K. Crop classification in a heterogeneous agricultural environment using ensemble classifiers and single-date Sentinel-2A imagery. *Geocarto Int.* **2021**, *36*, 2141–2159. [CrossRef]
25. Stahl, A.T.; Andrus, R.; Hicke, J.A.; Hudak, A.T.; Bright, B.C.; Meddens, A.J.H. Automated attribution of forest disturbance types from remote sensing data: A synthesis. *Remote Sens. Environ.* **2023**, *285*, 113416. [CrossRef]
26. Grabska, E.; Hostert, P.; Pflugmacher, D.; Ostapowicz, K. Forest stand species mapping using the Sentinel-2 time series. *Remote Sens.* **2019**, *11*, 1197. [CrossRef]
27. Nasiri, V.; Hawryło, P.; Janiec, P.; Socha, J. Comparing Object-Based and Pixel-Based Machine Learning Models for Tree-Cutting Detection with Planetscope Satellite Images: Exploring Model Generalization. *Int. J. Appl. Earth Obs. Geoinf.* **2023**, *125*, 103555. [CrossRef]
28. Abdollahnejad, A.; Panagiotidis, D.; Bílek, L. An Integrated GIS and Remote Sensing Approach for Monitoring Harvested Areas from Very High-Resolution, Low-Cost Satellite Images. *Remote Sens.* **2019**, *11*, 2539. [CrossRef]
29. Korea Forest Service (KFS). *Basic Statistics of Forest*; Korea Forest Service (KFS): Deajeon, Republic of Korea, 2023.
30. Ministry of Land Infrastructure and Transport (MLIT). *Cadastral Statistical Annual Report*; Ministry of Land Infrastructure and Transport (MLIT): Sejong, Republic of Korea, 2021.
31. Xue, J.; Su, B. Significant remote sensing vegetation indices: A review of developments and applications. *J. Sens.* **2017**, *2017*, 1353691. [CrossRef]
32. Hatfield, J.L.; Prueger, J.H.; Sauer, T.J.; Dold, C.; O'Brien, P.; Wacha, K. Applications of vegetative indices from remote sensing to agriculture: Past and future. *Inventions* **2019**, *4*, 71. [CrossRef]
33. Huete, A.R. A soil-adjusted vegetation index (SAVI). *Remote Sens. Environ.* **1988**, *25*, 295–309. [CrossRef]
34. Kaufman, Y.J.; Tanre, D. Atmospherically resistant vegetation index (ARVI) for EOS-MODIS. *IEEE Trans. Geosci. Remote Sens.* **1992**, *30*, 261–270. [CrossRef]
35. Huete, A.R.; Didan, K.; van Leeuwen, W.J.D.; Vermote, E.F. Global-scale analysis of vegetation indices for moderate resolution monitoring of terrestrial vegetation. *Remote Sens. Earth Sci. Ocean. Sea Ice Appl.* **1999**, *3863*, 141–151. [CrossRef]
36. Rikimaru, A.; Roy, P.S.; Miyatake, S. Tropical forest cover density mapping. *Trop. Ecol.* **2002**, *43*, 39–47.
37. Hu, Y.; Dong, Y.; Batunacun, B. An automatic approach for land-change detection and land updates based on integrated NDVI timing analysis and the CVAPS method with GEE support. *ISPRS J. Photogramm.* **2018**, *146*, 347–359. [CrossRef]
38. Gitelson, A.A.; Kaufman, Y.J.; Merzlyak, M.N. Use of a green channel in remote sensing of global vegetation from EOS-MODIS. *Remote Sens. Environ.* **1996**, *58*, 289–298. [CrossRef]
39. Pettorelli, N.; Vik, J.O.; Mysterud, A.; Gaillard, J.M.; Tucker, C.J.; Stenseth, N.C. Using the satellite-derived NDVI to assess ecological responses to environmental change. *Trends Ecol. Evol.* **2005**, *20*, 503–510. [CrossRef]
40. Lymburner, L.; Beggs, P.J.; Jacobson, C.R. Estimation of canopy-average surface-specific leaf area using landsat TM data. *Photogramm. Eng. Remote Sens.* **2000**, *66*, 183–192.
41. Korea Forest Service (KFS). *Statistical Yearbook of Forestry*; Korea Forest Service (KFS): Deajeon, Republic of Korea, 2020.
42. Ham, B.; Song, C.; Park, E.; Choi, S.-E.; Lee, W.-K. Development of forest activity data and forest management rate for national greenhouse gas inventory in the forest sector. *J. Clim. Chang. Res.* **2020**, *11*, 53–63. [CrossRef]
43. Kim, D.; Lim, C.H.; Lee, W.-K.; Song, C. A study on the improvement of evaluation system for implementation of national forest management. *J. Korean For. Soc.* **2015**, *104*, 640–648. [CrossRef]
44. Millard, K.; Richardson, M. On the importance of training data sample selection in random forest image classification: A case study in peatland ecosystem mapping. *Remote Sens.* **2015**, *7*, 8489–8515. [CrossRef]
45. Wibowo, A.; Salleh, K.O.; Frans, F.T.R.S.; Semedi, J.M. Spatial temporal land use change detection using google earth data. *IOP Conf. Ser. Earth Environ. Sci.* **2016**, *47*, 012031. [CrossRef]
46. Rwanga, S.S.; Ndambuki, J.M. Accuracy assessment of land use/land cover classification using remote sensing and GIS. *Int. J. Geosci.* **2017**, *08*, 611–622. [CrossRef]
47. Ayala-Izurieta, J.E.; Márquez, C.O.; García, V.J.; Recalde-Moreno, C.G.; Rodríguez-Llerena, M.V.; Damián-Carrión, D.A. Land cover classification in an Ecuadorian mountain geosystem using a random forest classifier, spectral vegetation indices, and ancillary geographic data. *Geosciences* **2017**, *7*, 34. [CrossRef]
48. Ehsan, S.; Kazem, D. Analysis of land use-land covers changes using Normalized Difference Vegetation Index (NDVI) differencing and classification methods. *Afr. J. Agric. Res.* **2013**, *8*, 4614–4622. [CrossRef]
49. Da Silva, V.S.; Salami, G.; da Silva, M.I.O.; Silva, E.A.; Monteiro Junior, J.J.; Alba, E. Methodological evaluation of vegetation indexes in land use and land cover (LULC) classification. *Geol. Ecol. Landsc.* **2020**, *4*, 159–169. [CrossRef]

50. Shishir, S.; Tsuyuzaki, S. Hierarchical classification of land use types using multiple vegetation indices to measure the effects of urbanization. *Environ. Monit. Assess.* **2018**, *190*, 342. [CrossRef] [PubMed]
51. Maschler, J.; Atzberger, C.; Immitzer, M. Individual tree crown segmentation and classification of 13 tree species using airborne hyperspectral data. *Remote Sens.* **2018**, *10*, 1218. [CrossRef]
52. Immitzer, M.; Neuwirth, M.; Böck, S.; Brenner, H.; Vuolo, F.; Atzberger, C. Optimal input features for tree species classification in Central Europe based on multi-temporal Sentinel-2 data. *Remote Sens.* **2019**, *11*, 2599. [CrossRef]
53. Morcillo-Pallarés, P.; Rivera-Caicedo, J.P.; Belda, S.; De Grave, C.; Burriel, H.; Moreno, J.; Verrelst, J. Quantifying the robustness of vegetation indices through global sensitivity analysis of homogeneous and forest leaf-canopy radiative transfer models. *Remote Sens.* **2019**, *11*, 2418. [CrossRef]
54. Trier, Ø.D.; Salberg, A.B.; Kermit, M.; Rudjord, Ø.; Gobakken, T.; Næsset, E.; Aarsten, D. Tree species classification in Norway from airborne hyperspectral and airborne laser scanning data. *Eur. J. Remote Sens.* **2018**, *51*, 336–351. [CrossRef]
55. Bolyn, C.; Michez, A.; Gaucher, P.; Lejeune, P.; Bonnet, S. Forest mapping and species composition using supervised per pixel classification of Sentinel-2 imagery. *Biotechnol. Agron. Soc. Environ.* **2018**, *22*, 172–187. [CrossRef]
56. Boonprong, S.; Cao, C.; Chen, W.; Bao, S. Random forest variable importance spectral indices scheme for burnt forest recovery monitoring—Multilevel RF-VIMP. *Remote Sens.* **2018**, *10*, 807. [CrossRef]
57. Khelifi, L.; Mignotte, M. Deep learning for change detection in remote sensing images: Comprehensive review and meta-analysis. *IEEE Access* **2020**, *8*, 126385–126400. [CrossRef]
58. Talukdar, S.; Singha, P.; Mahato, S.; Shahfahad; Pal, S.; Liou, Y.A.; Rahman, A. Land-Use Land-Cover Classification by Machine Learning Classifiers for Satellite Observations—A Review. *Remote Sens.* **2020**, *12*, 1135. [CrossRef]
59. Wilson, E.H.; Sader, S.A. Detection of forest harvest type using multiple dates of landsat TM imagery. *Remote Sens. Environ.* **2002**, *80*, 385–396. [CrossRef]
60. Franklin, S.E.; Moskal, L.M.; Lavigne, M.B.; Pugh, K. Interpretation and classification of partially harvested forest stands in the fundy model forest using multitemporal landsat TM digital data. *Can. J. Remote Sens.* **2000**, *26*, 318–333. [CrossRef]
61. Brooks, E.B.; Wynne, R.H.; Thomas, V.A.; Blinn, C.E.; Coulston, J.W. On-the-fly massively multitemporal change detection using statistical quality control charts and landsat data. *IEEE Trans. Geosci. Remote Sens.* **2014**, *52*, 3316–3332. [CrossRef]
62. Alonso, L.; Picos, J.; Armesto, J. Forest land cover mapping at a regional scale using multi-temporal Sentinel-2 imagery and RF models. *Remote Sens.* **2021**, *13*, 2237. [CrossRef]
63. Nink, S.; Hill, J.; Stoffels, J.; Buddenbaum, H.; Frantz, D.; Langshausen, J. Using landsat and Sentinel-2 data for the generation of continuously updated forest type information layers in a cross-border region. *Remote Sens.* **2019**, *11*, 2337. [CrossRef]
64. Salgueiro Romero, L.S.; Marcello, J.; Vilaplana, V. Super-resolution of Sentinel-2 imagery using generative adversarial networks. *Remote Sens.* **2020**, *12*, 2424. [CrossRef]
65. Feng, Q.; Liu, J.; Gong, J. UAV Remote Sensing for Urban Vegetation Mapping Using Random Forest and Texture Analysis. *Remote Sens.* **2015**, *7*, 1074–1094. [CrossRef]

forests

MDPI

Article

Changes in the Vegetation NPP of Mainland China under the Combined Actions of Climatic-Socioeconomic Factors

Yifeng Liu [1,†], Mei Xu [1,†], Bing Guo [1,*], Guang Yang [2,*], Jialin Li [1] and Yang Yu [1]

1 School of Civil Engineering and Geomatics, Shandong University of Technology, Zibo 255000, China; 21407010765@stumail.sdut.edu.cn (Y.L.); 22407010008@stumail.sdut.edu.cn (M.X.); 21507020776@stumail.sdut.edu.cn (J.L.); 21507020798@stumail.sdut.edu.cn (Y.Y.)

2 Research Institute of Aerospace Information, Chinese Academy of Sciences, Beijing 100101, China

* Correspondence: guobing@sdutedu.cn (B.G.); yangguang@radi.ac.cn (G.Y.); Tel.: +86-18766965987 (B.G.); +86-18510586581 (G.Y.)

† These authors contributed equally to this work.

Abstract: Under the combined impact of climatic, socioeconomic, and environmental factors, the vegetation NPP change process and its responses to drive factors in the sub-regions of Mainland China are not clear. This study analyzes the changing pattern of vegetation NPP in China from 2000 to 2022 from the perspective of zoning and clarifies its response mechanism to climate-human interaction based on the gravity center model, third-order partial correlation coefficient and geographical detector. The results showed that: (1) There was an overall decreasing trend of vegetation NPP in China from the southeast to the northwest; (2) The vegetation NPP gravity center in Northeast, Northwest, and North China migrated southwards, while that of Southwest, Central South, and East China showed northward migration.;(3) Human activities played a dominant role in zones with increasing vegetation NPP from 2000 to 2010, while climate change greatly contributed to the increase in vegetation NPP during 2011–2022; (4) Human activities, such as deforestation and overgrazing, in Northeast and North China should be reduced to prevent vegetation ecosystem degradation, and the negative impact of human activities should be reduced to maintain the growth of vegetation NPP. This study was conducted to support decision-making for the precise restoration of ecosystems.

Keywords: geographic demarcation; spatio-temporal variations; driving mechanisms; dominant factors; geographical detector

1. Introduction

Net Primary Productivity (NPP) refers to the total amount of organic matter accumulated by green plants per unit area per unit time [1]. Vegetation plays a major role in terrestrial ecosystems, which can better indicate the changes in regional ecological environments. To achieve the dual carbon goals, it is particularly important to clarify the spatial and temporal evolution patterns of vegetation NPP and its response mechanism to climate–human impacts [2].

Recently, research on vegetation NPP has been further conducted. Many scholars have conducted a series of studies on the estimation and analysis of vegetation NPP based on MODIS data [3,4]. Running et al. [5] estimated NPP by considering the difference between Gross Primary Productivity (GPP) and respiration and solving important ecological problems on time and space scales. Potter et al. [6] established a Carnegie–Ames–Stanford Approach (CASA) model to calculate vegetation NPP, aiming to control the seasonal production of terrestrial ecosystems and soil microbial respiration patterns associated with global climate and soil. Field et al. [7] proposed a simple modeling method for global NPP, which combines ecological principles with satellite data to obtain global estimates with an operational monitoring resolution. Mowll et al. [8] pointed out that precipitation played an important role in affecting vegetation NPP. Domestic scholars have conducted many studies on the influencing factors and driving mechanisms of vegetation NPP, such as spatio-temporal change patterns, dominant factors, and driving mechanisms, in different sub-regions of China [9–13]. Relevant studies have shown there is an increasing trend of vegetation in China, and the driving mechanisms of changes in NPP are different in different study areas [14–17].

Previous studies on vegetation NPP in China mainly focused on ecologically fragile areas. For example, Zhu et al. [18] applied the CASA model to analyze the change patterns of vegetation NPP in Inner Mongolia. Wang et al. [19] used the Hurst index to analyze NPP in the Qinling–Daba Mountains during 2000–2015. Yang et al. [20] analyzed the change patterns of vegetation NPP of Qinghai–Tibet Plateau utilizing a gravity center model [21]. Due to the vast territory of China, scholars have conducted many studies on different regions. For example, Xu et al. [22] investigated the evolution process and the influencing factors of different types of vegetation NPP in Southwest China. Li et al. [23] explored the change process of vegetation NPP in the Yangtze River Basin and its relationships with topographic factors. Based on the CASA model, Mao et al. [24] found that climate and land use are important factors affecting the change patterns of vegetation in Northeast China. Liu et al. [25] found that the NPP of vegetation in the Qinghai–Tibet Plateau is significantly positively correlated with temperature and precipitation.

The aims of this study are as follows: (1) To reveal the spatio-temporal evolution patterns of vegetation NPP and its differences in different sub-regions; (2) to systematically determine and clarify the differences in the dominant factors in different sub-regions from the perspective of partition; and (3) to explore the changes in the dominant factors during different historical periods in the same sub-region.

2. Materials and Methods

2.1. Study Area

China is located in the eastern part of Asia, east to the junction of Heilongjiang and the Ussuri (135°2′30″ E), west to the Pamirs (73°29′59.79″ E), south to the Cape Zengmuansha at the southern tip of the Nansha Archipelago (4°15′00″ N), north to the center line of the main channel of Heilongjiang (53°33′ N), with a total area of 9.6 million km^2 (Figure 1). Its terrain decreases from west to east, showing a three-stage ladder-like distribution. China has five types of territories based on relief: plateau, mountain, plain, hill, and basin. Due to its vast land area as well as its large latitude and longitude span, it has formed a variety of temperature and precipitation combinations, forming a complex and diverse climate [22]. There are five climate types in China, which include temperate monsoon, temperate continental, and alpine plateau climates. Yang et al. [26] studied the difference in plateau climate change in Tibet. Due to the large span of the north–south latitude, the north–south spatial heterogeneity of temperature is obvious [23]. There are a variety of vegetation types, which mainly include grasslands, tropical rainforests, and coniferous forests. The soil types (WRB) are complex and diverse and include 15 soil types, such as latosol, latosolic red, and reddish-yellow soils; their distribution varies based on geographical location and terrain height [27].

2.2. Data Source and Preprocessing

NPPs that were derived from the MODIS17A3H dataset were available at https://ladsweb.nascom.nasa.gov/ (accessed on 21 January 2023), with temporal and spatial resolution resolutions of 1 year and 500 m. The MRT (MODIS Reprojection Tool) tool was utilized to reproject and format the vegetation NPP data. The climate data came from the China Meteorological Data Network (http://data.cma.cn/ (accessed on 24 January 2023)), which included daily precipitation, daily average temperature, accumulated temperature (sum of temperatures greater than 10 °C), and sunshine data. Because kriging interpolation may cause interpolation errors compared with traditional methods, the difference result is more accurate. These datasets were interpolated into 1 km grids using ArcGIS 10.7 (Environmental Systems Research Institute, Inc., Redlands, CA, USA). Land use, socioeconomic (population density and GDP), and geographic partition data were retrieved from http://www.resdc.cn/ (accessed on 21 January 2023). The elevation and slope data were extracted from SRTM 90 m, which was downloaded at http://www.Gscloud.cn (accessed on 23 January 2023). Based on the Arcgis10.7 Clip tool for cutting, the abnormal values for the vegetation NPP data were removed, resampled to 1 km, and projected to a Krasovskii projection. The related driving factors and references are shown in Table 1. The soil distribution map, temperature and precipitation distribution map selected by driving factors are shown in Figures 2 and 3.

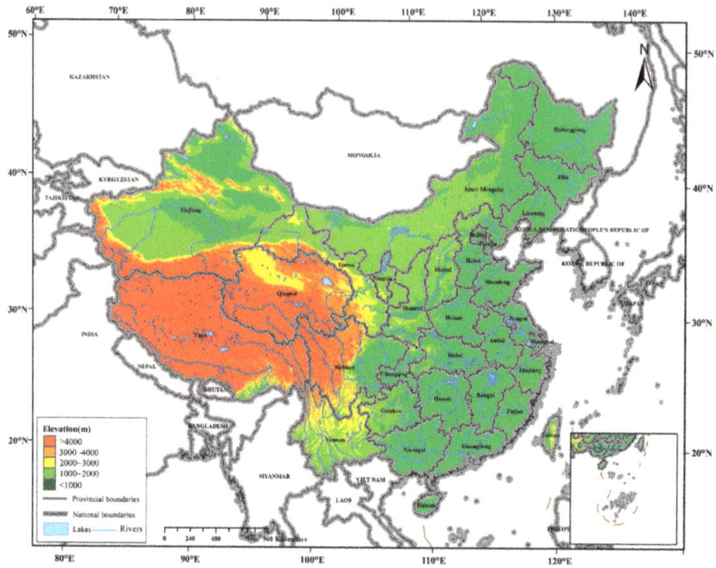

Figure 1. The location overview and elevation distribution map of the study area.

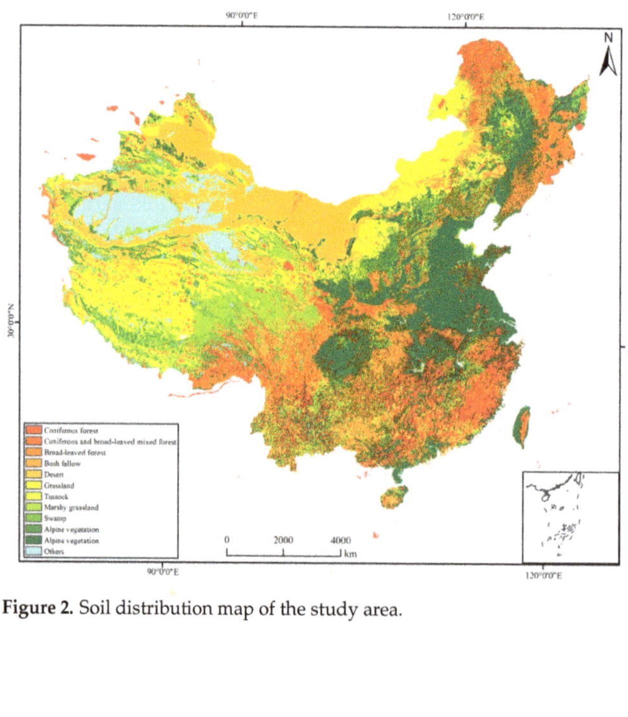

Figure 2. Soil distribution map of the study area.

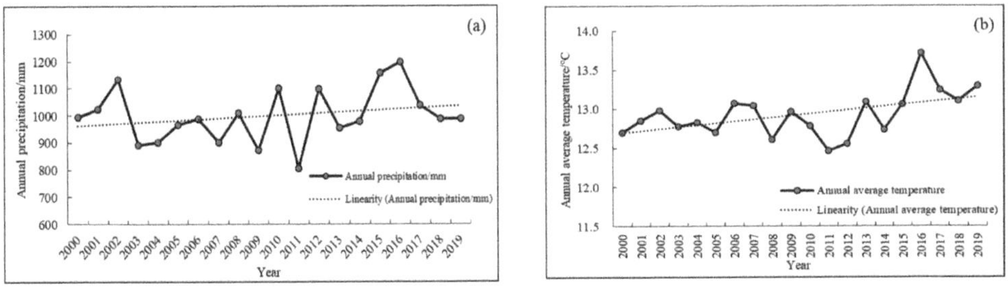

Figure 3. Driving factor line chart. (**a**) Annual precipitation; (**b**) Annual average temperature.

2.3. Methods

Most of the research is based on one partition, and few studies are based on different partitions. This study focuses on these differences, which are of great significance to protecting ecosystems according to local conditions. In addition, against the backdrop of global climate change, the dominant factors in the change process of vegetation NPP have also changed. It is not clear how the evolution mechanisms of vegetation NPP varied across different periods and ecological sub-regions. The related research is mostly based on the driving mechanisms of temperature, precipitation, population, and other factors [24,25,28,29]. Based on the research status, most relevant research has selected these data; in the size of the influence of the driving factors, these factors can well explain the reasons for the change. A series of ecological and environmental problems caused by climate change, such as air pollution and dust storms, are not only troubled. People's lives have also had a negative impact on industrial production, which has hindered the sustainable development of the social economy to a certain extent. Therefore, it is of scientific significance to understand and understand the impact of climate change on vegetation. Therefore, this study selects two climatic factors (temperature and precipitation), two social factors (population density and Gross Domestic Product), and two geographical factors (altitude and slope).

Moreover, previous studies mostly applied second-order partial correlation to explore the effects of different climatic factors on vegetation NPP but ignored the comprehensive and interactive effects of regional multi-type climatic factors. In the past, partial correlation analysis was mostly based on single factor and double factors. In this paper, third-order partial correlation analysis is used as an innovation point, which allows for comprehensively considering the complex effects of three factors on the evolution of vegetation, thereby more truly reflecting ecological processes.

Geodetectors detect the spatial stratification heterogeneity of elements and reveal the driving forces behind them [30], thus making it advantageous for exploring driving mechanisms across different partitions. In this study, we utilized a Geodetector, third-order partial correlation analysis, and a gravity center model to analyze the change process of vegetation NPP in China from the perspective of partitions. The relative role of climate–human interactions during different historical periods was quantitatively distinguished, and the dominant factors of changes in vegetation NPP across different sub-regions and historical periods were revealed. The effects of global change on Chinese vegetation NPP response mechanisms were clarified, which provides support for important decisions regarding the restoration of regional vegetation ecosystems.

The process was mainly divided into three parts: The first part was data preprocessing. The NPP, climate, land (land use, soil type), and socioeconomic (GDP, population density) datasets were preprocessed via kriging interpolation (Co-Kriging), re-projection, and clipping. The second part was the analysis of the spatio-temporal change process. Spatial distribution analysis was carried out from the perspective of mean distribution, different levels of vegetation NPP transfer scenarios, and mean comparative analysis. The analysis of temporal evolution was conducted from the perspective of gravity center migration. The third part was the analysis of the driving mechanisms. The driving factors were explored using a geographic detector. A flow chart of this study is presented in Figure 4.

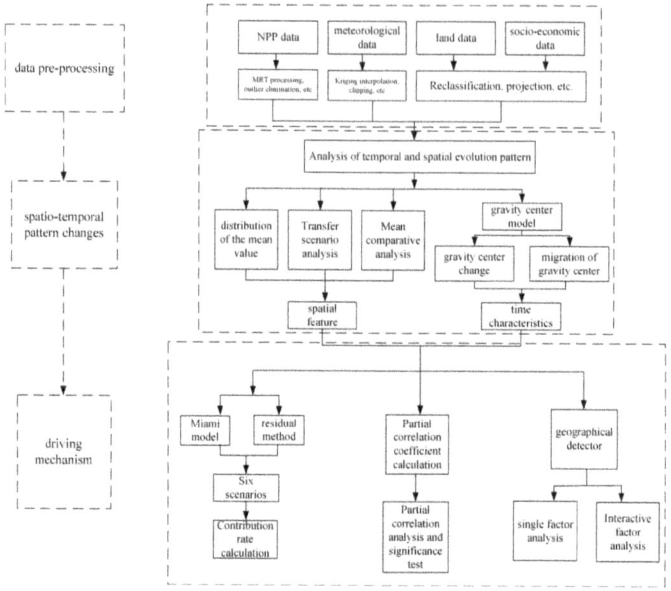

Figure 4. Flow chart of the research method.

2.3.1. Gravity Center Model

In physics, the gravity center refers to a point in a certain space where forces in all directions reach phase equilibrium [31]. Two common barycenter models are the quality gravity center model and the ecological gravity center model. The gravity mass center model is mainly used to describe the mass distribution, while the ecological gravity center model is suitable for describing the spatial distribution of biological communities or ecosystems. In addition, the gravity center model has also been widely applied in life, land use, and other fields. The migration trajectory and direction of the gravity center can directly indicate the change unevenness and the bias of certain phenomena from the perspective of spatial distribution [15,18,26]. Moreover, as typical geophysical and ecological indicators, there were obvious differences in change patterns for vegetation NPP across different periods and sub-regions. However, traditional statistical methods cannot reveal the laws governing changes in temporal and spatial distribution on a large scale [16]. Meanwhile, the migration distance and direction of vegetation NPP gravity center can indicate the increments and growth rate of vegetation NPP in different parts within the study region. The gravity center model can be applied to investigate the variation characteristics of vegetation NPP. The center of gravity model [32] is a mature tool in ArcGIS software with the following formulae:

$$\overline{x} = \frac{\sum_{i=1}^{n} z_i x_i}{\sum_{i=1}^{n} z_i} \tag{1}$$

$$\overline{y} = \frac{\sum_{i=1}^{n} z_i y_i}{\sum_{i=1}^{n} z_i} \tag{2}$$

where $(\overline{x}, \overline{y})$ is the barycenter coordinates, z_i is the attribute value of the *i*-th plane space element, and (x_i, y_i) is the attribute value of the i-th plane space element.

2.3.2. Third-Order Partial Correlation Coefficient and Significance Test

A correlation coefficient represents the degree of correlation between different elements [33]. The Pearson correlation coefficient was applied to analyze the correlation between vegetation NPP and typical climatic factors and their significance level. Because a first-order partial correlation coefficient can only control one variable, a high-order partial correlation coefficient can reflect the correlations between vegetation NPP and multiple variables. Therefore, this study utilized a third-order partial correlation coefficient to explore the correlations between climate variables and vegetation NPP at the $p < 0.05$ and $p < 0.01$ confidence levels. The formula can be expressed as follows:

$$r_{x,y} = \frac{\sum\limits_{i=1}^{n} (x_i - \overline{x})(y_i - \overline{y})}{\sqrt{\sum\limits_{i=1}^{n} (x_i - \overline{x})^2 \sum\limits_{i=1}^{n} (y_i - \overline{y})^2}} \quad (3)$$

In the formula, $r_{x,y}$ denotes the correlation coefficient (r_{xy_1}, r_{xy_2}, and $r_{y_1y_2}$ are similar) between variables x and y, where a positive value denotes a positive correlation; conversely, x_i and y_i represent the values of the two variables in year i. r_{xy_1,y_2} represents the partial correlation coefficient of x and y_1 in the case of the control variable y_2.

When controlling multiple variables, there are high-order partial correlation coefficients. The formula can be expressed as follows:

$$r_{xy,y_1y_2\ldots y_g} = \frac{r_{xy,y_1y_2\ldots y_{g-1}} - r_{xy_g,y_1y_2\ldots y_{g-1}} r_{yy_g,y_1y_2\ldots y_{g-1}}}{\sqrt{1 - r^2_{xy_g,y_1y_2\ldots y_{g-1}}}\sqrt{1 - r^2_{yy_g,y_1y_2\ldots y_{g-1}}}} \quad (4)$$

2.3.3. Climate–Human Interaction Distinction

The Miami model proposed by H. Lieth [34] is a widely used model to calculate potential vegetation NPP, and it can be expressed as follows:

$$\mathrm{NPP}_c = \min\{(1 + \frac{3000}{\exp(1.315 - 0.119T)}), 3000[1 - \exp(-0.000664P)]\} \quad (5)$$

In the formula, $\mathrm{NPP_c}$ refers to potential NPP ($\mathrm{gC \cdot m^{-2} \cdot a^{-1}}$); T refers to the average annual temperature (°C); and P refers to annual precipitation (mm).

Vegetation NPP is always influenced by the combined actions of climate–human activities. The residual method has been widely used to separate human-affected NPP from actual NPP [35]. Based on the residual method, NPP affected by human activities was calculated. Meanwhile, the change slope of the potential NPP was calculated to distinguish climate–human interactions according to the positive and negative slopes. This can be expressed as follows:

$$\mathrm{NPP_h} = \mathrm{NPP}_a - \mathrm{NPP}_c \quad (6)$$

In the formula, $\mathrm{NPP_h}$ is the human-influenced NPP, NPP_a is the actual NPP, and NPP_c is the potential NPP. If $\mathrm{NPP}_h < 0$, this indicates that human activities inhibit the increase in NPP. If $\mathrm{NPP}_h > 0$, this indicates that human activities promote an increase in NPP.

The formula to calculate the variation in vegetation NPP is as follows [36]:

$$\Delta\mathrm{NPP} = (n - 1) \times K_{\mathrm{slope}} \quad (7)$$

where n represents the study time scale, and K_{slope} is the change slope of NPP.

Based on the slope of NPP_a, NPP_c, and NPP_h, six scenarios were established to assess the relative contributions of climate-human to NPP change, as shown in Table 2:

Table 1. Different types of index references and contributions.

Index	Factor	References
Climate	Annual precipitation, average temperature, accumulated temperature, and sunshine	[8,9,25,30,32,36]
Terrain	Slope and altitude	
Social development	Population density and GDP	

Table 2. Six scenarios involving the contribution of human–climate interactions to the change in vegetation NPP.

NPP_a Status	Scenario	K_c	K_h	Climate Change Contribution Ratio	Human Activity Contribution Ratio
Increased NPP_a ($K_a > 0$)	Scenario 1	$K_c > 0$	$K_h < 0$	100	0
	Scenario 2	$K_c < 0$	$K_h > 0$	0	100
	Scenario 3	$K_c > 0$	$K_h > 0$	$\frac{\lvert\Delta NPP_c\rvert}{\lvert\Delta NPP_c\rvert+\lvert\Delta NPP_h\rvert}$	$\frac{\lvert\Delta NPP_h\rvert}{\lvert\Delta NPP_c\rvert+\lvert\Delta NPP_h\rvert}$
Reduced NPP_a ($K_a < 0$)	Scenario 4	$K_c < 0$	$K_h > 0$	100	0
	Scenario 5	$K_c > 0$	$K_h < 0$	0	100
	Scenario 6	$K_c < 0$	$K_h < 0$	$\frac{\lvert\Delta NPP_c\rvert}{\lvert\Delta NPP_c\rvert+\lvert\Delta NPP_h\rvert}$	$\frac{\lvert\Delta NPP_h\rvert}{\lvert\Delta NPP_c\rvert+\lvert\Delta NPP_h\rvert}$

In Table 2, K_i represents the slope of NPP_i, where i = a, b, h.

2.3.4. Geodetector

The Geodetector model can determine a factor's contribution rate to vegetation NPP, which is measured by its q-value [37]. The greater the value is, the greater the impact of this factor on NPP is, and vice versa. In this study, temperature, precipitation, population density, GDP, altitude, slope, land use, and soil type were selected, and the q-values of different factors were calculated via the Geodetector model to reveal the influences of different factors on vegetation NPP [28].

$$q = 1 - \frac{\sum_{h=1}^{L} N_h \sigma_h^2}{N\sigma^2} = 1 - \frac{SSW}{SST} \tag{8}$$

$$SSW = \sum_{h=1}^{L} N_h \sigma_h^2 \tag{9}$$

$$SST = N\sigma^2 \tag{10}$$

In the formula, h = 1, ..., L represents the classification of variable Y or factor X; N_h and N are the layer h and the number of pixels units in the whole region; σ_h the variance in the Y value produces the class h, which is the variance in the Y value of the whole region; SSW and SST are the sum of intra-layer variance and the total variance in the whole region, respectively.

2.3.5. Uncertainty Estimation Approaches

Uncertainty means that the potential and possible results of things cannot or cannot make full use of reasonable and known probability distribution to objectively analyze and characterize, and cannot subjectively analyze and estimate the results. The uncertainty of water resources caused by climate change is difficult to estimate and analyze based on the known probability distribution or subjectively. Therefore, quantitative estimation of the uncertainty of water resources caused by climate change from various aspects can help to understand the uncertainty of water resources.

3. Results

3.1. Spatial Distribution Characteristics of Vegetation NPP

The mean vegetation NPP during 2000–2022 was calculated to analyze its spatial distribution. Figure 5 showed that vegetation NPP in China ranged from 0 to 1933 gC·m^{-2}·a^{-1}. By using ArcGIS 10.7, the vegetation NPP was divided into five grades: level I (<200 gC·m^{-2}·a^{-1}), level II (200–400 gC·m^{-2}·a^{-1}), level III (400–600 gC·m^{-2}·a^{-1}), level IV (600–800 gC·m^{-2}·a^{-1}), and level V (>800 gC·m^{-2}·a^{-1}, as shown in Appendix A, Figure A1). The level V region accounted for the smallest proportion (only 12.54%), which was mostly located in the southern parts, including Guangdong Province, Guangxi, and Yunnan Province. The level IV region accounted for 13%, being mostly located in the central and southern regions, including Guizhou Province and Zhejiang Province. The level III region accounted for 25%, which was mostly located in the central parts, including Shaanxi Province and Anhui Province. The level II region accounted for 25.94%, being distributed in northeast Tibet. The level I region accounted for 25.00%, mainly distributed in the central and western regions, including the foothills of the Tianshan Mountains and Qinghai Province. Overall, the mean vegetation NPP gradually increased from north to south and from west to east.

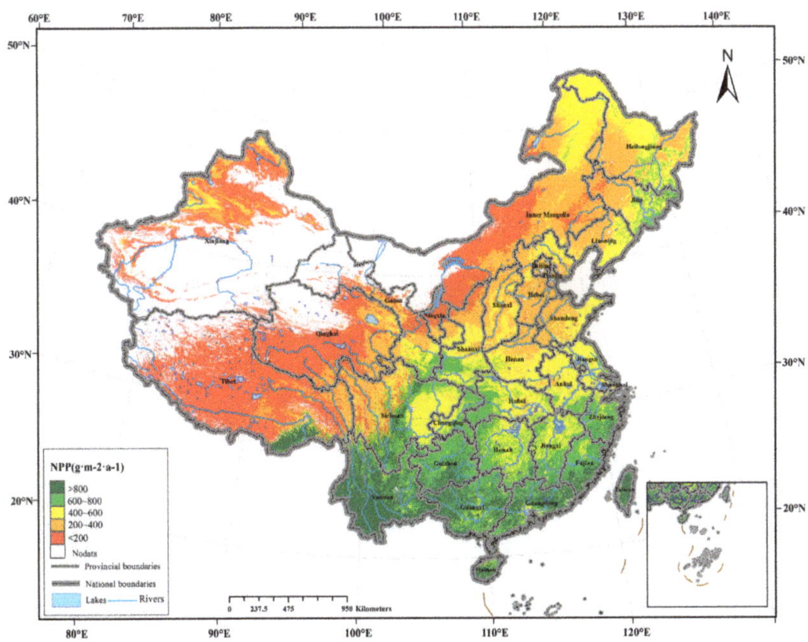

Figure 5. Annual average vegetation NPP in China over the last 23 years.

3.2. Spatial Distribution of Vegetation NPP

To reveal the changes in the different levels of vegetation NPP across different periods, we divided this study into two periods: 2000–2010 and 2011–2022, as shown in Appendix A Figure A2 and FigureA3.

Appendix A Figure A2a and FigureA3a showed that from 2000 to 2010, zones with levels II and III accounted for the largest proportion at 36.56%, which were mainly distributed across the Weihe Plain, Lesser Khingan, southern Henan, and northern Anhui. Secondly, zones with level I converted to zones with level II accounted for 28.06%, mostly concentrated in eastern Inner Mongolia, northern Shaanxi, and southern Qinghai Province. Areas comprising moderate to severe lushness accounted for 15.41%, being mainly distributed in southern Shaanxi, western Hunan, and southern Guangxi. During 2000–2010, zones with levels I and II to zones with levels III, IV, and V had the largest area, indicating that vegetation NPP showed an overall increasing trend.

Appendix A Figure A2b and FigureA3b showed that from 2011 to 2022, in the area where the vegetation NPP level changes, the area transferred from zones with level II to zones with level III accounted for the largest proportion, comprising 34.90%, which was mainly distributed in the Changbai Mountains, Shanxi Province, and western Shandong. The area transferred from zones with level III to zones with level IV accounted for 20.42%, mostly distributed in the Sichuan Basin and the Liaodong Peninsula. The area transferred from zones with level IV to zones with level V accounted for 16.45%, being mostly located in the western part of Guizhou, the central part of Hubei, and the junction of Heilongjiang and eastern Jilin. During 2011–2022, the area of zones with level II to zones with levels III, IV, and V accounted for the largest proportion, indicating that the overall NPP of vegetation showed an increasing trend. Overall, the vegetation NPP showed an increasing trend during 2000–2022.

Appendix A Figure A4 and FigureA5 showed that from 2000 to 2010, the average NPP of vegetation in China ranged from 0 to 1959 $gC{\cdot}m^{-2}{\cdot}a^{-1}$. The area proportion of level I was the largest, comprising 40.5%, which was mostly concentrated in the Qinghai–Tibet Plateau, Inner Mongolia, and other regions. Level V zones accounted for 3.2%, being mainly distributed in southern China, such as southern Fujian Province and the Guangxi Zhuang Autonomous Region. The area proportion of level III zones was 21.6%, being mostly concentrated in central and southern China, such as the Sichuan Basin and Hunan Province. Appendix A Figure A4 shows that the average annual NPP of vegetation in China from 2011 to 2022 had a similar trend to the average NPP of vegetation from 2000 to 2010. The difference is that the proportion of level I and level II zones decreased, and the proportion of level III zones and above had increased. When combined with our quantitative analysis in Appendix A Figure A5, in 2000–2010 and 2011–2022, we found that level I zones accounted for 40.5% and 38.4%, respectively, and the level V zones accounted for 3.2% and 3.6%, respectively.

3.3. Gravity Center Migration of Vegetation NPP

In order to explore the discreteness and bias of vegetation NPP distribution in the study area [38], the gravity center model was utilized to analyze the migration trend of vegetation NPP gravity center from 2000 to 2022 (Figure 6). In the polar coordinate system, the distance (polar diameter) and deflection angle (polar angle) of the gravity center of NPP from 2000 to 2022 were calculated to reflect its dynamic migration (Table 3).

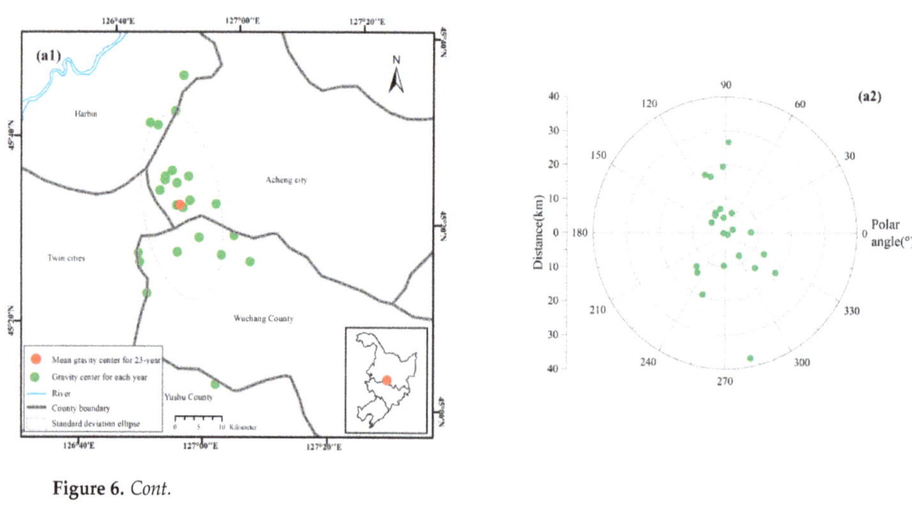

Figure 6. *Cont.*

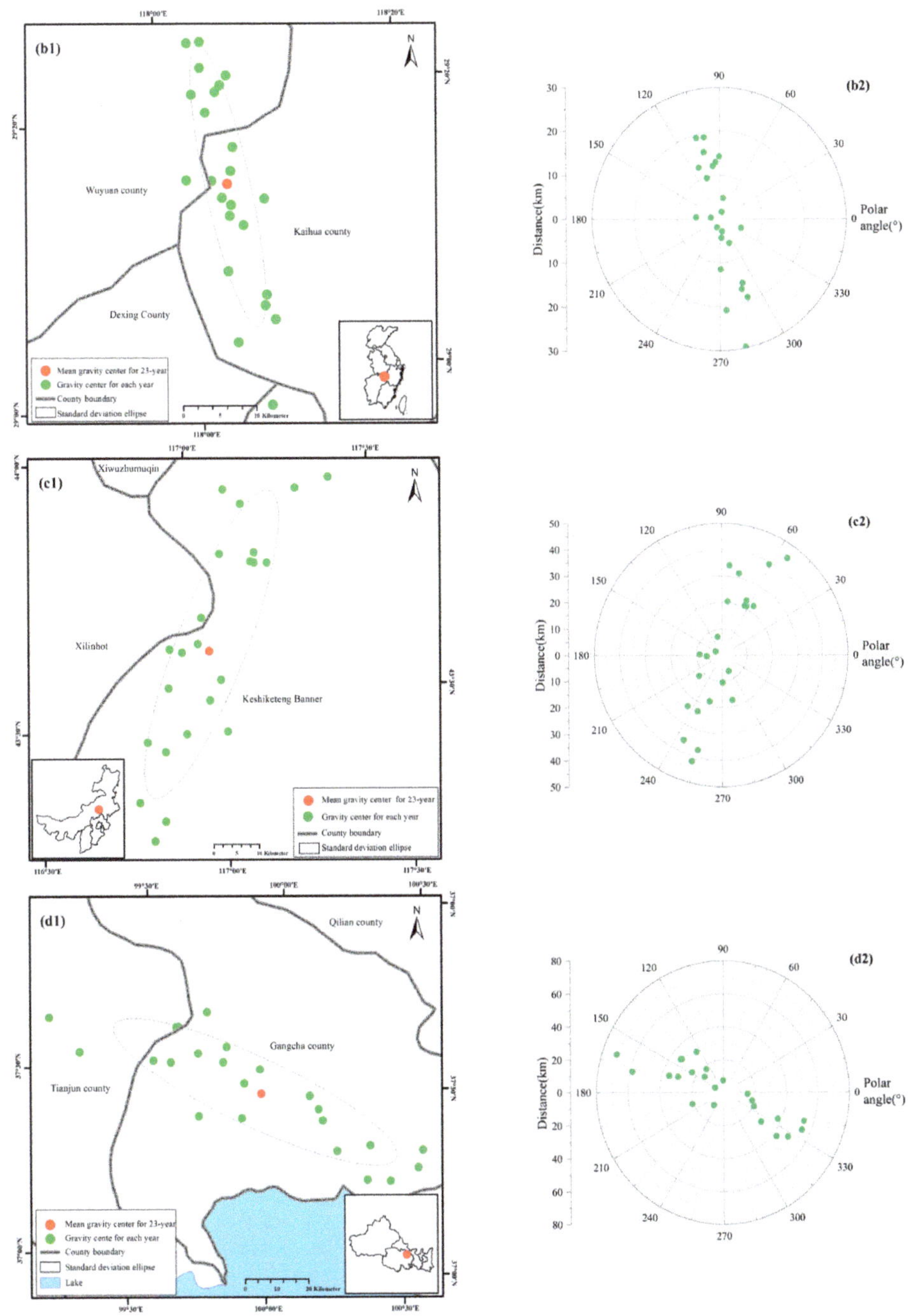

Figure 6. *Cont.*

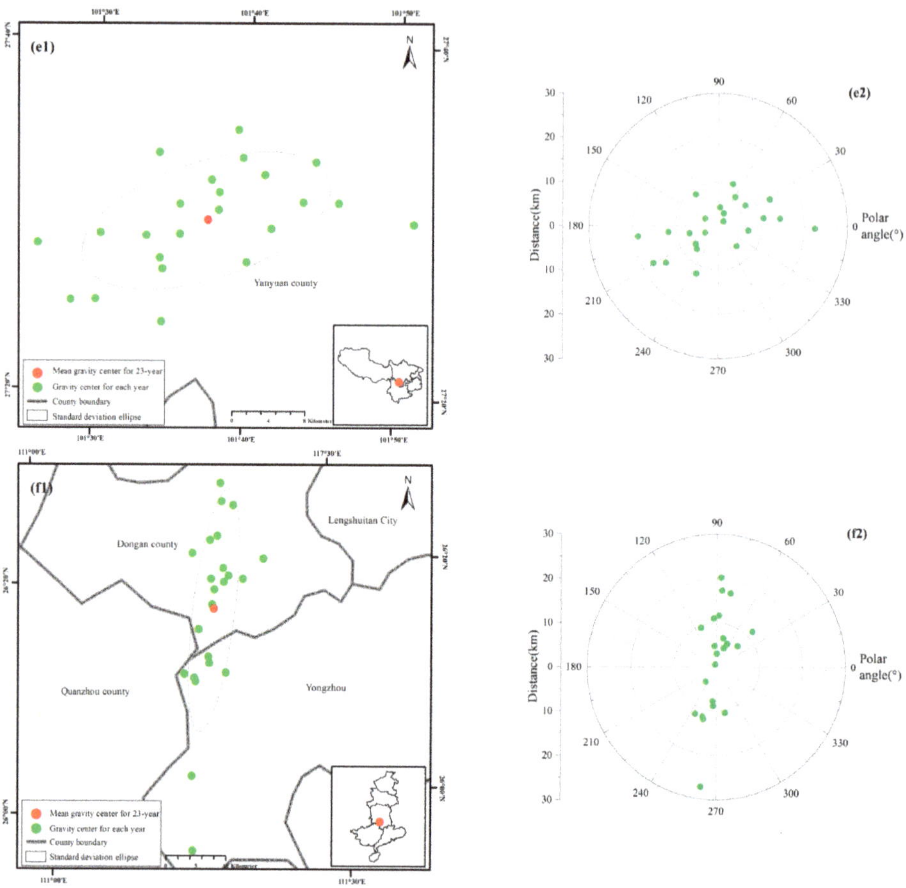

Figure 6. The vegetation NPP gravity centers in six regions of China. (**a1**) Northeast (space). (**a2**) Northeast (polar coordinates). (**b1**) East China (space). (**b2**) East China (polar coordinates). (**c1**) North China (space). (**c2**) North China (polar coordinates). (**d1**) Northwest China (space). (**d2**) Northwest China (polar coordinates). (**e1**) Southwest China (space). (**e2**) Southwest China (polar coordinates). (**f1**) Central South China (space). (**f2**) Central South China (polar coordinates).

Table 3. The standard deviation ellipse parameters of vegetation NPP in six sub-regions 2000 to 2022.

Standard Deviation Ellipse Parameters	Northeast China	East China	North China	Northwest China	Southwest China	Central South China
Rotating angle (°)	170.61	167.46	21.81	114.63	72.35	7.17
Standard deviation along x-axis (km)	8.317	3.068	7.702	48.940	14.128	3.532
Standard deviation along y-axis (km)	19.194	18.925	34.859	10.548	6.201	19.735
Ellipse area (km^2)	501.510	182.429	843.525	1621.792	275.257	219.001

The vegetation NPP gravity centers in Northeast China are mostly located in Acheng City and Heilongjiang Province (Figure 6(a1,a2)). The ellipse angle was 170.617°, and the gravity centers showed a northwest–southeast distribution pattern. The ellipse area was 501.510 km^2 (Table 3). Under the polar coordinates, the northern half of the gravity center (northwest and northeast quadrants) accounted for 56% of the total gravity centers, indicating that the vegetation NPP increments in the northern half were greater than that of the southern half. The western half of the gravity center (northwest and southwest quadrants) accounted for 56%, indicating that the vegetation NPP increments in the western half were greater than that of the eastern half.

The vegetation NPP gravity centers in East China are mostly distributed in Kaihua County, Quzhou City, and Zhejiang Province (Figure 6(b1,b2)). The standard deviation ellipse angle was 167.459°, which was significantly different from that of Northeast China (170.617°), showing a northwest–southeast distribution pattern. The standard deviation ellipse area of 182.429 km^2 was the smallest among the six distribution areas. The standard deviations were 3.068 km and 18.925 km along the x- and y-axes, respectively. In 2001, the gravity centers deviated the farthest (29.651 km) from the origin, distributed at the southern end, indicating that, in 2001, the vegetation NPP increments were more significant in the corresponding area.

The vegetation NPP gravity centers in North China were concentrated in Keshiketeng Banner, Chifeng City, and Inner Mongolia (Figure 6(c1,c2)). The ellipse angle was small (21.81°), showing a northeast–southwest distribution pattern. The standard deviation ellipse area was large, indicating that the gravity center distribution in North China was more discrete. Approximately 56% of the gravity centers were distributed in the north.

The vegetation NPP gravity centers in Northwest China were mainly distributed in Gangcha County and Haibei Tibetan Autonomous Prefecture, with a standard deviation ellipse angle of 114.625°, showing a northwest–southeast distribution pattern (Figure 6(d1,d2)). The area of standard deviation ellipse was the largest, at 1621.429 km^2. The distribution was the most discrete. The gravity center was the farthest from the origin in 2001, with a distance of 70.931 km, indicating that, in 2001, the increase in vegetation NPP was more significant in the corresponding area. In the polar coordinate system, the distribution of the gravity centers in the north–south direction was relatively uniform. Most of the gravity centers were distributed in the northwest quadrant, indicating that the vegetation NPP increments in the northwest were greater than those in the southeast.

The vegetation NPP gravity centers in Southwest China were concentrated in Yanyuan County and Liangshan Yi Autonomous Prefecture, with a deflection angle of 72.350° (Figure 6(e1,e2)). The area of standard deviation ellipse was 275.257 km^2, and the dispersion degree was not large. According to the polar coordinate image, 2006 and 2015 had the farthest distances from the origin, at 18.857 km and 22.432 km, respectively, indicating that the vegetation NPP changed considerably in the corresponding area. Moreover, 2009 and 2012 had the smallest distances from the origin, at 3.412 km and 1.595 km, respectively, indicating that the vegetation NPP changed slightly in the corresponding area.

The vegetation NPP gravity centers in the central and southern regions were concentrated in Dong'an County and Yongzhou City, with a north–south spatial distribution (Figure 6(f1,f2)). The standard ellipse angle was 7.17°, and the standard ellipse area was small, indicating that its dispersion was small. The long axis of the standard deviation ellipse was 19.735 km in the north–south direction, and the short axis was 3.532 km in the east–west direction. The ratio was 5.59, indicating that the dispersion in the north–south direction was greater than that in the east–west direction. From the polar coordinates, it can be observed that the gravity centers in the northern quadrant had the largest proportion of 65%, indicating that the vegetation NPP increments in the northern half were higher than that in the southern quadrant. From the perspective of distance, in 2000 and 2001, it was far from the origin, and the distances were 27.237 km and 39.186 km, respectively, indicating that the vegetation NPP increased significantly in the corresponding areas in these two years. In 2006 and 2007, it was close to the origin, and the distances were 3.08 km and 0.69 km, respectively, indicating that the vegetation NPP did not fluctuate much.

Figure 7a shows that the vegetation NPP gravity centers in Northeast China showed a migration trend of southeast–northwest–northeast–southwest from 2000 to 2022. From 2001 to 2005, the gravity center shifted to the southeast compared with that of 2000, indicating that the vegetation NPP increments in the south were greater than those in the north. The vegetation NPP gravity center from 2006 to 2010 and 2011 to 2015 shifted to the northwest compared with that of 2001–2005, indicating that the increments of vegetation NPP in the northwest were greater than those in the southeast during the study periods. From 2016 to 2020, the gravity center shifted slightly to the northeast, indicating that the growth and increments of vegetation NPP in the northeast were greater than those in the southwest. From 2021 to 2022, the vegetation NPP gravity center shifted significantly to the southwest, indicating that the growth and increments of vegetation NPP in the southwest were greater than those in the northeast. On the whole, the vegetation NPP gravity center showed a trend of rotating from north to south to north and then moving south. However, the distance moving south was greater than that moving north, which indicates that the growth rate and increments in the south of Northeast China in the past 23 years were greater than those in the north.

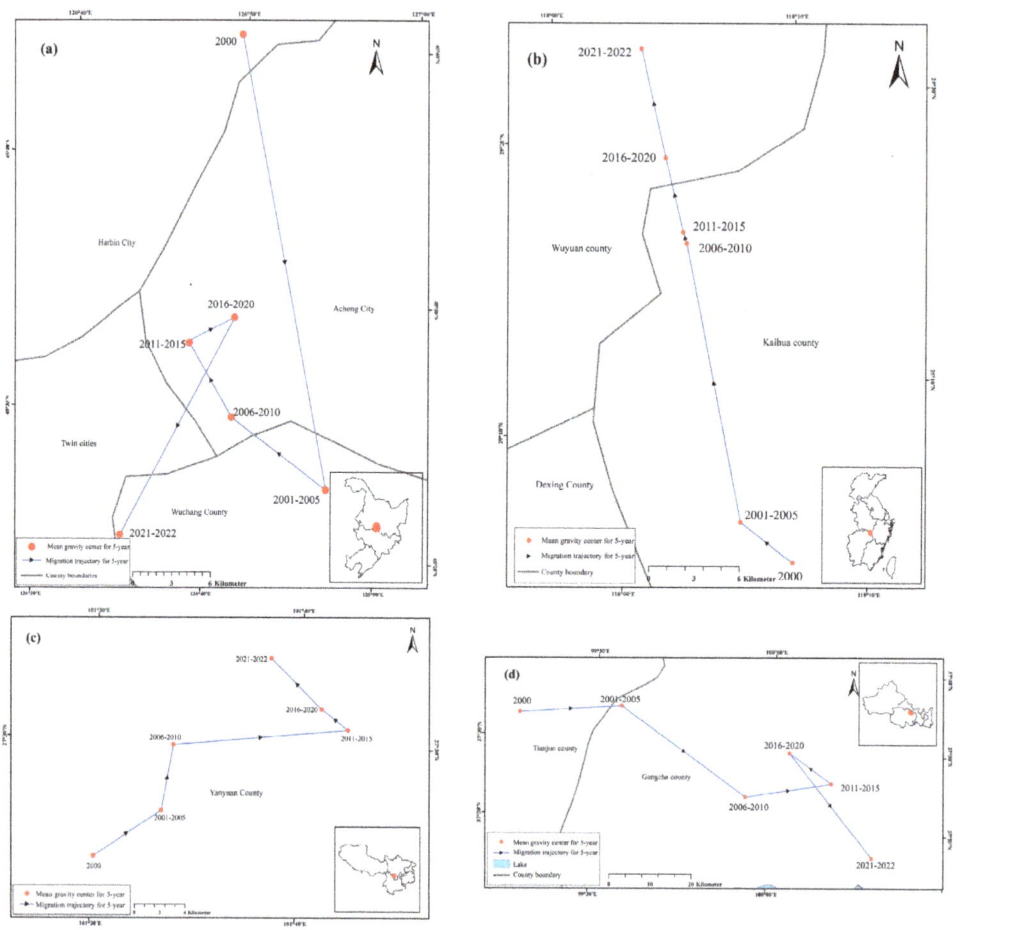

Figure 7. *Cont.*

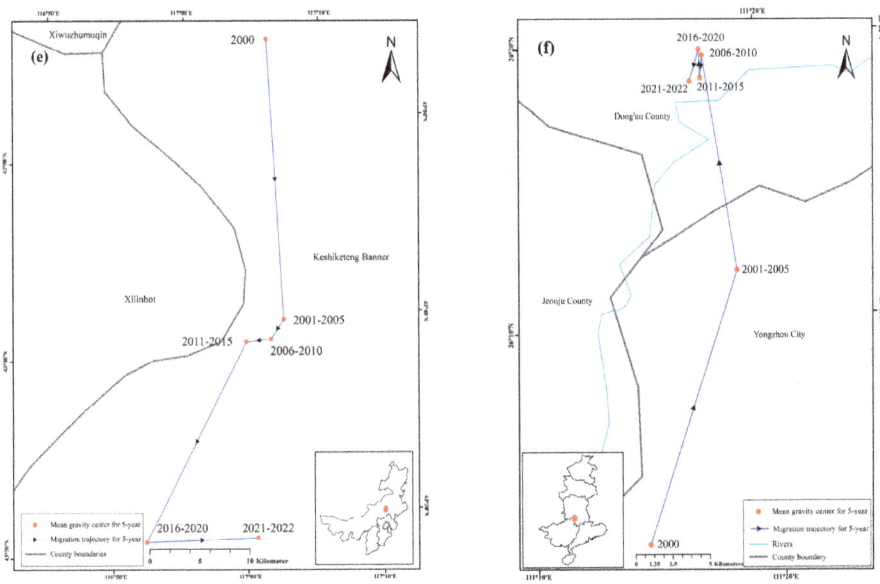

Figure 7. The migration trajectory of vegetation NPP center of gravity in six regions of China (5-year interval). (**a**) Northeast China. (**b**) East China. (**c**) Southwest China. (**d**) Northwest China. (**e**) North China. (**f**) Central South China.

Figure 7b shows that the gravity center migration trajectory of vegetation NPP in East China showed a trend of moving northwestward. During 2001–2005 and 2006–2010, the northwest migration distances were the largest, at 18.312 km, indicating that the growth rate and increments of vegetation NPP in the northwest direction were the largest during this period. During 2006–2010 and 2011–2015, the migration distances of the gravity center of vegetation NPP were the smallest, at 0.768 km, indicating that the growth rate and increments of vegetation NPP in the northwest direction were the smallest during this period.

Figure 7c shows that the vegetation NPP gravity centers in Southwest China moved northeastward in the first three periods and then moved northwestward in the last two periods. The five periods are 2000→2001–2005, 2001–2005→2006–2010, 2006–2010→2011–2015, 2011–2015, and 2016–2020→2021–2022. During 2006–2010 and 2011–2015, the migration distances were the largest, at 13.914 km, but the migration angle to the northeast was smaller, indicating that the growth rate and increments of vegetation NPP in the northeast direction were greater than those in the southwest direction during this period. During 2011–2015 and 2016–2020, the migration distances were the smallest, at 2.615 km. Overall, the gravity centers of vegetation NPP in Southwest China moved to the northeast, indicating that in the past 23 years, the increments of vegetation NPP in the northeast parts were greater than those in the southwest parts.

Figure 7d shows that the migration trajectory of the gravity centers in Northwest China was relatively complex. In 2001–2005, the vegetation NPP gravity center moved to the northeast. During 2001–2005 and 2006–2010, the gravity centers moved to the southeast, with the largest migration distance of 37.198 km, indicating that the vegetation NPP increments in the southeast parts were more significant. During 2011–2015 and 2016–2020, the gravity centers shifted to the northwest, and the migration distance was the shortest at 12.663 km, indicating that the increments and growth rate of vegetation NPP in the northwest parts were larger than those of other regions. Overall, the vegetation NPP gravity center moved to the southeast, indicating that the vegetation NPP increments and growth rate in the southeast parts were greater than those in the northwest parts.

Figure 7e shows that in 2001–2005, the gravity center of North China migrated to the southeast, with the largest migration distance of 26.884 km, indicating that the vegetation NPP increments in the southeast parts were higher than those in other parts. During 2001–2005 and 2006–2010, the gravity centers migrated to the southwest, with the smallest migration distance of 2.261 km, indicating that the vegetation NPP increments in the southwest parts during 2006–2010 were smaller than those in other years. Overall, the gravity center of vegetation NPP in North China showed a trend of moving southward.

Figure 7f showed that the change in vegetation NPP in the central and southern regions had a strong circuitry and complex change state. During 2000–2005, the gravity centers of vegetation NPP migrated to the northeast, with the largest migration distance of 18.837 km, and during 2001–2005 and 2006–2010, it moved to the northwest. During 2006–2010 and 2011–2015, 2011–2015 and 2016–2020, as well as 2016–2020 and 2021–2022, there was a trend of circuitous change. During 2006–2010 and 2011–2015, the gravity centers moved to the southwest, with the smallest migration distance of 1.488 km. During 2011–2015 and 2016–2020, the gravity center moved to the northwest. During 2016–2020 and 2021–2022, the gravity centers continued to move to the southwest. During the period from 2000 to 2022, the migration trajectory of the gravity center showed a trend of northeast–northwest–southwest–northwest–southwest. Overall, the vegetation NPP gravity center moved to the northeast.

3.4. *Quantitative Discrimination of Climate–Human Effect on Vegetation NPP*

The combined effects of climate-human interactions on vegetation NPP are examined through six scenarios. These six development scenarios were defined as areas with an increase in vegetation NPP via climatic change, human activities, and combined effects, as well as areas with a decrease in vegetation NPP via climatic change, human activities, and combined effects (Tables 3 and 4; Appendix A Figure A6).

Table 4. The area proportion of zones affected by climate–human interactions on vegetation NPP.

Relative Action Zone	Percentage (%) 2000–2010	Percentage (%) 2011–2022
NPP increased (climatic change)	23.87	36.03
NPP increased (human activities)	39.92	26.99
NPP increased (combined effects)	16.11	11.88
NPP decreased (climatic change)	12.89	9.09
NPP decreased (human activities)	5.00	14.29
NPP decreased (combined effects)	2.21	1.72

From 2000 to 2010, zones with human activities that promoted an increase in vegetation NPP had the largest proportion, comprising 39.92%, which was mainly distributed in Ningxia Hui Autonomous Region, Henan Province, and the junction of Heilongjiang Province and eastern Jilin Province. Zones with a reduction in vegetation NPP caused by combined effects had the smallest area, accounting for 2.21%, which was mainly distributed at the junction of Guangdong Province, Zhejiang Province, and Anhui Province. Overall, during 2000–2010, the proportion of areas with increased vegetation NPP was greater than that of areas with decreased vegetation NPP. From 2011 to 2022, zones with climate-change-promoted increases in vegetation NPP accounted for 36.03%, being mostly concentrated in Heilongjiang Province, Yunnan Province, Gansu Province, and southern Qinghai Province. Zones with a reduction in vegetation NPP caused by combined effects had the smallest area, accounting for 1.72%.

In contrast, among areas with an increase in vegetation NPP, zones with climate-change-promoted increases in vegetation NPP expanded significantly, while areas that underwent human-activity-promoted increases in vegetation NPP decreased significantly. These change areas were mainly distributed in Sichuan Province, Yunnan Province, Guizhou

Province, and Heilongjiang Province. For areas with a reduction in vegetation NPP, the proportion of zones with a reduction caused by human activities increased significantly.

We divided the climate-human contribution rate to the increase in vegetation NPP in the study area into five equal intervals: 0%–20%, 20%–40%, 40%–60%, 60%–80%, and 80%–100%. Firstly, 0%–20% indicates that the change in vegetation NPP (increase or decrease) was almost unaffected by human activities or climate change. Secondly, 20%–40% indicates that the change in vegetation NPP was slightly affected by human activities or climate change. Thirdly, 40%–60% indicates that the change in vegetation NPP was moderately affected by climate change or human activities. Fourthly, 60%–80% indicates that the change in vegetation NPP was strongly affected by climate change or human activities. Finally, 80%–100% indicates that the change in vegetation NPP was almost entirely affected by climate change or human activities (Appendix A Figure A7).

Figure 8 shows the region wherein climate change promoted an increase in vegetation NPP in 2000–2010, which was mainly distributed in North China (Figure 8a,b). In terms of climate change, the proportion of area with contribution rates greater than 80% was the largest, comprising 67.57%, which was mainly distributed in Heilongjiang Province, western Jilin Province, Shandong Province, Hebei Province, Anhui Province, eastern Qinghai Province, and northern Shanxi Province. Zones with a contribution rate between 20% and 40% had the smallest area, accounting for only 7.84%. Figure 8b and Appendix A Figure A7 showed that during 2000–2010, zones with an increase in vegetation NPP promoted by human activities were mainly distributed in southern China, and zones with contribution rates greater than 80% accounted for the highest proportion (77.29%), which were mainly in Guizhou Province, Sichuan Province, Henan Province, Yunnan Nasheng, Hubei Province, and eastern Inner Mongolia. Zones with contribution rates of 0%–20% and 60%–80% accounted for only 5.60%.

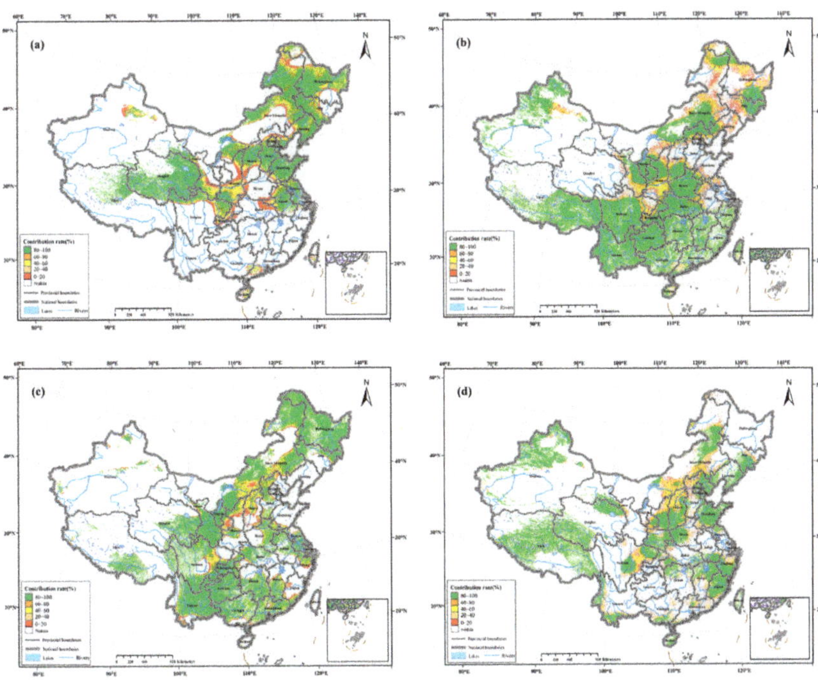

Figure 8. The spatial distribution of the contribution rate of climate–human interactions to the increase in vegetation NPP across different historical periods. (**a**) Climate change (2000–2010); (**b**) human activities (2000–2010); (**c**) climate change (2011–2022); (**d**) human activities (2011–2022).

From 2011 to 2022, climate change contributed 36.03% of the increase in vegetation NPP (Figure 8c,d). In terms of climate (Figure 8c), zones with contribution rates of 80–100% accounted for the largest proportion, comprising 79.78%, which were mainly distributed in Heilongjiang, Yunnan, Guangdong, Jiangsu, and Ningxia. Zones with contribution rates of 0%–20% had the smallest area proportion, comprising 4.22%. In terms of human activities (Figure 8d), zones with increases in vegetation NPP promoted by human activities were distributed in inland China, such as the patch areas formed by Shandong Province, Fujian Province, and Zhejiang Province; the Qinghai–Tibet Plateau; and the junction of Qinghai Province and Ningxia Hui Autonomous Region. From the perspective of contribution levels, zones with contribution rates of 80%–100% accounted for the largest proportion, at 74.65%, which were mainly located in each patch area, followed by zones with contribution rates of 40%–60%, accounting for 7.12%, which were mainly distributed at the junction of Shanxi Province and Inner Mongolia Autonomous Region.

Areas with a reduction in vegetation NPP caused by climate–human interaction are less extensive than areas with an increase in vegetation NPP, accounting for only 20.1% (Figure 9 and Appendix A Figure A8). In terms of climate change (Figure 9a), zones with contribution rates of 80%–100% had the largest area proportion, comprising 88.17%, which was mostly located in the southern part. The reduction in vegetation NPP caused by human activities was mostly located in the middle of the study area, such as the junction of Anhui Province and Zhejiang Province, southern Qinghai Province, and central and western Tibet (Figure 9b). Zones with contribution rates of 80%–100% accounted for the highest proportion, at 74.54%, which was mainly distributed in Guangdong and Guangxi. Zones with vegetation NPP decreases induced by climate change had larger areas than those caused by human activities. Our comprehensive analysis showed that the decrease in vegetation NPP during 2000–2010 was mainly related to climate change.

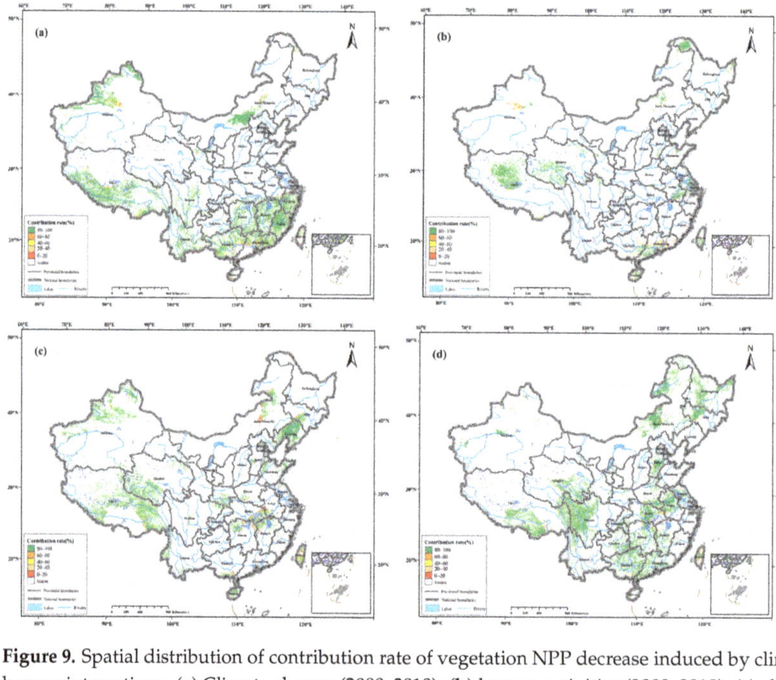

Figure 9. Spatial distribution of contribution rate of vegetation NPP decrease induced by climate–human interactions. (**a**) Climate change (2000–2010); (**b**) human activities (2000–2010); (**c**) climate change from (2011–2022); (**d**) human activities (2011–2022).

Areas with reduced vegetation NPP accounted for 25.1%, which was smaller than areas with increased vegetation NPP (Figure 9c,d). In terms of climate change (Figure 9c), zones with contribution rates of 80%–100% accounted for the largest proportion, at 86.81%, being mostly located in the southern part of the study area, such as the Tibet Autonomous Region, Jiangxi Province, and Hubei Province. Zones with contribution rates of 60%–80% were mostly located in the northern parts, accounting for only 2.80%. In terms of human activities (Figure 9d and Appendix A Figure A8), zones with contribution rates of 80%–100% accounted for the largest proportion, at 92.05%, which were mainly distributed across all provinces and cities. Zones with contribution rates of 0%–20% accounted for the smallest proportion, comprising only 1.78%.

3.5. Dominant Factors Influencing the Evolution of Vegetation NPP in Different Sub-Regions with Time Changes

3.5.1. Third-Order Partial Correlation Analysis

To further explore the correlations between vegetation NPP and climatic factors (accumulated temperature, sunshine, precipitation, and temperature) from 2000 to 2022, a third-order partial correlation coefficient was used to explore the relationships between vegetation NPP and climatic factors (Figure 10).

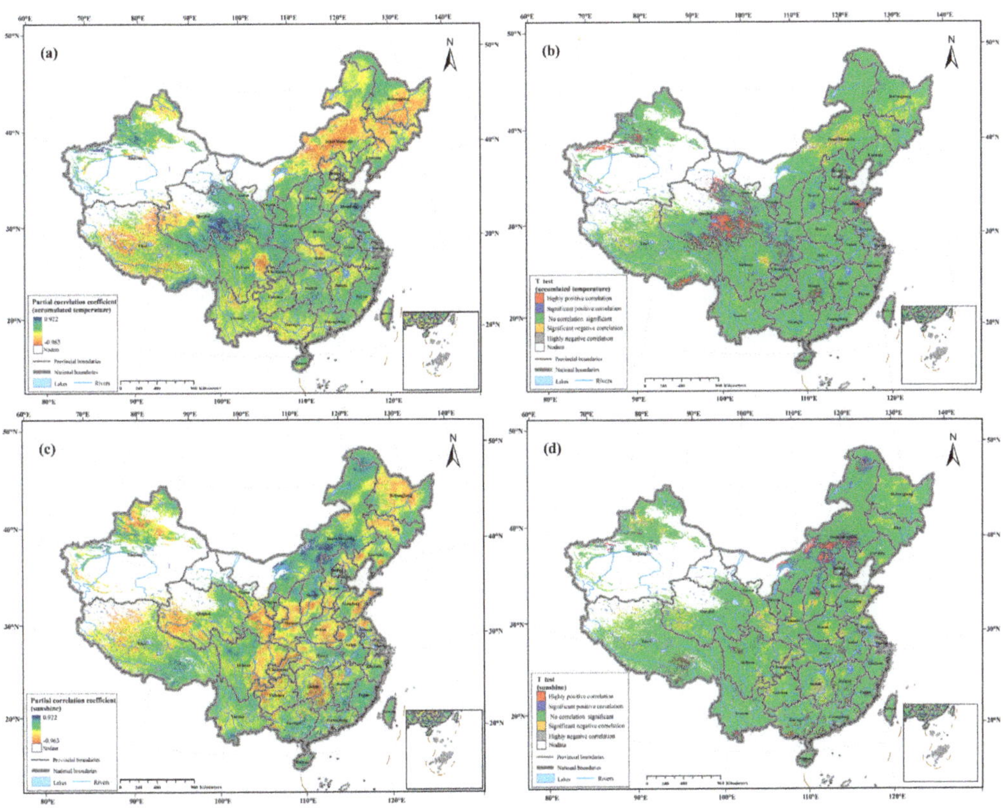

Figure 10. *Cont.*

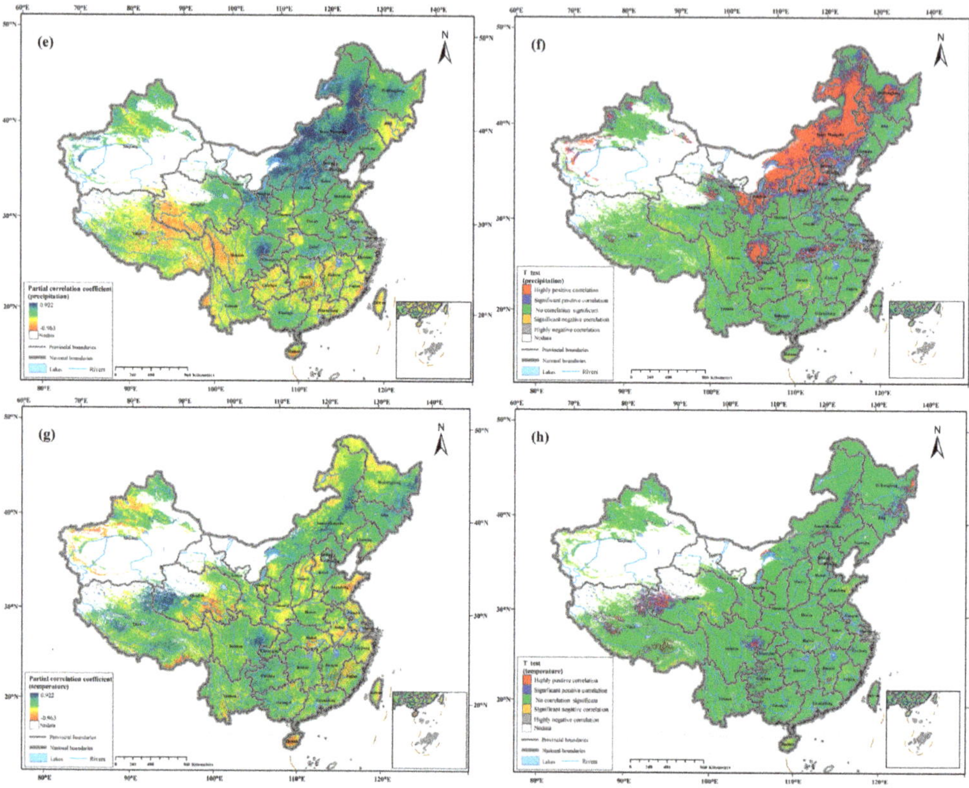

Figure 10. A partial correlation analysis distribution map and a significance test distribution map of vegetation NPP and climatic factors. (**a**) Vegetation NPP is partially correlated with accumulated temperature; (**b**) a significance test of vegetation NPP and accumulated temperature; (**c**) vegetation NPP is correlated with sunshine; (**d**) a significance test of vegetation NPP and sunshine; (**e**) vegetation NPP is partially correlated with precipitation; (**f**) a significance test of vegetation NPP and precipitation; (**g**) vegetation NPP is partially correlated with air temperature; (**h**) a significance test of vegetation NPP and temperature.

The third-order partial correlation coefficient between vegetation NPP and accumulated temperature ranged from −0.916 to 0.967. The positive correlation area accounts for 56.93%, which was mainly distributed south of the Yellow River and north of the Yangtze River, such as Henan Province, Shandong Province, and other regions (Figure 10a). Zones with negative correlation accounted for 43.06%, being mainly distributed in Jilin, Liaoning, and Tibet Autonomous Region. Figure 10c showed that the partial correlation coefficient between vegetation NPP and sunshine ranged from −0.953 to 0.940, of which the positive correlation area accounted for 47.76%, which was mostly concentrated in the northern part, such as Inner Mongolia, Loess Plateau, and northern Heilongjiang Province. Zones with negative correlation accounted for 52.24%, mainly concentrated in the southern part, such as Hunan Province and Guizhou Province. As shown in Figure 10e, the partial correlation coefficient between vegetation NPP and precipitation ranged from −0.954 to 0.974, of which the positive correlation area accounted for 71.67%, being mostly located in the northern parts, such as Inner Mongolia Autonomous Region, Heilongjiang Province, Shandong Province, and other regions. Zones with negative correlations accounted for 28.33%, which were mainly distributed in southern China, such as Fujian Province, Jiangxi Province, Hunan Province, and Guizhou Province. Figure 10g showed that the partial correlation coefficient between vegetation NPP and temperature was between −0.962 and 0.922, of which the positive correlation area accounted for 64.45%, being mainly distributed in the northern parts. Zones with negative correlations accounted for 35.55%, which were mainly distributed in southern China, such as Anhui, Zhejiang, Jiangsu, Fujian, and Hainan provinces.

Areas with extremely significant positive correlations between vegetation NPP and accumulated temperature accounted for 3.34%, and most of these areas were located in the Three-River Source Region. Areas with significant positive correlations accounted for 6.47%, most of which were located in the surrounding areas with significant positive correlations (Figure 10b). Zones with extremely significant negative correlations ($p < 0.01$, $R < 0$) accounted for only 1.15%, being mostly concentrated in the central and eastern parts of Inner Mongolia. Zones with significant negative correlations ($p < 0.05$, $R < 0$) accounted for 4.16%, which were mostly distributed in the Sichuan Basin, Qinghai–Tibet Plateau, and Songhua River–Nenjiang–Mudanjiang River Basin. According to Figure 10d, zones with extremely significant positive correlations in the partial correlation coefficient between vegetation NPP and sunshine accounted for 2.21%, which were mainly concentrated in Inner Mongolia. Zones with significant positive correlations accounted for 5.63%, and surrounding areas had extremely significant positive correlations. Zones with extremely significant negative correlation ($p < 0.01$, $R < 0$) had the smallest area proportion, at 1.97%, which were mainly distributed in internal areas with significant negative correlations. Zones with extremely significant positive correlation between vegetation NPP and precipitation partial correlation coefficient accounted for the largest proportion, comprising 16.94% in the significance level test (Figure 10f), being mainly distributed in North China, such as eastern Inner Mongolia, Loess Plateau, the Songhua River–Nenjiang River Basin, northern Hebei, and northwestern Shanxi. Zones with significant positive correlations accounted for 12.64%, mostly located in the surrounding areas with extremely significant positive correlations. Areas with significant negative correlations ($p < 0.05$, $R < 0$) accounted for 2.63%, which were scattered in southern China. Zones with extremely significant negative correlations ($p < 0.05$, $R < 0$) had the smallest area, accounting for only 0.96%. The zones with extremely significant positive correlations between vegetation NPP and temperature partial correlation coefficient accounted for 2.32% (Figure 10h), being widely distributed in northern Inner Mongolia, eastern Heilongjiang, Sichuan Basin, and the junction of Tibet Autonomous Region and Qinghai Province. Zones with significant positive correlations accounted for 7.74% and were distributed in the surrounding areas with extremely significant positive correlations. Zones with significant negative correlations ($p < 0.05$, $R < 0$) accounted for 2.04% and were scattered.

3.5.2. Dominant Factor Analysis

Single Factor

Factor detection can quantitatively determine the contribution rate of geographic factors on the spatial distribution difference of an index value. A factor detector of a geographic detector is used to analyze the q-value of each factor to obtain a single factor weight table.

As shown in Appendix A Figure A9a and Table 5, precipitation, soil type, and land use played a dominant role in vegetation NPP change in Northeast China in 2000, with q-values of 0.515, 0.461, and 0.443, respectively. The explanatory power of population density and GDP on vegetation NPP was smaller, with both less than 0.02. Appendix A Figure A9b and Table 5 show that, compared with 2000, the explanatory power changed to a certain extent. The dominant factors were land use, soil type, and altitude, with q-values of 0.456, 0.435, and 0.412, respectively. The influences of GDP and population density on vegetation NPP increased to a certain extent. However, the explanatory power of vegetation NPP was still small, with a q-value of no more than 0.1. Appendix A Figure A9c and Table 5 showed that land use, soil type, and slope were the dominant factors of vegetation NPP in Northeast China in 2022, with q-values of 0.440, 0.409, and 0.373, respectively. GDP and population density were the two factors with the smallest influence on vegetation NPP in Northeast China, with q-values of less than 0.3. It can be concluded that land use and soil type were the main factors affecting the changes in vegetation NPP in Northeast China.

Table 5. Q values for different regions in 2000, 2010, and 2020.

Region	Year	TEM	PRE	POP	GDP	ELE	GRA	LAND	SOIL
NEC	2000	0.181	0.515	0.007	0.018	0.439	0.346	0.443	0.461
	2010	0.102	0.205	0.011	0.071	0.412	0.384	0.456	0.435
	2020	0.097	0.281	0.02	0.029	0.332	0.373	0.44	0.409
NC	2000	0.485	0.584	0.005	0.004	0.160	0.237	0.580	0.714
	2010	0.43	0.705	0.005	0.011	0.19	0.249	0.649	0.736
	2020	0.344	0.673	0.004	0.015	0.175	0.265	0.621	0.732
EC	2000	0.441	0.372	0.094	0.053	0.341	0.303	0.347	0.39
	2010	0.393	0.256	0.08	0.164	0.307	0.276	0.349	0.347
	2020	0.356	0.237	0.056	0.058	0.298	0.273	0.326	0.323
NWC	2000	0.206	0.648	0.013	0.038	0.058	0.136	0.528	0.763
	2010	0.173	0.691	0.014	0.152	0.05	0.118	0.55	0.783
	2020	0.156	0.682	0.011	0.203	0.056	0.111	0.535	0.798
SWC	2000	0.719	0.776	0.137	0.011	0.713	0.03	0.454	0.753
	2010	0.729	0.706	0.003	0.099	0.736	0.029	0.486	0.773
	2020	0.766	0.744	0.002	0.078	0.77	0.023	0.508	0.801
CSC	2000	0.397	0.321	0.03	0.125	0.164	0.177	0.27	0.378
	2010	0.35	0.233	0.128	0.121	0.172	0.179	0.29	0.347
	2020	0.317	0.255	0.111	0.09	0.174	0.171	0.27	0.313

NEC—Northeast China; NC—North China; EC—East China; NWC—Northwest China; SWC—Southwest China; CSC—Central South China; TEM—temperature; PRE—precipitation; POP—population density; ELE—elevation; GRA—gradient; LAND—land utilization; SOIL—soil type.

Precipitation, land use, and soil type were the dominant factors affecting changes in vegetation NPP in North China in 2000, with q-values of 0.714, 0.584, and 0.580, respectively (Appendix A Figure A9a and Table 5). Slope and altitude exhibited little influence on changes in vegetation NPP, with q-values of 0.237 and 0.160, respectively. Similarly, population density and GDP showed little influence on vegetation NPP, with q-values less than 0.01. Appendix A Figure A9b and Table 5 showed that the dominant factors in 2010 were still soil type, precipitation, and land use, with q-values of 0.736, 0.705, and 0.649, respectively. Compared with 2000, the explanatory power of the dominant factors increased. The influence of altitude on vegetation NPP was insignificant, with a q-value of 0.190. The contribution rate of GDP and population density on vegetation NPP was still small, with q-values of 0.011 and 0.005, respectively. Appendix A Figure A9c and Table 5 showed that, compared with 2010, the dominant factors in 2022 did not change. The influence of altitude on vegetation NPP was still insignificant, with a q-value of 0.175. The influence of GDP and population density on vegetation NPP was still insignificant, with q-values of 0.015 and 0.004, respectively. We can conclude that soil type was the dominant factor affecting changes in vegetation NPP in North China.

The temperature in 2000, 2010, and 2022 was the main factor affecting changes in vegetation NPP, with q-values of 0.441, 0.393, and 0.356, respectively (Appendix A Figure A9). Compared with 2010, the explanatory power of precipitation decreased significantly, with a q-value reduction from 0.372 to 0.256. In 2020, the q-values of each factor decreased to varying degrees. The q-values of land use and soil types were roughly the same, with both around 0.32. GDP and population density had little impact on changes in vegetation NPP, with q-values less than 0.06. Temperature was the dominant factor influencing the variations in vegetation NPP in East China.

The dominant factor in 2000, 2010, and 2022 was soil type, with q-values of 0.763, 0.783, and 0.798, respectively (Appendix A Figure A9a and Table 5). The q-values of GDP increased significantly, indicating that human activities gradually became the dominant factors influencing vegetation NPP. Soil type, precipitation, and land use were the dominant factors influencing changes in vegetation NPP in Northwest China.

In 2000, the q-values of precipitation, soil type, temperature, and altitude affecting the changes in vegetation NPP were greater than 0.7 in Southwest China (Appendix A Figure A9a and Table 5). The q-values of slope and GDP were the smallest at 0.030 and 0.011, respectively, indicating an insignificant influence on vegetation NPP. Appendix A Figure A9b and Table 5 show that the dominant factors were soil type, altitude, temperature, and precipitation. Appendix A Figure A9c and Table 5 showed that the dominant factor did not change compared with 2010, and the q-values of the dominant factors increased to 0.801, 0.770, 0.766, and 0.744. The explanatory power of population density on the change in vegetation NPP was still the smallest, with a q-value of only 0.002. Soil type, altitude, temperature, and precipitation were the dominant factors influencing changes in vegetation NPP in Southwest China.

According to Appendix A Figure A9a and Table 5, temperature, soil type, and precipitation were the dominant factors in 2000 in central and southern regions, with q-values of 0.397, 0.378, and 0.321, respectively. Population density had the smallest explanatory power for vegetation NPP change, with a q-value of 0.03. Appendix A Figure A9b and Table 5 showed that, in 2010, the dominant factors were temperature and soil type, with q-values of 0.350 and 0.347, respectively. Appendix A Figure A9c and Table 5 showed that the dominant factors in 2022 were temperature and soil type, with q-values of 0.31, 0.317, and 0.313. Therefore, temperature and soil type were the dominant factors influencing changes in vegetation NPP in central and southern regions.

Interactive Factor

Interactive detection is a kind of geographical detector. The interaction between factors can be calculated using Formula (9). Interaction factor detection can be utilized to analyze the influence of interactions between different factors on changes in vegetation NPP (Figure 11). Using the origin to draw a color-scale diagram, the higher the value, the stronger the interaction between the representative factors, and vice versa.

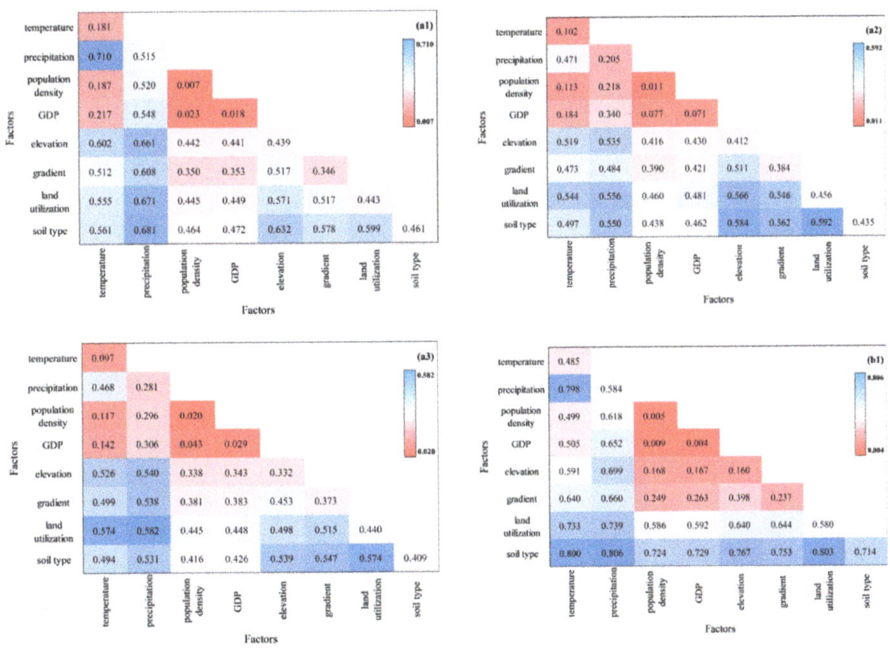

Figure 11. *Cont.*

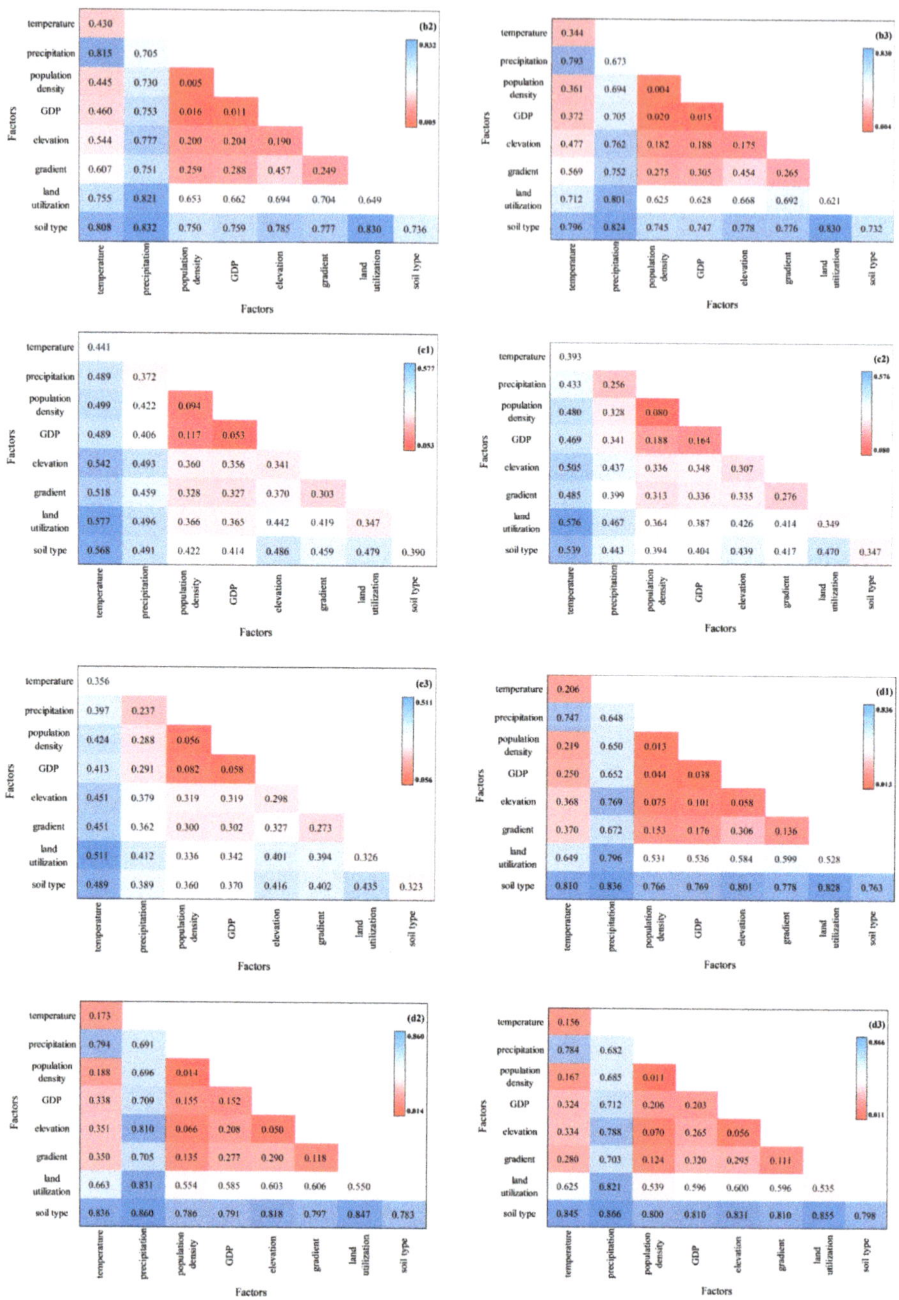

Figure 11. *Cont.*

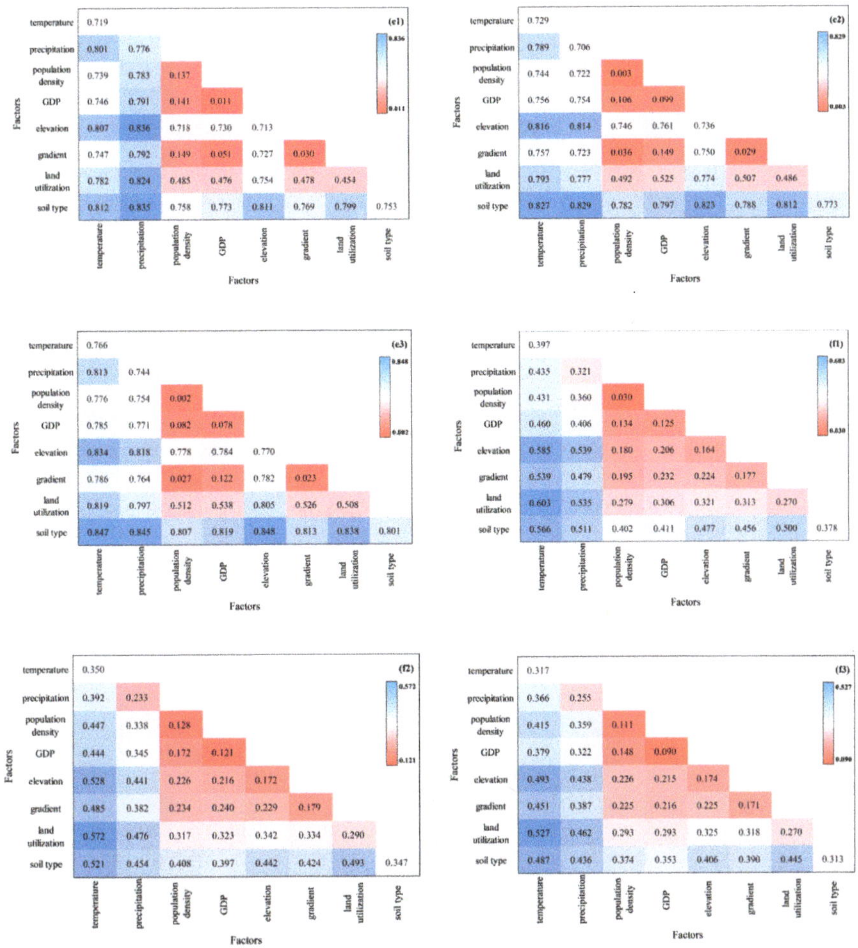

Figure 11. The interactive dominant factors affect vegetation NPP in different regions in different periods. (**a1**) Northeast China (2000). (**a2**) Northeast China (2010). (**a3**) Northeast China (2022). (**b1**) North China (2000). (**b2**) North China (2010). (**b3**) North China (2022). (**c1**) East China (2000). (**c2**) East China (2010). (**c3**) East China (2022). (**d1**) Northwest China (2000). (**d2**) Northwest China (2010). (**d3**) Northwest China (2022). (**e1**) Southwest China (2000). (**e2**) Southwest China (2010). (**e3**) Southwest China (2022). (**f1**) Central South China (2000). (**f2**) Central South China (2010). (**f3**) Central South China (2022).

Figure 11(a1–a3) shows that the interaction between the two factors in Northeast China is mostly mutually reinforcing. In 2000, the dominant interactive factor was temperature ∩ precipitation (0.710), while in 2010, land use ∩ soil type (0.592) became the dominant factor. In 2022, the dominant interactive factor was precipitation ∩ land use (0.582). As shown in Figure 11(b1–b3), in 2000 and 2010, the dominant interaction factor was precipitation ∩ soil type (0.806 and 0.832). In 2022, the dominant interactive factor was land use ∩ soil type (0.830). As shown in Figure 11(c1–c3), in 2000, 2010, and 2022, the dominant interactive factor was temperature ∩ land use (0.577, 0.576, and 0.511, respectively). As shown in Figure 11(d1–d3), in 2000, 2010, and 2022, the dominant interactive factor was precipitation ∩ soil type (0.836, 0.860, and 0.866, respectively). In general, the interaction between precipitation and soil type was the main factor affecting the change in vegetation

NPP in Northwest China. Figure 11(e1–e3) showed that in 2000, the dominant interactive factor was precipitation ∩ elevation (0.836). The dominant interactive factor in 2010 was precipitation ∩ soil (0.829), which changed significantly compared with that in 2000. The dominant interactive factor in 2022 was altitude ∩ soil type (0.848). Overall, the change in vegetation NPP in Southwest China was greatly affected by the interaction between altitude and precipitation, as well as other factors. As shown in Figure 9(f1–f3), in 2000, 2010, and 2022, the dominant interactive factor was temperature ∩ land use (0.603, 0.572, and 0.527, respectively). The interaction between GDP, population density, land use, and other factors showed a trend of strengthening first and then weakening. Overall, the interaction between temperature and other factors was the dominant factor affecting the change in vegetation NPP, and the influence of human activities on the change in vegetation NPP showed a trend of increasing first and then decreasing.

4. Discussion

4.1. The Reasons for the Spatio-Temporal Changes in Vegetation NPP

The spatial distribution of vegetation NPP during 2000–2022 was roughly the same as that in 2000–2010 and 2011–2022, showing a decreasing trend from south to north and from east to west, which is consistent with the research results of Shi et al. [39]. From the north to the south, the climate in southern China was hot and humid, and the precipitation was higher, which is conducive to vegetation growth. In the northern region, the climate type is mainly temperate continental. It was hot and rainy in summer and cold and dry in winter, and the vegetation growth environment was worse than that in the south, thus forming a spatial distribution pattern of high in the south and low in the north [18]. The western region of China was dominated by deserts and plateaus. The Taklimakan Desert and other regions had high temperatures all year round, scarce precipitation, and almost no vegetation coverage. Moreover, the Qinghai–Tibet plateau, with many glaciers and permafrost, had high altitude, low temperature, and scarce precipitation [16]. The vegetation type was dominated by alpine meadows, resulting in low levels of vegetation NPP. The eastern region was close to the Pacific and Indian Oceans. Due to the influence of ocean currents, the precipitation was richer than in the western region, and the vegetation types were more abundant. There were a small number of high-value areas in the northeast region and the southeast of the Qinghai–Tibet Plateau, which is consistent with the research results of Xue et al. [40]. Specifically, Northeast China had rich forest resources, high-quality black soil, and organic matter, which is conducive to the accumulation of vegetation NPP. At the same time, in 1998, China implemented the "Natural Forest Resource Protection Project", which further helped to improve vegetation NPP, resulting in a small number of high-value NPP areas in Northeast China. The southeast of the Qinghai–Tibet Plateau, with more precipitation than other parts, was affected by a Pacific monsoon, and its vegetation types were diverse.

In terms of time transfer from different grades, the transfer area from low to high accounts for a large proportion, and vegetation NPP showed an increasing trend, which is consistent with the research results obtained by Xu et al. [41] and Lou et al. [42]. This may be related to China's implementation of returning farmland to forest and grassland as part of its protection policy for ecological areas. NPP is a key factor in determining the ecological process of regulating carbon sources within ecosystems. It plays an important role in global change and carbon balance, providing insights for scientific decision-making in ecological management and climate change mitigation.

4.2. The Changes in Dominant Factors Influencing the Evolution of Vegetation NPP

Precipitation affects spatio-temporal changes of vegetation NPP in both dry and wet environments, temperature affects the rates of evaporation, and the synchronization of water and heat affects the growth state of the vegetation itself, thus affecting vegetation NPP. Compared with 2000–2010, human activities led to an obvious increase in areas with less vegetation NPP and a significant decrease in areas that promoted increases in vegetation NPP in 2011–2022 compared with 2000–2010. This might be due to the increase in human activities, including vegetation deforestation, corresponding with an improvement in living standards [39]. From the perspective of partial correlation, precipitation was mainly positively correlated in the north and negatively correlated in the south, which is related to the good hydrothermal conditions in the south and the cold and dry environment in the north. This is consistent with the research results of Guo et al. [26], which found a positive relationship between NPP and precipitation in Northeast China. In terms of temperature, zones with positive correlations were mainly located in the Qinghai–Tibet Plateau and Northeast China. Liu et al. [25] found that vegetation NPP of the Qinghai-Tibet Plateau was significantly positively affected by climate. Warm and humid climate change is an important driving force for the significant increase in vegetation NPP, which is consistent with the results of this study. Zones with negative correlations were mainly concentrated in the south, which is mainly related to the high latitude in Northeast China and the high altitude in the Qinghai–Tibet Plateau with low temperatures. An increase in temperature is conducive to plant photosynthesis and promotes the increase in vegetation NPP. In addition, precipitation and land use were the main factors affecting vegetation NPP in Northeast China, which is related to the high latitude and low precipitation in local areas. Due to the different research periods and driving factors, this is different from the research results obtained by Mao et al. [43]. In North China and Northwest China, precipitation and soil type were the main factors affecting vegetation NPP (the maximum q-values were 0.832 and 0.866, respectively). The main reason is that the soil type in North China is dominated by the river alluvial yellow dry farming type, which is consistent with the research results of Huang et al. [44]. The northwest region was mainly dominated by desertification, and precipitation was scarce, which was not suitable for vegetation growth. Precipitation became the dominant factor of changes in vegetation NPP in the northwest region. This is consistent with the research conclusion of Tong et al. [45]. The imbalance and uncertainty of the spatial and temporal distribution of water resources will lead to the contradiction between the ecological environment and economic development in the arid area of Northwest China.

The dominant factors in Southwest China were altitude and precipitation (the maximum q-value was 0.836). Xu et al. [30] found that precipitation had a strong enhancing effect on NPP in Southwest China, which is consistent with the results of this study. The local plateau area accounted for a large proportion, with high altitude, cold and dry climate, and mainly alpine meadow vegetation. The photosynthesis of vegetation was weak, which is not conducive to increases in NPP [46,47]. In Central South and East China, temperature and land use were dominant (the maximum q-values were 0.577 and 0.603, respectively). Although local precipitation and temperature levels were high, the temperature could not reach the dynamic balance of temperature and precipitation during vegetation photosynthesis, resulting in temperature being the dominant factor of vegetation NPP in Central South and East China.

From analyzing the impact of human activities, we determined that land use had more obvious impacts on changes in vegetation NPP in East China, Central and South China, and North China than other factors (the maximum q-values were 0.349, 0.290, and 0.649, respectively). In East China and Central South China, the rapid expansion of the population and urbanization had a great influence on the change in vegetation NPP. In North China, due to overgrazing, zones dominated by grassland were reduced, and desertification was aggravated, which is not conducive to an increase in vegetation NPP. Human activities had negative effects on vegetation NPP changes. Soil type had the greatest impact on the northwest region (the maximum q-value was 0.798). This region has the largest desert in China, which is not suitable for an increase in vegetation NPP [48–50]. The increase in human activities was conducive to the change in soil types and the increase in vegetation NPP. The influence of land use and population density on the change in vegetation NPP in Northeast China gradually increased. As the primary heavy industrial base in Northeast China, the region experienced rapid economic development, and human activities had negative effects on changes in vegetation NPP. Mao et al. [24] found that NPP changed greatly with the change in land use in Northeast China, which indicates that human activities have a significant influence on NPP. This is consistent with the results of this study. The influence of land use, GDP, and population density on changes in vegetation NPP in Southwest China was low (the maximum q-values were 0.508, 0.099, and 0.137, respectively), which was mainly due to the weak local economic foundations and the influence of human activities on vegetation NPP [51,52].

5. Conclusions

In this study, Geodetector, a gravity center model, and third-order partial correlation analysis methods were used to analyze the change patterns of vegetation NPP and its driving mechanisms in different sub-regions of China from a partition perspective.

(1) Vegetation NPP in China showed a decreasing trend from southeast to northwest. The gravity centers of Northeast, Northwest, and North China showed a trend of southward migration, indicating that the increments of vegetation NPP in the south of the corresponding region were greater than those in the north. The gravity centers of Southwest, Central South, and East China showed a trend of northward migration, indicating that the increments of vegetation NPP in the north of the corresponding region were greater than those in the south.

(2) During 2000–2010, human activities contributed greatly to the vegetation NPP increase and during 2011–2022, climate change was the dominant factor for the increase in vegetation NPP.

(3) Zones with significant positive correlations between vegetation NPP and accumulated temperature were mostly located in the southern part of Qinghai Province. Zones with significant positive correlations with precipitation were mostly concentrated in Inner Mongolia and other regions. Zones with significant positive correlations with temperature were widely distributed in the junction of Tibet and Qinghai Province and the northeast region. Zones with significant positive correlations with sunshine were mainly distributed in the central and eastern regions of Inner Mongolia.

(4) Precipitation and land use were the dominant factors influencing changes in vegetation NPP in Northeast China, while precipitation and soil types played an important role in the vegetation NPP changes in North China. Temperature was the dominant factor influencing the change in vegetation NPP in East China, while precipitation and soil types were the main factors affecting the vegetation NPP changes in Northwest China. The explanatory power of human activities on the change in vegetation NPP in Northwest China gradually increased. Altitude and precipitation contributed considerably to the change in vegetation NPP in Southwest China. Provided that urbanization is ensured, reducing land use can effectively promote an increase in local vegetation NPP.

Author Contributions: Conceptualization, methodology, writing—original draft preparation, Y.L. and M.X.; investigation, supervision, project administration, and funding acquisition, B.G. and G.Y. Methodology, investigation, J.L. and Y.Y. All authors have read and agreed to the published version of the manuscript.

Funding: This research was funded by the Natural Science Foundation of Shandong Province (grant number: ZR2021MD047); the Scientific Innovation Project for Young Scientists in Shandong Provincial Universities (grant number: 2022KJ224); the National Natural Science Foundation of China (grant numbers: 42101306 and 62006096); the Natural Science Foundation of Fujian Province (grant number: 2020J05146); and a grant from the State Key Laboratory of Resources and Environmental Information System.

Data Availability Statement: Data are contained within the article.

Conflicts of Interest: The authors declare no conflict of interest.

Appendix A

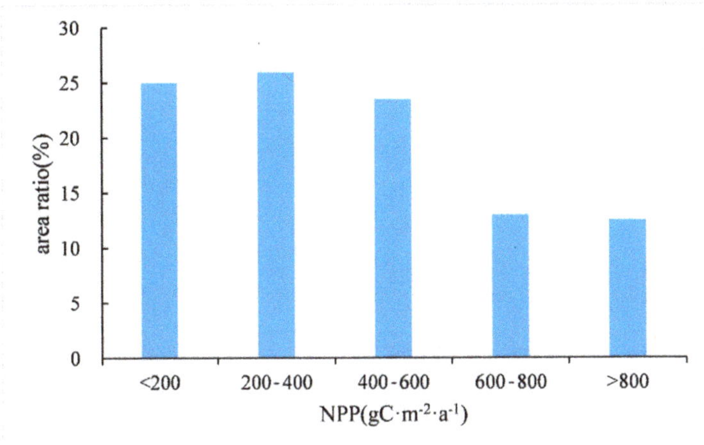

Figure A1. The proportion of different grades of vegetation NPP areas.

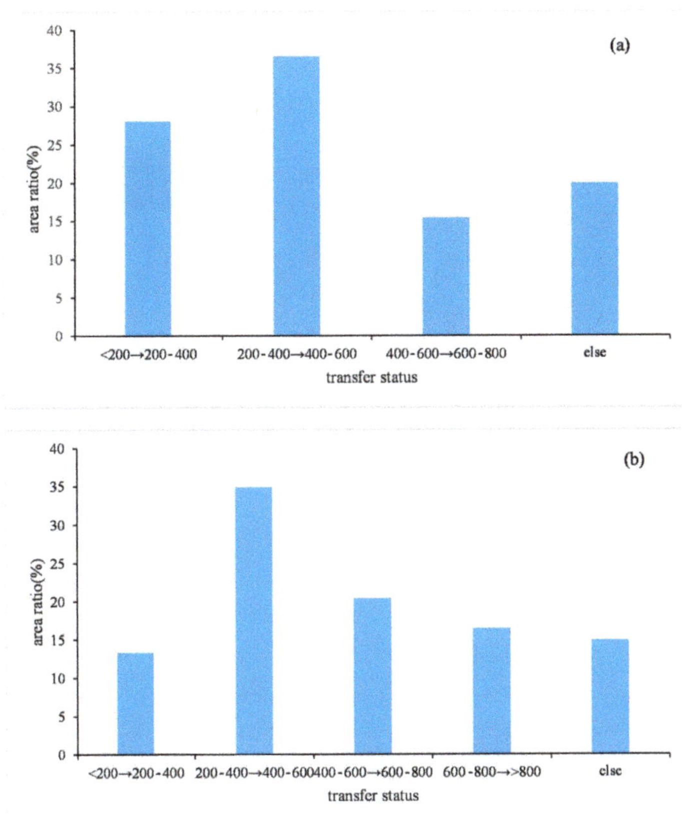

Figure A2. The proportion of different grades of vegetation NPP areas. (**a**) 2000–2010; (**b**) 2011–2022.

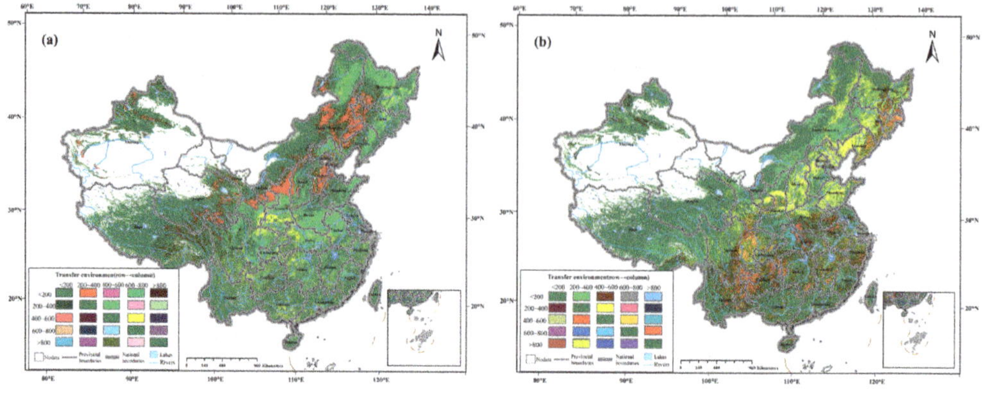

Figure A3. Conversion among different levels of vegetation NPP transfer at different levels. (**a**) 2000–2010; (**b**) 2011–2022.

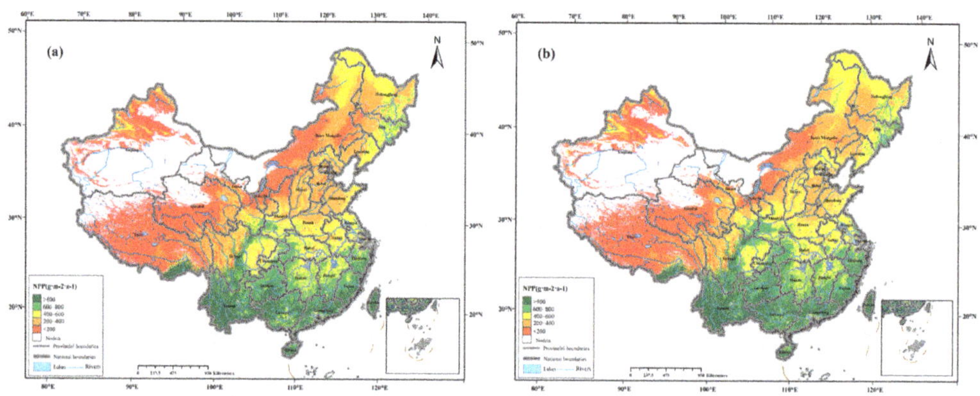

Figure A4. The mean distribution map of NPP at different levels during different historical periods. (**a**) 2000–2010; (**b**) 2011–2022.

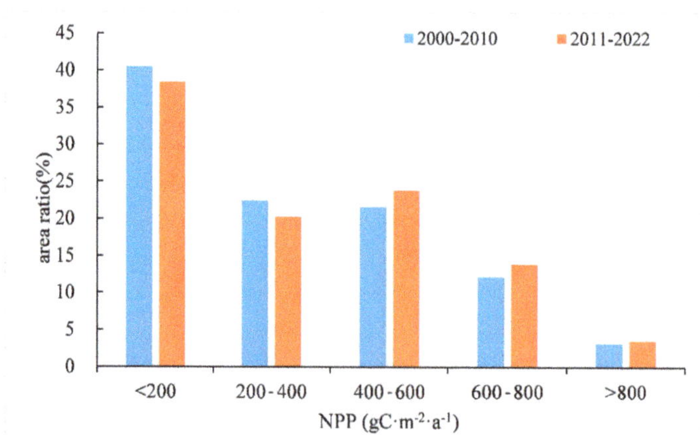

Figure A5. The area of different NPP levels during different historical periods.

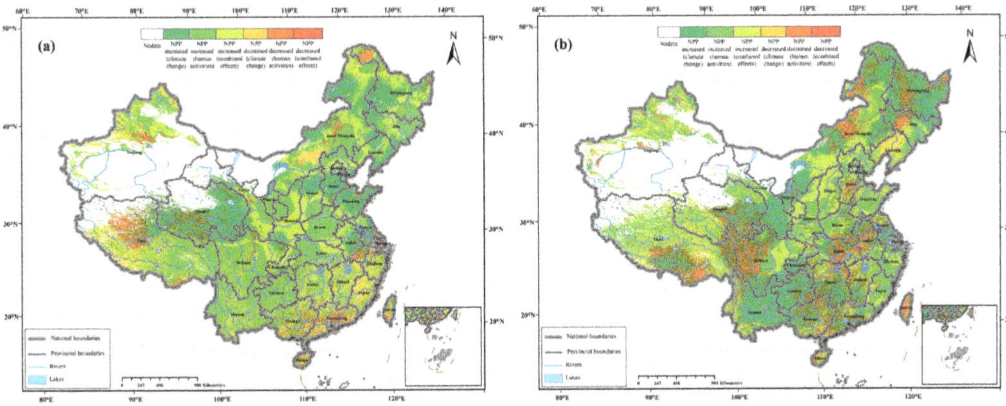

Figure A6. The spatial distributions of climate–human impacts on vegetation NPP. (**a**) 2000–2010; (**b**) 2011–2022.

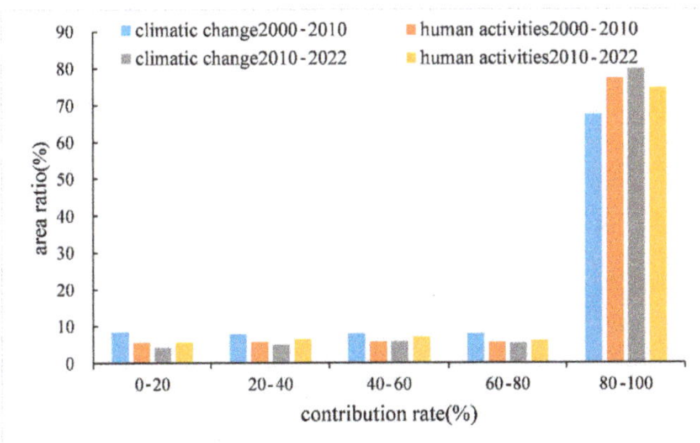

Figure A7. The area ratio of contribution rates for different grades during different historical periods (increase).

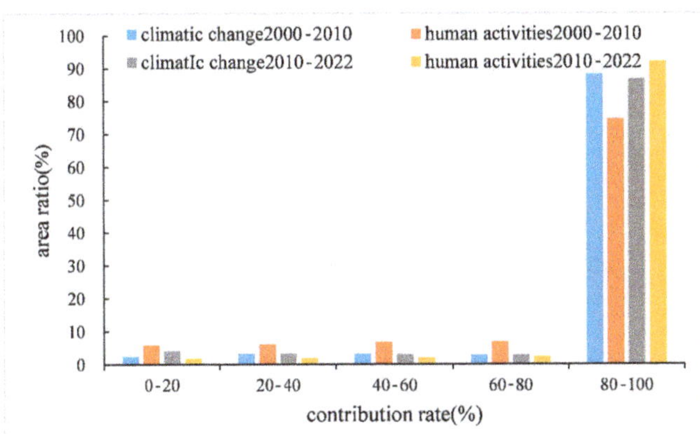

Figure A8. The area ratio of contribution rates for different grades during different historical periods (decrease).

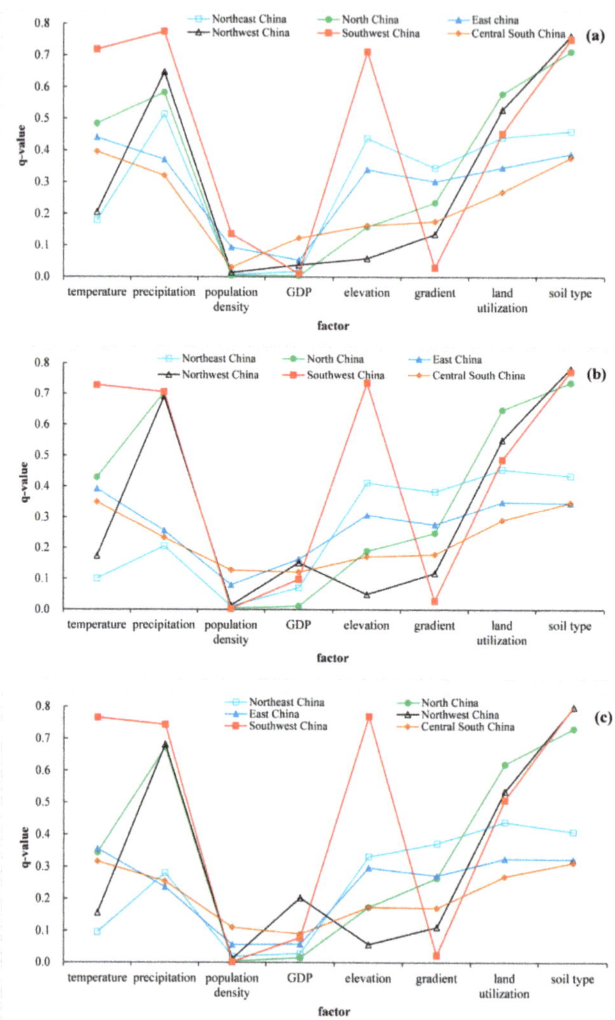

Figure A9. The explanatory power of each factor across different regions and years (q-value). (**a**) 2000; (**b**) 2010; (**c**) 2022.

References

1. Mustafa, A.; Szydowski, A. The Impact of Spatiotemporal Changes in Land Development (1984–2019) on the Increase in the Runoff Coefficient in Erbil, Kurdistan Region of Iraq. *Remote Sens.* **2020**, *12*, 1302. [CrossRef]
2. Xia, C.; Liu, G.; Xia, H.; Jiang, F.; Meng, Y. Influence of saline intrusion on the wetland ecosystem revealed by isotopic and hydrochemical indicators in the Yellow River Delta, China. *Ecol. Indic.* **2021**, *133*, 108422. [CrossRef]
3. Dong, S.Y.; Liu, W.X. Implement the new development concept to help carbon peak carbon neutrality. *People's Dly.* **2023**, *10*, 25–31.
4. Amen, A.R.M.; Mustafa, A.; Kareen, D.A.; Hameed, H.M.; Mirza, A.A.; Szydtowski, M.; Saleem, B.K.M. Mapping of Flood-Prone Areas Utilizing GIS Techniques and Remote Sensing: A Case Study of Duhok, Kurdistan Region of Iraq. *Remote Sens.* **2023**, *15*, 1102. [CrossRef]
5. Fatah, K.K.; Mustafa, Y.T.; Hassan, M.O. Flood Susceptibility Mapping Using an Analytic Hierarchy Process Model Based on Remote Sensing and GIS Approaches in Akre District, Kurdistan Region, Iraq. *Iraqi Geol. J.* **2022**, *55*, 2414–6064. [CrossRef]
6. Running, S.W.; Hunt, E.R., Jr. 8-Generalization of a Forest Ecosystem Process Model for Other Biomes, BIOME-BGC, and an Ap-plication for Global-Scale Models. *Scaling Physiol. Process.* **1993**, *6*, 141–158.

7. Potter, C.S.; Randerson, J.T.; Field, C.B.; Matson, P.A.; Klooster, S.A. Terrestrial ecosystem production: A process model based on global satellite and surface data. *Glob. Biogeochem. Cycles* **1993**, *7*, 811–841. [CrossRef]
8. Field, C.B.; Randerson, J.T.; Malmstrom, C.M. Global net Primary Production: Combining ecology and remote sensing. *Remote Sens. Environ.* **1995**, *51*, 74–88. [CrossRef]
9. Mowll, W.; Blumenthal, D.M.; Cherwin, K.; Smith, A.; Symstad, A.J.; Vermeire, L.T.; Collins, S.L.; Smith, M.D.; Knapp, A.K. Climatic controls of aboveground net primary production in semi-arid grasslands along a latitudinal gradient portendlow sensitivity to warming. *Oecologia* **2015**, *177*, 959–969. [CrossRef]
10. Seino, H.; Uchijima, Z. Agroclimatic Evaluation of Net Primary Productivity of Natural Vegetation: Assessment of Total Net Primary Production in Japan. *J. Agric. Meteorol.* **1985**, *41*, 139–144. [CrossRef]
11. Kang, M.C.; Zhu, L.P.; Xu, X.; Cha, T.G.; Zhang, Z.Q. Study on the response of carbon and water fluxes to climate change in northern poplar plantations based on the Biome-BGC model. *Ecology* **2019**, *39*, 2378–2390.
12. Xu, F.Y.; Jiao, H.B.; Ding, X.D.; Ye, Q.; Chen, F.S.; Lin, Y.L. Temporal and spatial characteristics of vegetation net primary productivity in subtropical evergreen broad-leaved forest. *J. Northwest For. Univ.* **2019**, *34*, 62–68.
13. Mao, D.H.; Luo, L.; Wang, Z.M.; Zhang, C.H.; Ren, C.Y. Variations in net primary productivity and its relationships with warming climate in the permafrost zone of the Tibetan Plateau. *J. Geogr. Sci.* **2015**, *25*, 967–977. [CrossRef]
14. Zhu, Y.Y.; Han, L.; Zhao, Y.H.; Ao, Y.; Li, J.J.; Xu, K.B.; Liu, B.; Ge, Y.Y. NPP simulation and its spatial and temporal pattern in Northwest China. *J. Ecol.* **2019**, *38*, 1861–1871.
15. Han, W.Y.; Zhang, C.; Zeng, Y.; Liu, G.H. Spatio-temporal changes and driving factors of NPP in Lhasa River Basin from 2000 to 2015. *Ecol. J.* **2018**, *38*, 8787–8798.
16. Tian, Z.H.; Zhang, D.D.; He, X.H.; Guo, H.L.; Wei, H.T. Spatio-temporal variation characteristics and driving factors of veg-etation net primary productivity in the Yellow River Basin from 2000 to 2015. *Int. Soil Water Conserv.* **2019**, *26*, 255–262.
17. Luo, Y.; Zhang, S.L. Spatial and temporal distribution characteristics and driving factors of vegetation NPP in Shandong Province. *Plants Guangxi* **2019**, *39*, 690–700.
18. Ji, P.P.; Gao, M.H.; Yang, X.D. Analysis of NPP driving force in arid areas of northwest China: A case study of the Ili River Valley and parts of the Tianshan Mountains in Xinjiang. *Ecol. J.* **2019**, *39*, 1–11.
19. Zhu, W.Q.; Chen, Y.H.; Xu, D.; Li, J. Advances in calculation models of net primary productivity of terrestrial vegetation. *J. Ecol.* **2005**, *11*, 296–300.
20. Wang, Y.B.; Zhao, Y.H.; Han, L.; Ao, Y. Spatial and temporal changes of vegetation net primary productivity and its driving factors in Qinba Mountains from 2000 to 2015. *Appl. Ecol.* **2018**, *29*, 2373–2381.
21. Yang, X.; Guo, B.; Han, B.M.; Chen, S.T.; Yang, F.; Fan, Y.W.; He, T.L.; Liu, Y. Analysis of spatial and temporal evolution pattern and driving mechanism of NPP in Qinghai–Tibet Plateau. *Resour. Environ. Yangtze River Basin* **2019**, *28*, 3038–3050.
22. Lv, X.; Wang, J.L.; Kang, H.J.; Zhao, Q.; Han, X.H.; Wang, Y.J. Study on the temporal and spatial variation of grass yield in the source region of the Three Rivers from 2006 to 2015 based on MODIS NPP. *J. Nat. Resour.* **2017**, *32*, 1857–1868.
23. Wan, H.W.; Li, H.X.; Gao, J.X.; Sun, C.X.; Li, G.Y.; Gao, Y.H.; Li, Q. Spatial pattern analysis of carbon sequestration potential of vegetation ecosystem in China. *Acta Ecol. Sin.* **2022**, *42*, 8568–8580.
24. Li, S.F.; Ge, W.Y.; Wang, F. Characteristics of drought events and their impacts on vegetation in northern China from 1982–2019. *Res. Soil Water Conserv.* **2023**, *30*, 259–521.
25. Mao, D.H.; Wang, Z.M.; Han, J.X.; Ren, C.Y. Spatio-temporal Pattern of Net Primary Productivity and Its Driven Factors in Northeast China in 1982–2010. *Sci. Geogr. Sin.* **2012**, *32*, 1106–1111.
26. Wang, H.; Gao, J.B.; Hou, W.J. Quantitative attribution analysis of soil erosion in different morphological types of geomorphology in Karst areas: Based on the geographical detector method. *J. Geogr. Sci.* **2019**, *29*, 271–286. [CrossRef]
27. Yang, X. Research on Key Technology of Vulnerability Quantitative Evalution for Multi-Type Ecological Functional Areas. Master's Thesis, Shandong University of Technology, Zibo, China, 2022.
28. Liu, J.; Ji, Y.H.; Zhou, G.S.; Zhou, L.; Lv, X.M.; Zhou, M.Z. Temporal and spatial variations of net primary productivity (NPP) and its climate driving effect in the Qinghai-Tibet Plateau, China from 2000 to 2020. *Chin. J. Appl. Ecol.* **2022**, *33*, 1533–1538.
29. Guo, Z.X.; Wang, Z.M.; Zhang, B.; Liu, D.W.; Yang, G.; Song, K.S.; Li, F. Analysis of Tempora-l Spatial Characteristics and Factors Influencing Vegetation NPP in Northeast China from 2000 to 2006. *Resour. Sci.* **2008**, *30*, 1226–1235.
30. Xu, Y.; Huang, H.Y.; Dai, Q.Y.; Guo, Z.D.; Zheng, Z.W.; Pan, Y.C. Spatial-temporal Variation in Net Primary Productivity in Terrestrial Vegetation Ecosystems and Its Driving Forces in Southwest China. *Environ. Sci.* **2023**, *44*, 2704–2714.
31. Liu, Y.; Liu, H.H.; Chen, Y.; Gang, C.C. Spatial-temporal dynamic characteristics and driving factors of vegetation optical depth in China from 2000 to 2018. *Acta Geogr. Sin.* **2023**, *78*, 729–745.
32. He, K.D.; Sun, J.; Chen, Q.J. Effects of climatic factors and soil texture on grassland net primary productivity and precipitation use efficiency on the Tibetan Plateau. *Pratacultural Sci.* **2019**, *36*, 1053–1065.
33. Liu, H.W.; Wang, C.L.; Lun, M.H.; Chen, S.W. Study on Gravity Model in Optimization of Logistics Facilities Location and Its Solution. *Logist. Technol.* **2013**, *32*, 287–290.
34. Zhang, X.L.; Wang, Y.C.; Xiao, W.H.; Yang, R.X.; Wang, Y.; Zhu, L.S. The response of NPP to climate change in the Shiyang River Basin. *J. Ecol.* **2018**, *37*, 3110–3118.
35. Lieth, H.; Whittaker, R.H. Primary productivity of the biosphere. *ResearchGate* **1975**, *32*, 274.

36. Xu, H.J.; Wang, X.P.; Zhang, X.X. Alpine grasslands response to climatic factors and anthropogenic activities on the Tibetan Plateau from 2000 to 2012. *Ecol. Eng.* **2016**, *92*, 251–259. [CrossRef]
37. Zhou, Y.Y.; Zhu, M.X.; Guo, X.J.; Li, K.; Miao, J.X.; Guo, J.J.; Xu, X.F.; Yue, D.X. Evaluation of relative impact of climate change and human activities on vegetation NPP in Shule River Basin. *Ecology* **2019**, *39*, 5127–5137.
38. Qian, S.; Cui, X.J.; Jiang, Y.Q. Interpretation of QX/T 494-2019. Land Vegetation Meteorological and Ecological Quality In-spection and Evaluation Grade. *Stand. Sci.* **2022**, *7*, 125–134.
39. Shi, Z.Y.; Wang, Y.T.; Zhao, Q.; Zhang, L.P.; Zhu, C.M. Spatio-temporal variation and driving mechanism of vegetation net primary productivity in China from 2001 to 2020. *Ecol. Environ. Sci.* **2022**, *31*, 2111–2123.
40. Xue, M. Spatio-Temporal Distribution and Driving Forces of Vegetation Productivity in Northeast China. Master's Thesis, Fuzhou University, Fuzhou, China, 2018.
41. Xu, Y.; Zheng, Z.W.; Meng, Y.C.; Pan, Y.C.; Guo, Z.D.; Zhang, Y. Spatio-temporal variation of Net Primary Produvtivity of different vegetation types and its influencing factors exploration in Southwest China. *Environ. Sci.* **2023**, *15*, 58–64.
42. Lou, Y. The Temporal-Spatial Changes of Vegetation NPP in the Main Forest Areas of Henan Province Based on CASA Model. Ph.D. Thesis, Zhengzhou University, Zhengzhou, China, 2020.
43. Yan, J.N.; Wang, L.Z.; He, H.X.; Liang, D.; Song, W.J.; Han, W. Large area land cover changes monitoring with time-series remote sensing images using transferable deep models. *IEEE Trans. Geosci. Remote* **2022**, *60*, 4409917. [CrossRef]
44. Huang, Y.Y. Spatial-Temporal Distribution and Driving Factors of Vegetation Net Primary Productivity in North China Plain from 2000 to 2018. Ph.D. Thesis, Northwest Normal University, Lanzhou, China, 2020.
45. Tong, L.J.; Liu, Y.Y.; Wang, Q.; Yang, Y.; Li, J.L. Spatial and temporal dynamics of Net Primary Productivity and its driving factors in Northwest China. *Res. Soil Water Conserv.* **2019**, *26*, 367–374.
46. Wang, S.; Zhang, S.W.; Lin, X.; Li, X.Y.; Li, R.S.; Zhao, X.Y.; Liu, M.M. Response of soil water and carbon storage to short-term grazing prohibition in arid and semi-arid grasslands of China. *J. Arid. Environ.* **2022**, *202*, 104754. [CrossRef]
47. Zhang, L.H.; Wang, J.F.; Zhao, R.F.; Guo, Y.F.; Hao, L.Y. Aboveground net primary productivity and soil respiration display different responses to precipitation changes in desert grassland. *J. Plant Ecol.* **2022**, *15*, 57–70. [CrossRef]
48. Guo, Y.; He, Y.J.; Wu, P.; Wu, B.L.; He, M.H.; Han, X.; Xia, T.T.; Shen, K.P.; Kang, L.L.; Tan, Q.Y.; et al. The interspecifc competition presents greater nutrient facilitation compared with intraspecifc competition through AM fungi interacting with litter for two host plants in karst soil. *J. Plant Ecol.* **2022**, *15*, 399–412. [CrossRef]
49. Guo, B.; Liu, Y.F.; Fan, J.F.; Lu, M.; Zang, W.Q.; Liu, C.; Wang, B.Y.; Huang, X.Z.; Lai, J.B.; Wu, H.W. The salinization process and its response to the combined processes of climate change-human activity in the Yellow River Delta between 1984 and 2022. *Catena* **2023**, *231*, 107301. [CrossRef]
50. Chen, G.P.; Cai, Q.; Fang, W.J.; Feng, Y.H.; Zhu, J.J.; Ji, C.J.; Tang, Z.Y.; Fang, J.Y. The structural characteristics and climatic and human impacts of deciduous oak forests in China. *J. Plant Ecol.* **2022**, *15*, 265–267. [CrossRef]
51. Wang, S.; Tuya, H.; Zhang, S.W.; Zhao, X.Y.; Liu, Z.Q.; Li, R.S.; Lin, X. Random forest method for analysis of remote sensing inversion of aboveground biomass and grazing intensity of grasslands in Inner Mongolia, China. *Int. J. Remote Sens.* **2023**, *44*, 2867–2884. [CrossRef]
52. Luo, M.; Zhang, S.W.; Huang, L.; Liu, Z.Q.; Yang, L.; Li, R.S.; Lin, X. Temporal and Spatial Changes of Ecological Environment Quality Based on RSEI: A Case Study in Ulan Mulun River Basin, China. *Sustainability* **2022**, *14*, 13232. [CrossRef]

forests

Article

Tree Skeletonization with DBSCAN Clustering Using Terrestrial Laser Scanning Data

Lei You [1,2,3,4,*], Yian Sun [1,2], Yong Liu [1,2], Xiaosa Chang [1,2], Jun Jiang [5], Yan Feng [1,2] and Xinyu Song [6]

1 School of Computer and Information Technology, Xinyang Normal University, Xinyang 464000, China; sya@xynu.edu.cn (Y.S.); liuyong@xynu.edu.cn (Y.L.); xiaosachang@126.com (X.C.); yfeng@xynu.edu.cn (Y.F.)
2 Henan Engineering Research Center of Internet of Things and Smart Security, Xinyang 464000, China
3 Henan Dabieshan National Field Observation and Research Station of Forest Ecosystem, Zhengzhou 450046, China
4 Xinyang Academy of Ecological Research, Xinyang 464000, China
5 College of Forestry, Beijing Forestry University, Beijing 100083, China; jiang@bjfu.edu.cn
6 School of Mathematics and Statistics, Xinyang Normal University, Xinyang 464000, China; xysong88@163.com
* Correspondence: leiyou@xynu.edu.cn

Abstract: A tree skeleton reflects the geometric and structural characteristics of a tree. Terrestrial laser scanning (TLS) provides an effective means to construct tree skeletons that can capture the surface and topological features of trees. However, it is difficult to construct a tree skeleton located at the geometric centre of the tree with a detailed hierarchy of branches because of the natural intricate topology of the tree and the defects of the tree point cloud scanned by TLS. In this study, we presented a tree-skeletonization method based on density-based spatial clustering of applications with noise (DBSCAN) using TLS data. First, outliers were removed using DBSCAN, and the point-traversal order of each point was recorded. Next, a tree point cloud was divided into several tree slices using contour planes, and several tree segments were obtained by applying DBSCAN to each tree slice. Tree skeleton points were retrieved from each tree segment after the point-inversion transformation. Then, the adjacent relationship between skeleton points and the flow weight of each skeleton point was calculated based on the point-traversal order. After that, the skeleton points were classified into stems and different levels of branch points using the flow weights of the skeleton points, and the branch hierarchies were identified. Finally, the tree skeleton was optimized using the angle consistency. The positional deviation d_p and directivity deviation d_d were presented in this study to verify the validity of the constructed tree skeleton. From the visualization results, the constructed tree skeleton effectively reflected the geometrical structure and branch hierarchies of the tree. The averages of d_p and d_d were 0.418 cm and 8.474 degrees, respectively. The results demonstrated the validity of the presented method. It will help improve the visualization and accurate measurement of trees.

Keywords: TLS; tree skeleton; DBSCAN; stem; branch; branch hierarchy

1. Introduction

1.1. Background

Trees are common and vital objects with complex geometric structures in the natural environment. Tree geometry modelling is beneficial to related fields, such as tree modelling for parameter retrieval in forest inventory [1,2], tree morphology representation in computer graphics, and digital urban planning management [3,4]. A tree skeleton is a kind of tree geometry model that is usually represented by points and line segments connecting the points. It represents the topology and geometry of a tree [5] and contains the essential data for three-dimensional (3D) tree model reconstruction with interconnected 3D geometries [6]. Tree skeletonization has attracted the attention of many researchers because of its simplicity and effectiveness. However, it is a challenge to construct a tree skeleton consistent with the actual geometry and composition of the tree, as the tree and its branches have naturally complex topologies [4,7].

In recent years, terrestrial laser scanning (TLS) can be used to obtain 3D geometric data of an object's surface and can store the data as a point cloud. As a millimetre-level measurement tool, TLS can provide an accurate approach for nondestructively acquiring 3D tree structural attributes [8,9], which opens up the possibility of constructing detailed tree skeleton models. However, due to the complexity of the tree geometry structure and the defects of the tree point cloud scanned with TLS, most of the existing methods focus on the geometric topological structure of the tree and do not consider the composition characteristics of the tree. In this study, we focus on these two aspects and present a tree-skeletonization method that can construct an accurate tree skeleton with detailed composition information.

1.2. Related Work

Many studies have focused on the construction of tree skeletons using TLS data. Ref. [10] optimized a branch-structure graph similar to an input point cloud to represent a tree skeleton. Ref. [11] segmented the cylindrical components and ramified regions based on the variational k-means clustering algorithm and then extracted a skeleton based on an adjacency graph built from the neighbourhood information of the components. Finally, a B-spline surface representation was used to represent the tree. Ref. [12] reduced the tree graph extracted from an octree organization of a tree point cloud to the tree skeleton. Ref. [13] segmented the tree branches and estimated the tree skeleton vertices based on the estimated neighbour domain and normal vector and then used a single-source path to construct a skeleton graph of tree skeleton vertices. The smoothing tree curve skeleton was obtained by utilizing the cubic Hermite curves. Ref. [14] trained a neural network to generate a series of cylinders to represent the tree geometry under the assumption that the local shape of branches is cylindrical. Ref. [4] first contracted 3D tree-shaped point clouds to a graph as an initial skeleton by minimizing the sum of the graph geodesic distances of the k-nearest geodesic neighbours and a topology term for preventing point clouds from shrinking in the local principal direction. They then connected the skeleton points using the confidence of the distance and the included angle of the principal directions. Ref. [5] filtered noise and identified bifurcations using density-based spatial clustering of applications with noise (DBSCAN) and then extracted a tree skeleton using space colonization. Ref. [15] presented a fast tree skeleton extraction method that improved the skeleton connectivity and completeness with a breakpoint connection after extracting a raw tree skeleton using tree voxel thinning. The above studies achieved satisfactory results; however, there are some issues worthy of attention, such as the centre location of skeleton points, the category of skeleton points, and branch-level category determination.

The difficulty of tree skeletonization lies not only in the complexity of tree topology but also in the defects of the tree point clouds scanned with TLS. The mutual occlusion of branches and leaves makes the tree point cloud uneven, incomplete, and noisy. The complexity of the tree topology and the defects of the tree point cloud are exactly where the advantages of the DBSCAN algorithm can be exerted. In this study, we present a tree skeletonization method using DBSCAN to extract tree skeletons, which are represented by points and line segments with category labels. The specific objectives are (1) to accurately calculate tree skeleton points using the inner points of a tree, (2) to classify skeleton points and to identify the branch hierarchies of a tree, (3) to construct a tree skeleton located at the geometrical centre with detailed branch hierarchy information, and (4) to present the positional deviation and directivity deviation for evaluating a tree skeleton in the absence of ground truth.

2. Materials and Methods

2.1. Field Data Collection and Processing

Scanning was performed in a planted *poplar* plantation plot in Luoshan County, Henan Province, China. The trees were scanned from nine TLS stations (stand centre, four corners, and centre of each border) using FARO X330. The scan quality was 4× (a scanning parameter of the FARO TLS scanner), the scan speed was 122 kpts/s, the point spacing was 6.136 mm at 10 m, and reference spheres were used for registration. Point cloud registration was performed using FARO Scene software (Figure 1). Afterwards, the tree points were extracted according to the continuity of the stem in the vertical direction [16]. Ten trees were extracted from the scanned plot, and their average stem diameter at breast height was 19.29 cm.

Figure 1. 2D and 3D views in FARO Scene software of a scanning station using FARO X330. (**a**) 2D view. (**b**) 3D view.

The presented method was implemented using C++ with the Point Cloud Library [17], Qt and VTK. The experimental results were obtained on a computer with Intel i7-8700K, 3.70 GHz CPU, 32 GB memory, and a 64-bit Windows 10 operating system.

2.2. DBSCAN

The core idea of DBSCAN [18] is that a point belongs to a cluster if it is close to many points of that cluster. Two points are considered neighbours if the distance between them is less than or equal to *eps*, and *minPts* specifies the minimum number of points to define a cluster. A core point is a point that has at least *minPts* points (including itself) in its neighbourhood based on *eps*. A border point is a point reachable from a core point and that has fewer than *minPts* points in its neighbourhood. A point is noise if it is not a core point and not reachable from any core points.

Each point needs to be traversed to calculate the number of neighbouring points in the process of clustering. After a point has been traversed, its neighbourhood points can be traversed to complete the discovery of a cluster. In this study, the point-traversal order is calculated as follows. Once a point is directly traversed, that is, not from the neighbours of other points, its predecessor is itself. Moreover, the predecessor of its neighbour point whose predecessor has not been set is the current point. For example, in Figure 2, if point A is first traversed, that is, directly traversed, the predecessor of A is A, and the predecessor of C, D, and E is A. When D is traversed, the predecessor of C, A, and E remains unchanged, and the predecessor of F is D.

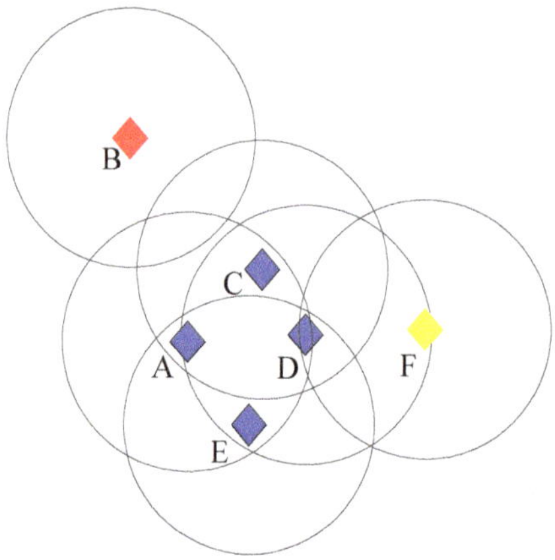

Figure 2. The diagram of DBSCAN clustering. The blue points are the core points. The yellow point is a border point. The red point is noise. The black circles represent the neighbourhoods of points. Here, $minPts$ is 3.

2.3. *Tree-Skeletonization Method*

2.3.1. Outlier Removal and Recorded Traversal Order

Due to mutual occlusion and breeze disturbances, a tree point cloud scanned with TLS has the following defects: the density of the tree point cloud is very uneven, outliers are distributed throughout the tree point cloud, and some discontinuous points exist in the tree point cloud. To eliminate these effects, outliers and discontinuous points were first removed using DBSCAN.

The tree point cloud was clustered using DBSCAN with the parameters eps and $minPts$ using a bottom-up approach. Clustering started from the lowest point of the Z axis and gradually expanded in ascending order of the Z axis. In the clustering process, the point-traversal order of each point was recorded. The cluster with the largest number of points was chosen as the major tree point cloud and the other clusters and noise were removed.

2.3.2. Tree Slice and Segment Generation

In this study, the tree skeleton points were mainly obtained from tree segments, which were the clustering result from tree slicing applied to the major tree point cloud.

Tree slicing is similar to the stem slice partitioning in [19]. In this study, each tree slice is formed by two contour planes, and the distance between the two planes is h cm. Then, m (Formula (1)) tree slices are obtained (Figure 3a), and each tree slice contains one region or several unconnected regions (Figure 3b). Then, each tree slice is clustered using DBSCAN with parameters eps and $minPts$. In the clustering result, a slice consists of several clusters and noise, and a cluster is a tree segment (Figure 3c).

$$m = \lceil \frac{z_{max} - z_{min}}{h} \rceil \tag{1}$$

where z_{max} and z_{min} are the maximum and minimum values of the z-axis of the major tree point cloud, respectively.

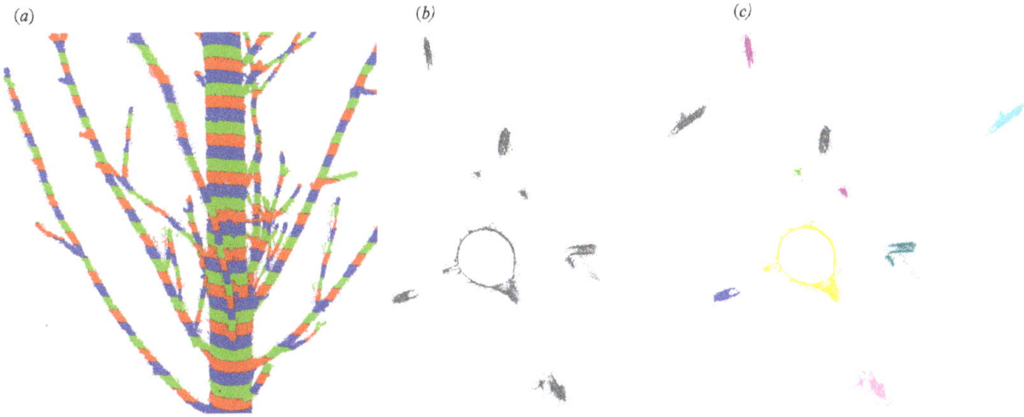

Figure 3. Diagrams of the tree-slice partitioning and DBSCAN results for a tree slice. (**a**) Partial view of a major tree point after tree slice partitioning; adjacent slices are shown in different colours. (**b**) Top view of a tree slice. (**c**) Top view of the clustering result of the tree slice in (**b**) from DBSCAN; different clusters are shown in different colours, and noise is shown in red.

2.3.3. Tree Skeleton Point Calculation with Point-Inversion Transformations

Each tree segment represents a component of the tree. Therefore, each tree segment is abstracted to a tree skeleton point. Similar to the geometric central point calculation in [20], a tree skeleton point of each tree segment is the centroid of the convex polygon that is formed from the planar projection point set, which is the resultant point set of the tree segment points projecting onto its lower contour plane. However, the irregular shape of the tree segment at the joint part of the stem and the branch causes a positional deviation of the centroid point of the polygon. The tree segment in Figure 4b is the segment coloured yellow in Figure 3c. The centroid of the stem segment in Figure 4b is not in the centre of the stem due to the influence of branch points.

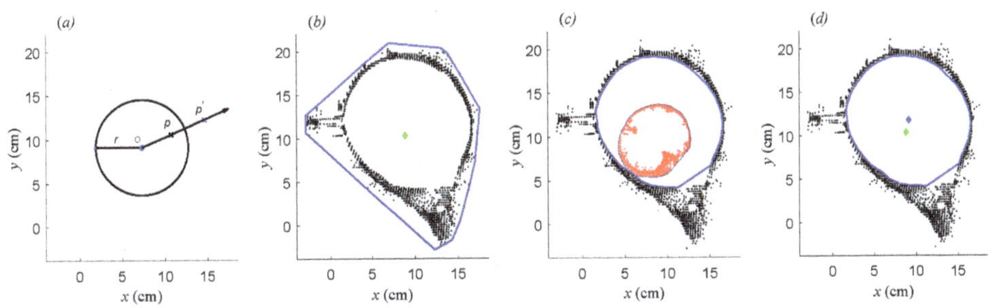

Figure 4. Diagrams of the point-inversion transformation and the centroid calculation before and after the point-inversion transformation. The black points represent tree segment points, the red points represent points found using the inversion transformation, the blue lines represent polygons, and the green and blue diamond points represent the centroid of the polygon before and after the point-inversion transformation, respectively. (**a**) Diagram of the point-inversion transformation. p' is the inversion-transformed point of point p and the centre and radius of inversion are o and r, respectively. (**b**) Centroid calculation directly based on the convex polygon formed by tree segment points. (**c**) Tree segment points, corresponding inversion-transformed points, and their corresponding polygons. (**d**) Tree segment points and the polygon of tree segment points corresponding to the convex hull points of transformed points.

Naturally, a point closer to the centre of the tree segment is preferred as a skeleton point. To ensure that a point close to the centre is used, the point-inversion transformation is used in this study. For a tree segment stem point set P, the centre of inversion o is the centroid of the convex polygon of the tree segment stem point set, and the radius of inversion r is the radius of the tree segment calculated by the closure Bézier curve fitting with global convexity [21]. The inverse of point p is the unique point p' on the ray starting from o and passing through p and it satisfies Formula (2).

$$op \cdot op' = r^2 \tag{2}$$

Then, the inverse point set P' of P is obtained (denoted by red points in Figure 4c), and the convex hull points of the inverse point set P' are calculated. According to the corresponding relationship between P and P', several points in P corresponding to the convex hull points of P' are chosen and used to form a polygon (denoted by the blue polygon in Figure 4d). The centroid of the polygon is the tree skeleton point in this study (denoted by the blue diamond point in Figure 4d). It is obvious that the centroid after the point-inversion transformation is closer to the centre of the segment than the centroid before the point-inversion transformation.

2.3.4. Adjacent Relationship Construction

According to the point-traversal order of the major tree point set extracted in Section 2.3.1, the adjacent relationship between tree skeleton points can be easily extracted by voting. If the predecessor points of most points in tree segment S_2 are in tree segment S_1, the parent of S_2 is S_1, that is, the parent of the skeleton point of S_2 is the skeleton point of S_1. Based on this, the adjacent relationships between tree skeleton points, that is, the parent–child relationships, are constructed. Obviously, a skeleton point has only one parent and at least one child. The bottom-up approach of major tree point set extraction ensures the integrity and correctness of the constructed adjacent relationship.

2.3.5. Skeleton Point Classification and Branch Hierarchy Identification

After adjacent relationship construction, the flow weights of the tree skeleton points are calculated as follows. The flow weight of a skeleton point is the sum of the number of its children and the sum of the flow weights of all its children. If a skeleton point has no child, the flow weight of the skeleton point is zero (as shown in Figure 5).

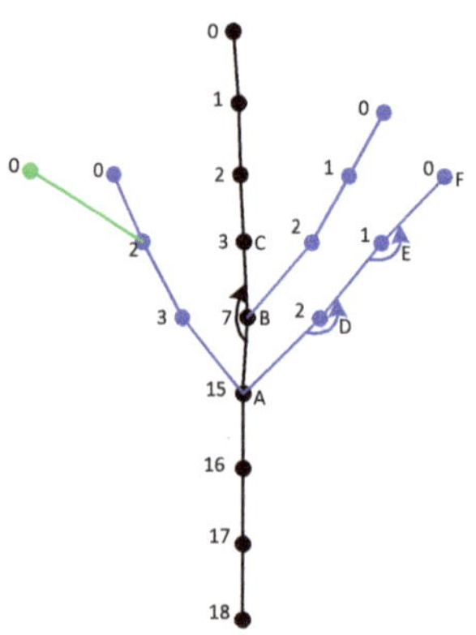

Figure 5. Diagram of skeleton-point classification, branch hierarchy recognition, and the included angle of the skeleton points. The numbers are the flow weights of the points. Black, blue, and green represent stem skeleton points and first-level and second-level branch points, respectively. The black, blue, and green lines represent the stem skeleton and first-level and second-level branches, respectively. The included angle at point B is the angle formed by points A, B, and C.

Obviously, the point with the largest flow weight must be the lowest tree skeleton point, i.e., the root point of the constructed tree skeleton. From the root, the point type of a point, i.e., a stem skeleton point or branch skeleton point, is classified as follows. If a point has only one child, the type of the child is equal to the type of the point. If a point has more than one child, the type of the child whose weight is closest to the weight of the point is equal to the type of the point, and the types of all the other children are equal to the point's type plus one (the type of the stem skeleton point plus one is the type of the first-level branch point, the type of the first-level branch point plus one is the type of the secondary branch point, and so on). If the weights of the two children points are the same and both are closest to the parent's weight, a child is selected so that the angle formed by the child's growth direction and the parent's growth direction is the smallest, the selected child's point type is equal to its parent's point type, and the point types of all the other children are equal to its parent's point type plus one. The growth direction of a tree skeleton point here is a vector formed by its coordinate minus its parental coordinate.

Based on the point type of a point, the procedure of branch hierarchy identification is as follows. The skeleton lines connected with stem skeleton points are the stem skeleton, the skeleton lines connected with a stem skeleton point and a first-level branch point or connected with first-level branch points are the first-level branch, and so on (Figure 5).

2.3.6. Tree Skeleton Optimization

Affected by missing TLS data, especially in the middle and upper regions of the tree, the calculated skeleton point may deviate from the centre of the stem segment (as shown by the black skeleton lines of the tree-slice index of 166 and 167 in Figure 6a). Furthermore, the joints of the stem and first-level branches are usually in the same cluster. This causes the direction between the stem and first-level branches to be inaccurate (denoted by the yellow skeleton lines of the slice index of 164, 167, 169 and 172 in Figure 6a). Therefore, some tree skeleton points need to be optimized to ensure angle consistency. In this study, the optimization consists of three processes in turn: position optimization of the stem skeleton point, position optimization of the joint point between the stem and first-level branches, and position optimization of the branch skeleton point.

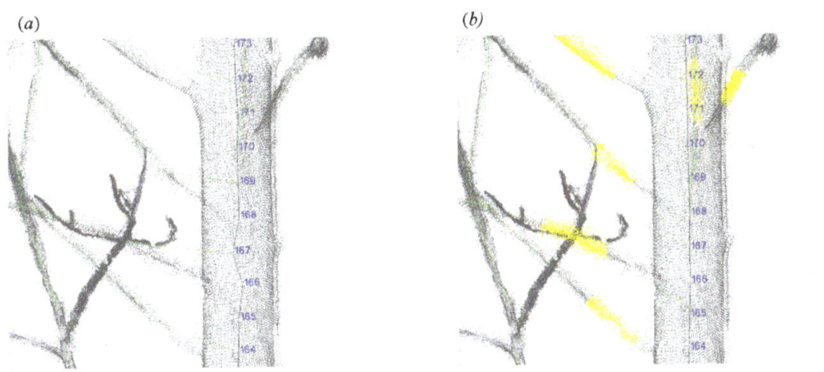

Figure 6. Diagrams of tree skeleton optimization. The lines represent the constructed tree skeleton; different branch hierarchies are represented by different colours. (**a**) The unoptimized tree skeleton. (**b**) The optimized tree skeleton. The yellow points represent the selected points for the position optimization of the joint point.

The aim of position optimization of the stem skeleton point is for the stem growth direction calculated by the constructed stem skeleton to be consistent with the actual stem growth direction. If the included angle at a stem skeleton point (the three points involved in the calculation are the stem skeleton points, as denoted by $\angle ABC$ in Figure 5) is smaller than 165 degrees, the stem skeleton point needs to be repositioned to match the actual growth direction. According to the relative stability of the growth direction of the stem for a small length (no larger than 30 cm in this study), the new position was along the local stem growth direction and at a distance h cm from its parent stem skeleton point. The local stem growth direction is calculated from the k consecutive ancestors of the repositioned stem skeleton point using principal component analysis (PCA) [19]. This optimization approach is a bottom-up approach, which can ensure the correctness of the stem growth direction.

The aim of position optimization of the joint point is to make the angle between the stem and a first-level branch from the constructed tree skeleton consistent with the actual angle between the stem points and points of the first-level branch. Based on the adjacent relationship construction in Section 2.3.4, the joint point between the stem and a first-level branch is a stem skeleton point. The growth direction of a first-level branch is the eigenvector corresponding to the largest eigenvalue using PCA for selected points with a distance of h cm from the first point of the first-level branch (represented by yellow points in Figure 6b). The new position of the joint point is in the stem skeleton for which the angle between the growth direction of the first-level branch and the vector (the first point of the first-level branch minus the repositioned joint point) is the smallest. This optimization approach is a bottom-up approach and is executed after position optimization of the stem skeleton point.

The aim of position optimization of the branch skeleton point is the same as that of stem skeleton point optimization. Branches have more complex geometries than stems, and branches may have relatively large bends in a short segment, especially in the last-level branches. For a branch skeleton point, if its included angle (denoted by $\angle DEF$ in Figure 5) is less than 120 degrees and the absolute value of the difference between its included angle and its parent's included angle is larger than 60 degrees, the position of the branch skeleton point is optimized, and the optimization approach is the same as the position-optimization approach for the stem skeleton point.

Based on the above, a flowchart of the presented tree-skeletonization method is shown in Figure 7.

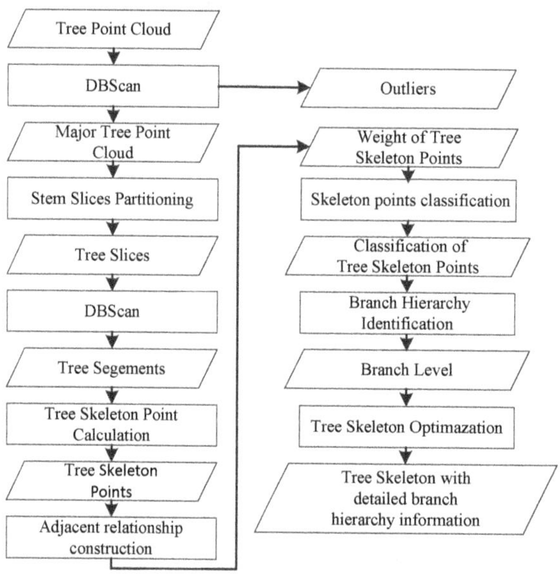

Figure 7. Flowchart of the presented tree-skeletonization method.

2.4. Assessment Indices

For artificially synthetic tree point clouds with a ground-truth skeleton, the Hausdorff distance [22] and the Chamfer distance [23], which are calculated between the extracted tree skeleton and the ground-truth tree skeleton, can be used to perform quantitative assessments. However, the ground-truth tree skeleton is unknown, as the tree point cloud is scanned using TLS in this study. The root mean square error ($RMSE$) and R^2 of the branch angle, tree skeleton length, and branch length are used to quantitatively assess the accuracy of the skeleton [5]. However, the height of the branch is more than 10 m in this study, and it is difficult to measure the branch angle and branch length at this height for standing trees.

The extracted tree skeleton should be located at the centre of the tree point cloud, and the direction between two adjacent tree skeleton points should be consistent with the growth direction of the current tree part: stem or branch. Therefore, the two indices are presented to evaluate the accuracy of the extracted tree skeleton in this study. (a) The positional deviation d_p of the tree skeleton (Formula (3)) is the distance between the median point of two adjacent tree skeleton points p_i and p_{i+1} and the centroid of the constructed tree segment $p_{i,c}$, which is between the two planes formed by the two adjacent tree skeleton points with its direction (denoted by the zone between the two dotted lines in Figure 8). (b) The directivity deviation d_d of the tree skeleton (Formula (4)) is the included angle $p_{i+1} - p_i$ between the direction of the two adjacent tree skeleton points p_i and p_{i+1} (denoted by the black arrows in Figure 8) and the growth direction of the constructed tree segment $p_{i,g}$. It is obvious that the smaller the location deviation and the direction deviation are, the higher the accuracy of the extracted tree skeleton.

$$d_p = \mid \frac{p_i + p_{i+1}}{2} - p_{i,c} \mid \quad (3)$$

$$d_d = < p_{i+1} - p_i, p_{i,g} > \quad (4)$$

In this study, the difference d_H between the tree skeleton height and the tree point cloud height and the maximum, minimum, and average of d_p and d_d are used as quantitative assessments for a constructed tree skeleton. Then, the maximum max_{d_p}, minimum min_{d_p}, and average avg_{d_p} of d_p and the maximum max_{d_d}, minimum min_{d_d}, and average avg_{d_d} of d_d for all constructed tree skeletons are used to evaluate the performance of the presented tree-skeletonization method in this study. The growth direction of the constructed tree segment is calculated using PCA, which is closely related to the point distribution. Therefore, the two indices d_p and d_d are calculated based on the tree skeleton points that have only one child skeleton point.

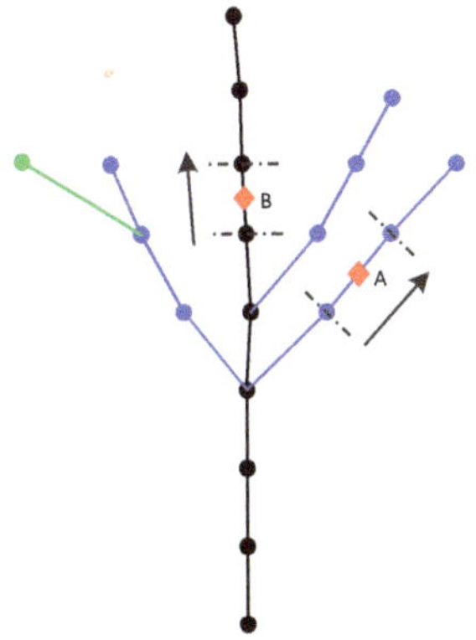

Figure 8. Diagrams of the accuracy evaluation. The round points represent tree skeleton points. The diamond points A and B represent the median points of two adjacent tree skeleton points. The zone between the two dotted lines around the diamond points represents the constructed tree segment. The black arrows represent the direction calculated by the two adjacent tree skeleton points.

3. Results

3.1. Results of the Constructed Tree Skeleton with Four Groups of Parameter Values

There are four parameters in the presented method: *eps* and *minPts* for the DBSCAN method, *h* for tree slicing, and *k* for local stem growth direction calculation. The more tree skeleton points there are, the finer the tree skeleton. In this study, a tree segment generates a skeleton point. In other words, the smaller the value of *h* is, the denser the skeleton points. Tree skeleton optimization is based on the local growth direction of the tree using PCA, which requires at least three skeleton points. Furthermore, the calculation of the local growth direction should be calculated on a tree segment whose length is not greater than 30 cm. This means that $k \geqslant 3$ and the value of $k \times h$ should be less than 30 cm in this study. The purpose of the DBSCAN algorithm is to cluster and determine the order relationship between points. For a tree point cloud, the values of *eps* and *minPts* should maximize the number of major tree point clouds and should be constant. In this study, four group parameters of *h* and *k* were used for the experiments. The numerical results of the assessment indices are shown in Table 1. The average values of the assessment indices are shown in Table 2. The visualization effects of the four groups of parameters of a tree point cloud are shown in Figures 9 and 10.

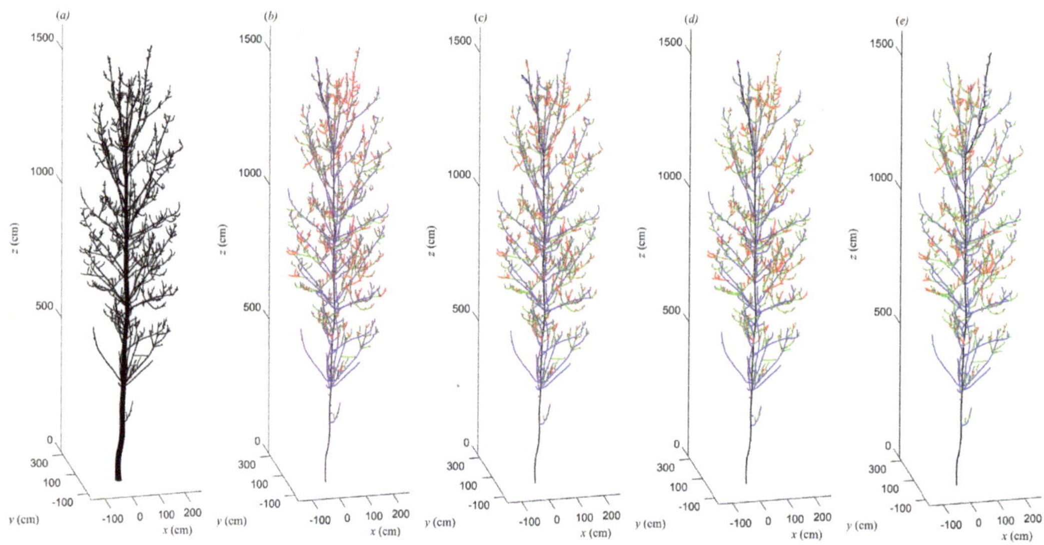

Figure 9. Diagrams of a tree skeleton constructed from tree point cloud 0213 with *eps* = 1 cm and *minPts* = 6. (**a**) The tree point cloud. (**b**) The tree skeleton with *h* = 2 cm and *k* = 10. (**c**) The tree skeleton with *h* = 3 cm and *k* = 8. (**d**) The tree skeleton with *h* = 5 cm and *k* = 5. (**e**) The tree skeleton with *h* = 8 cm and *k* = 3. The magenta points represent stem skeleton points. The black, blue, green, and red lines represent the stem and first-, second-, and third-level branches, respectively.

According to Tables 1 and 2, the number of skeleton points increases as *h* decreases, i.e., the parameter *h* can control the fineness of the tree skeleton. The height error between the tree height and skeleton height decreases as *h* decreases. The maximum, minimum, and average values of the height error are 0.119, 0.016, and 0.053 metres, respectively. This demonstrates that the height of the constructed tree skeleton is close to the height of the tree point cloud. The assessment index values based on d_d demonstrate that most of the tree skeleton points are very close to the centre points, there is indeed a certain degree of deviation between the skeleton point and the geometric centre point of the tree, and the deviation is not significantly affected by the parameters, which are related to the geometric topology of the tree. The assessment index values based on d_p demonstrate that there is an angular deviation between the local growth direction calculated from the skeleton and the local growth direction calculated from the tree point cloud, and the angular deviation tends to increase as h decreases. The values of max_{d_p} are approximately 50 degrees, and most of the values of avg_{d_p} are below 10 degrees, which means that there are few large angular deviations. The average height error and avg_{d_d} increase as h increases, and there is no obvious relationship between the other indices and h. Based on the above, although the maximum values of d_d and d_p are slightly high, the average values of d_d and d_p are satisfactory.

From Figure 9, the constructed tree skeletons with different parameters are almost replicas of the tree point cloud and can effectively reflect the geometric structure characteristics of the tree point cloud. The difference is that the smaller h is, the denser the skeleton points and the richer the details.

From Figure 10, the partially enlarged views clearly show the effect of different parameters on skeleton construction. The rich details are reflected not only in the additional branch skeletons, but also in the more delicate shapes of the branches. The number of skeleton points decreases as h increases, and the complexity of the displayed tree structure is weakened. Notably, the more skeleton points there are, the less smooth the tree skeleton, and the larger the included angle. This is consistent with the numerical results that the angle deviation tends to increase as h decreases. Based on the above results, the presented tree-skeletonization method can control the density of skeleton points through parameters, thereby controlling the output quality of the skeleton. However, this does not mean that the more skeleton points there are, the better the tree skeleton quality.

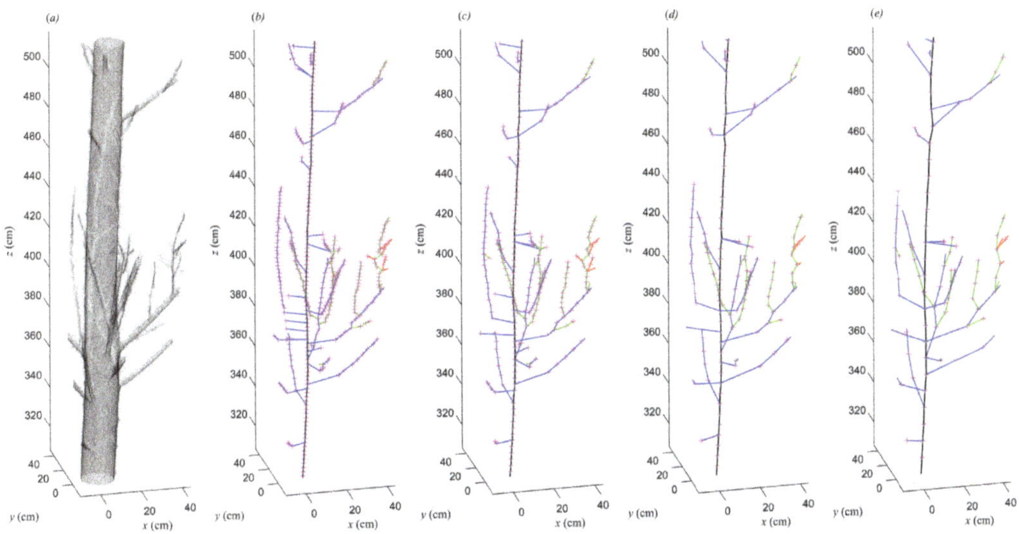

Figure 10. Partial enlarged views of the constructed tree skeleton in Figure 9. (**a**) Partial enlarged tree point cloud. (**b**) Partial enlarged tree skeleton with h = 2 cm and k = 10. (**c**) Partial enlarged tree skeleton with h = 3 cm and k = 8. (**d**) Partial enlarged tree skeleton with h = 5 cm and k = 5. (**e**) Partial enlarged tree skeleton with h = 8 cm and k = 3. The magenta points represent stem skeleton points. The black, blue, green, and red lines represent the stem and first-, second-, and third-level branches, respectively.

Table 1. Numerical result for ten trees with four group parameters.

Tree ID	Tree Height (m)	Point Number	*eps* (cm)	*minPts*	*h* (cm)	*k*	Skeleton Height (m)	Skeleton Point Number	Height Error (m)	max_{d_d} (cm)	min_{d_d} (cm)	avg_{d_d} (cm)	max_{d_p} (o)	min_{d_p} (o)	avg_{d_p} (o)
0213	15.927	2,289,875	1.0	6	8	3	15.829	3473	0.099	4.211	0.017	0.429	50.346	0.121	4.548
					5	5	15.880	5574	0.048	7.766	0.006	0.320	50.089	0.055	6.130
					3	8	15.891	9467	0.037	9.983	0.005	0.310	50.206	0.012	7.831
					2	10	15.896	14445	0.031	6.431	0.008	0.288	54.609	0.015	11.169
0111	15.922	1,349,179	1.0	6	8	3	15.840	1983	0.083	8.177	0.009	0.423	47.506	0.090	4.025
					5	5	15.884	3183	0.038	8.177	0.009	0.365	48.969	0.068	5.479
					3	8	15.894	5434	0.028	8.177	0.007	0.339	48.969	0.063	6.802
					2	10	15.898	8290	0.024	8.538	0.005	0.354	54.232	0.045	8.860
0510	10.108	833,913	1.0	6	8	3	10.049	992	0.058	7.226	0.016	0.424	49.142	0.041	6.817
					5	5	10.050	1613	0.058	7.226	0.016	0.409	45.146	0.041	7.149
					3	8	10.075	2838	0.033	11.367	0.013	0.525	50.526	0.022	8.763
					2	10	10.079	4444	0.029	6.711	0.008	0.430	54.415	0.018	13.802
0406	17.827	3,027,495	1.5	6	8	3	17.746	3910	0.081	15.290	0.009	0.543	51.734	0.028	7.239
					5	5	17.751	6186	0.077	15.290	0.009	0.512	51.734	0.033	7.551
					3	8	17.759	10543	0.081	15.290	0.008	0.476	53.679	0.016	9.380
					2	10	17.778	16035	0.049	15.290	0.008	0.444	53.679	0.007	12.311
0305	17.128	3,378,374	1.5	6	8	3	17.038	5107	0.090	12.168	0.010	0.464	51.928	0.033	6.954
					5	5	17.085	8285	0.043	12.168	0.010	0.444	48.858	0.046	7.443
					3	8	17.095	14107	0.033	12.168	0.008	0.415	52.264	0.022	9.324
					2	10	17.097	21521	0.030	17.298	0.006	0.400	53.596	0.011	14.309
0407	14.350	2,192,439	1.5	6	8	3	14.231	2989	0.119	13.366	0.006	0.426	53.806	0.019	11.750
					5	5	14.301	4830	0.049	7.672	0.010	0.469	51.360	0.131	7.564
					3	8	14.310	8197	0.040	9.514	0.006	0.417	53.806	0.019	12.114
					2	10	14.320	12446	0.030	7.672	0.006	0.398	53.806	0.019	13.061
0511	15.674	1,218,069	1.0	6	8	3	15.596	1929	0.078	8.605	0.015	0.512	53.294	0.081	6.632
					5	5	15.633	3175	0.041	8.605	0.015	0.512	53.294	0.150	7.692
					3	8	15.891	5406	0.037	9.983	0.005	0.310	50.206	0.012	7.831
					2	10	15.896	8456	0.031	9.983	0.005	0.296	54.609	0.012	9.959

Table 1. *Cont.*

Tree ID	Tree Height (m)	Point Number	*eps* (cm)	*minPts*	*h* (cm)	*k*	Skeleton Height (m)	Skeleton Point Number	Height Error (m)	max_{d_d} (cm)	min_{d_d} (cm)	avg_{d_d} (cm)	max_{d_p} (o)	min_{d_p} (o)	avg_{d_p} (o)
0617	15.551	2,825,789	1.0	6	8	3	15.437	4252	0.114	8.709	0.007	0.409	53.984	0.046	6.253
					5	5	15.495	6859	0.056	4.990	0.007	0.383	52.103	0.046	6.549
					3	8	15.530	11826	0.021	13.255	0.005	0.363	53.984	0.025	0.025
					2	10	15.535	18109	0.016	13.255	0.004	0.343	54.107	0.009	9.957
0514	13.193	1,203,242	1.0	6	8	3	13.107	1913	0.087	17.298	0.008	0.417	50.521	0.040	6.876
					5	5	13.098	3191	0.095	17.298	0.008	0.401	50.521	0.040	6.979
					3	8	13.137	5591	0.057	17.298	0.008	0.465	54.493	0.026	9.050
					2	10	13.160	8610	0.033	17.298	0.008	0.468	53.514	0.011	13.004
0519	14.171	1,246,086	1.0	6	8	3	14.074	1436	0.096	9.448	0.006	0.444	48.738	0.097	7.199
					5	5	14.1354	2394	0.035	9.448	0.010	0.451	48.738	0.097	7.947
					3	8	14.149	4108	0.021	12.644	0.005	0.507	49.850	0.027	9.817
					2	10	14.154	6375	0.016	7.860	0.008	0.397	52.283	0.031	12.820

Table 2. Average values of different groups' parameters values.

h (cm)	*k*	Average of height error (m)	Average of max_{d_d} (cm)	Average of min_{d_d} (cm)	Average of avg_{d_d} (cm)	Average of max_{d_p} (o)	Average of min_{d_p} (o)	Average of avg_{d_p} (o)
2	10	0.029	11.034	0.007	0.382	53.885	0.018	11.925
3	8	0.039	11.968	0.007	0.413	51.798	0.024	8.094
5	5	0.054	9.864	0.010	0.427	50.081	0.071	7.048
8	3	0.090	10.450	0.010	0.449	51.100	0.060	6.829
- [a]	- [a]	0.053	10.829	0.008	0.418	51.716	0.043	8.474

[a] all the values of *h* and *k*.

3.2. Visualizations of Four Constructed Tree Skeletons with a Group of Parameter Values

To further demonstrate the visualization effect of tree skeleton construction, four constructed tree skeletons from four trees with $h = 5$ cm and $k = 5$ are shown in Figure 11.

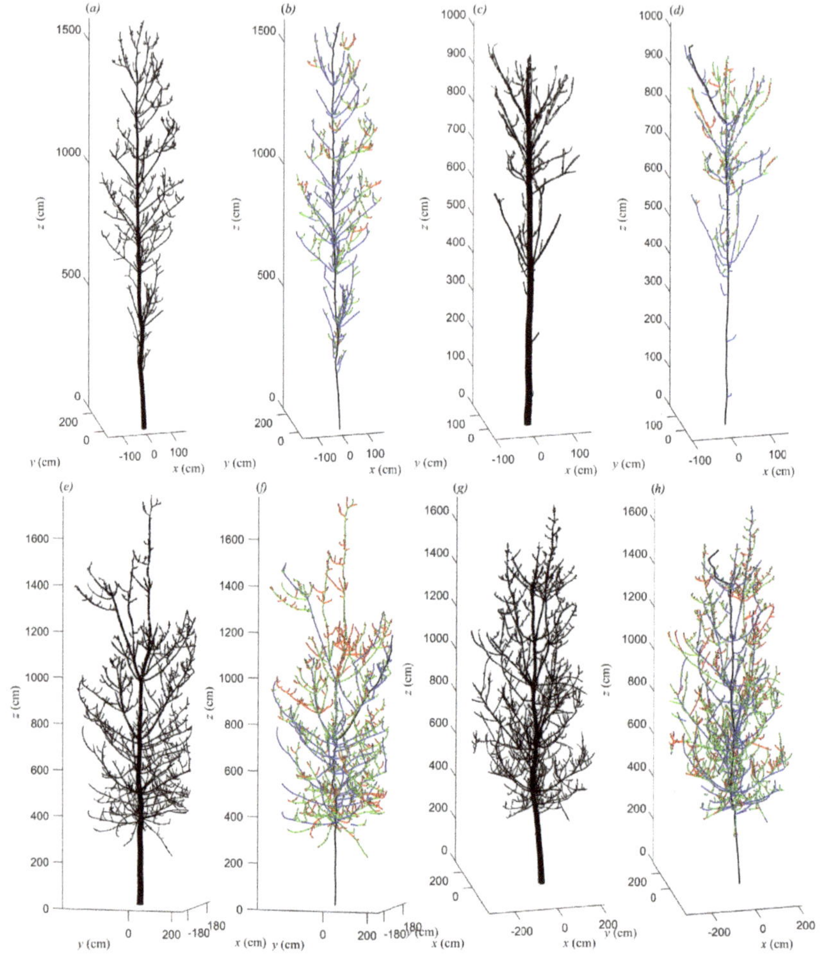

Figure 11. Diagrams of four tree skeletons constructed from four tree point clouds. (**a**) Tree point cloud of 0111. (**b**) Tree skeleton of 0111. (**c**) Tree point cloud of 0510. (**d**) Tree skeleton of 0510. (**e**) Tree point cloud of 0406. (**f**) Tree skeleton of 0406. (**g**) Tree point cloud of 0305. (**h**) Tree skeleton of 0305. The black, blue, green, and red lines represent the stem and first-, second-, and third-level branches, respectively.

The four trees in Figure 11 have different morphological characteristics. The height of tree 0510 is the lowest; trees 0406 and 0305 are tall and have very complex geometries with many scattered branches. The structures of the four skeletons are all consistent with the structures of the four trees. According to the colours of the stems or branches, the vast majority of the stem or branch levels are identified correctly, and only the upper stem of tree 0406 is identified incorrectly.

4. Discussion

A tree skeleton is an efficient graphical tool to express the geometric structure of a tree. However, it is a challenge to construct a tree skeleton that is consistent with the tree geometry. In this study, we present tree skeletonization with DBSCAN using TLS data. The constructed stem skeleton can effectively reflect the geometrical structure of a tree, i.e., the stem and branch hierarchy information. The heights of the study materials in [5] were not more than 5 m (the height value is not explicitly given; it was reflected in figures), and the first-level branches were not more than eight. The study materials in [4] included artificially generated trees and raw-scan trees. It is difficult to create synthetic trees that have geometries similar to those of actual trees. For raw-scan trees, the level of branch complexity is based on the number of branching points, and the number of points for the most complex branches ranges from 71 to 85. The complex geometric structure of a tree does not appear in either of the above cases. In this study, the minimum number of tree point clouds is 833,913, which is much higher than that of [4]. Compared with the study in [4,5], the heights of our studied trees were much higher (Table 1) and the geometry structure was much more complex (Figure 11). Even so, our constructed tree skeleton can accurately output the tree geometry structure. Compared with the trees studied in [15], our constructed tree skeletons are continuous, and the stems and branches can be distinguished with detailed branch hierarchy information, which has not been achieved by the existing research. Therefore, the presented method has clear advantages over the recent existing methods.

DBSCAN is a density-based clustering algorithm that can discover clusters of any shape without interference from noise. Trees are diverse and have natural geometric topological complexity. Affected by wind disturbance and other factors, a tree point cloud has noise. These unfavourable factors of the tree point cloud are exactly the advantages of the DBSCAN algorithm. In this study, we make full use of the connectivity characteristics of the DBSCAN algorithm to perform outlier removal, determine the point-traversal order, create connections between skeleton points, and calculate the weight of skeleton points. This is one of the advantages of our algorithm.

Accurate calculation of the tree skeleton points is another advantage of our algorithm. For a tree section, there are some factors that prevent accurate calculation of the section's centre point, such as adjacent branch points (Figure 4b) and points in the joint part between branches. In this study, a point-inversion transformation is used to calculate the tree skeleton points. Points on the inside of the tree section are used to calculate the skeleton position (Figure 4d) rather than points on the outside. Therefore, the aforementioned disadvantages are overcome and the skeleton points can be accurately calculated.

How can the effectiveness of a tree skeleton be quantitatively evaluated? This question is more challenging when the ground-truth skeleton is unknown. A tree skeleton point should locate the centre position of the tree, and the direction between adjacent and connected tree skeleton points should be consistent with the growth direction of the tree. It is obvious that a constructed tree skeleton should meet these conditions. Based on this, the positional deviation and directivity deviation are presented to quantitatively evaluate the effectiveness of a tree skeleton in this study. The experimental results preliminarily verify the effectiveness of these two parameters in this study; however, these two parameters still need to be further verified.

According to the numerical results, max_{d_d} is 53.885 degrees; according to the visualization results, there are few large angular deviations. This means that the method in this study can be further improved. In this study, k consecutive skeleton points are used to calculate the local stem growth direction with PCA in the tree-skeleton optimization. When the number of points is insufficient, optimization will fail. This is also one of the reasons for the large values of max_{d_d}. Based on the geometric structure characteristics of trees, improving the optimization method of skeleton points is one of the future research topics.

With the improvement in data accuracy, tree point clouds already have millimetre-level data accuracy. The constructed tree skeleton can not only be applied to the field of tree visualization, but also should be able to be applied to the precise retrieval of tree parameters. Skeleton visualization shows that the stem and branch hierarchies can be distinguished. Then, the included angle between branches, the height and the length of branches, and the crown width in multiple directions can be calculated based on the constructed tree skeleton. Accurately retrieving the above parameters using a constructed tree skeleton is one of the future research topics.

It is worth noting that the presented method is based on the tree point cloud from TLS, which does not mean that the method can only use TLS as the data source. Other data sources are also possible, such as the tree point cloud acquired by a backpack Lidar scanner, as long as the preprocessed tree point cloud can reflect the geometric structure of the tree.

5. Conclusions

We presented a tree-skeletonization method based on the advantages of the DBSCAN algorithm in this study. The visualization results showed that the constructed tree skeleton can accurately represent the stem, and the branch hierarchy of a tree and is consistent with the geometric characteristics of the tree. Experiments with ten trees showed that the avg_{d_p} of d_p was 0.01 cm and the avg_{d_d} of d_d was 7.07 degrees, and the numerical results demonstrated that most line segments of the constructed tree skeleton were in the centre of the tree.

The presented tree-skeletonization method achieved satisfactory results, and this method will facilitate the accurate extraction of tree parameters, tree modelling, and 3D tree canopy modelling and analysis. In addition, the positional deviation and directivity deviation were presented to evaluate the validity of the tree skeleton. This provides a feasible way to verify the validity of the tree skeleton in the absence of a ground-truth skeleton.

Author Contributions: Conceptualization, software, and writing—original draft preparation, L.Y.; data processing, software, validation, and investigation, Y.S., Y.L. and J.J.; investigation and data processing, X.C.; investigation and writing—review and editing, Y.F.; writing—review and editing, X.S. All authors have read and agreed to the published version of the manuscript.

Funding: This research was funded in part by the National Natural Science Foundation of China (No. 31872704), the Foundation for Distinguished Young Talents in Higher Education of Henan (No. 2020GGJS157), the Xinyang Academy of Ecological Research Open Foundation (No. 2023XYMS12), the Beijing Shisanling Forest Farm Coniferous Plantation Multifunctional Management Monitoring Project (No. 2022SSL110), and the Nanhu Scholars Program for Young Scholars of XYNU.

Data Availability Statement: The data presented in this study are available on request from the corresponding author.

Conflicts of Interest: The authors declare no conflict of interest.

References

1. Raumonen, P.; Kaasalainen, M.; Åkerblom, M.; Kaasalainen, S.; Kaartinen, H.; Vastaranta, M.; Holopainen, M.; Disney, M.; Lewis, P. Fast Automatic Precision Tree Models from Terrestrial Laser Scanner Data. *Remote Sens.* **2013**, *5*, 491. [CrossRef]
2. Abegg, M.; Bösch, R.; Kükenbrink, D.; Morsdorf, F. Tree volume estimation with terrestrial laser scanning-Testing for bias in a 3D virtual environment. *Agric. For. Meteorol.* **2023**, *331*, 109348. . [CrossRef]
3. Huang, H.; Wu, S.; Cohen-Or, D.; Gong, M.; Zhang, H.; Li, G.; Chen, B. L1-Medial Skeleton of Point Cloud. *ACM Trans. Graph.* **2013**, *32*, 1–8. [CrossRef]
4. Jiang, A.; Liu, J.; Zhou, J.; Zhang, M. Skeleton Extraction from Point Clouds of Trees with Complex Branches via Graph Contraction. *Vis. Comput.* **2021**, *27*, 2235–2251.
5. Xu, Y.; Hu, C.; Xie, Y. An improved space colonization algorithm with DBSCAN clustering for a single tree skeleton extraction. *Int. J. Remote Sens.* **2022**, *43*, 3692–3713. [CrossRef]
6. Li, J.; Wu, H.; Xiao, Z.; Lu, H. 3D modeling of laser-scanned trees based on skeleton refined extraction. *Int. J. Appl. Earth Obs. Geoinf.* **2022**, *112*, 102943. [CrossRef]
7. Su, Z.; Zhao, Y.; Zhao, C.; Guo, X.; Li, Z. Skeleton extraction for tree models. *Math. Comput. Model.* **2011**, *54*, 1115–1120. [CrossRef]
8. Pueschel, P. The influence of scanner parameters on the extraction of tree metrics from FARO Photon 120 terrestrial laser scans. *ISPRS J. Photogramm. Remote Sens.* **2013**, *78*, 58–68. [CrossRef]

9. Liang, X.; Hyyppä, J.; Kaartinen, H.; Lehtomäki, M.; Pyörälä, J.; Pfeifer, N.; Holopainen, M.; Brolly, G.; Francesco, P.; Hackenberg, J.; et al. International benchmarking of terrestrial laser scanning approaches for forest inventories. *ISPRS J. Photogramm. Remote Sens.* **2018**, *144*, 137–179. [CrossRef]
10. Livny, Y.; Yan, F.; Olson, M.; Chen, B.; Zhang, H.; El-Sana, J. Automatic reconstruction of tree skeletal structures from point clouds. *ACM Trans. Graph. (TOG)* **2010**, *29*, 151. [CrossRef]
11. Yan, D.M.; Wintz, J.; Mourrain, B.; Wang, W.; Boudon, F.; Godin, C. Efficient and robust reconstruction of botanical branching structure from laser scanned points. In Proceedings of the 2009 11th IEEE International Conference on Computer-Aided Design and Computer Graphics, Huangshan, China, 19–21 August 2009; pp. 572–575. [CrossRef]
12. Bucksch, A.; Lindenbergh, R.; Menenti, M. SkelTre - Robust skeleton extraction from imperfect point clouds. *Vis. Comput.* **2010**, *26*, 1283–1300. [CrossRef]
13. He, G.; Yang, J.; Behnke, S. Research on geometric features and point cloud properties for tree skeleton extraction. *Pers. Ubiquitous Comput.* **2018**, *22*, 903–910. [CrossRef]
14. Liu, Y.; Guo, J.; Benes, B.; Deussen, O.; Zhang, X.; Huang, H. TreePartNet Neural Decomposition of Point Clouds for 3D Tree Reconstruction. *ACM Trans. Graph.* **2021**, *40*, 232:1–232:16. [CrossRef]
15. Sun, J.; Wang, P.; Li, R.; Zhou, M.; Wu, Y. Fast Tree Skeleton Extraction Using Voxel Thinning Based on Tree Point Cloud. *Remote Sens.* **2022**, *14*, 2558. [CrossRef]
16. Liu, L.; Pan, g.Y.; Li, Z. Individual Tree DBH and Height Estimation Using Terrestrial Laser Scanning (TLS) in A Subtropical Forest. *Sci. Silvae Sin.* **2016**, *52*, 26–37. [CrossRef]
17. Rusu, R.B.; Cousins, S. 3d is here: Point cloud library (pcl). In Proceedings of the 2011 IEEE International Conference on Robotics and Automation, Shanghai, China, 9–13 May 2011; pp. 1–4.
18. Ester, M.; Kriegel, H.P.; Sander, J.; Xiaowei, X. *A Density-Based Algorithm for Discovering Clusters in Large Spatial Databases with Noise*; AAAI Press: Washington, DC, USA, 1996; pp. 226–231.
19. You, L.; Guo, J.; Pang, Y.; Tang, S.; Song, X.; Zhang, X. 3D stem model construction with geometry consistency using terrestrial laser scanning data. *Int. J. Remote Sens.* **2021**, *42*, 714–737. [CrossRef]
20. You, L.; Tang, S.; Song, X.; Lei, Y.; Zang, H.; Lou, M.; Zhuang, C. Precise Measurement of Stem Diameter by Simulating the Path of Diameter Tape from Terrestrial Laser Scanning Data. *Remote Sens.* **2016**, *8*, 717. [CrossRef]
21. You, L.; Wei, J.; Liang, X.; Lou, M.; Pang, Y.; Song, X. Comparison of Numerical Calculation Methods for Stem Diameter Retrieval Using Terrestrial Laser Data. *Remote Sens.* **2021**, *13*, 1780. [CrossRef]
22. Qin, H.; Han, J.; Li, N.; Huang, H.; Chen, B. Mass-Driven Topology-Aware Curve Skeleton Extraction from Incomplete Point Clouds. *IEEE Trans. Vis. Comput. Graph.* **2020**, *26*, 2805–2817. [CrossRef] [PubMed]
23. Yifan, W.; Wu, S.; Huang, H.; Cohen-Or, D.; Sorkine-Hornung, O. Patch-based progressive 3d point set upsampling. In Proceedings of the IEEE/CVF Conference on Computer Vision and Pattern Recognition, Long Beach, CA, USA, 15–20 June 2019; pp. 5958–5967.

forests

MDPI

Article

Forest Canopy Water Content Monitoring Using Radiative Transfer Models and Machine Learning

Liang Liu [1], Shaoda Li [1,*], Wunian Yang [1,*], Xiao Wang [2], Xinrui Luo [1], Peilian Ran [1] and Helin Zhang [3,4]

1 College of Earth Science, Chengdu University of Technology, Chengdu 610059, China; liuliangcdut@gmail.com (L.L.); 2019010040@stu.ceut.edu.cn (X.L.); ranpeilian@stu.cdut.edu.cn (P.R.)
2 School of Architecture and Civil Engineering, Chengdu University, Chengdu 610106, China; wangxiao@cdu.edu.cn
3 State Key Laboratory of Remote Sensing Science, Jointly Sponsored by Beijing Normal University and Institute of Remote Sensing and Digital Earth of Chinese Academy of Sciences, Beijing 100875, China; zhanghelinrs@gmail.com
4 Beijing Engineering Research Center for Global Land Remote Sensing Products, Institute of Remote Sensing Science and Engineering, Faculty of Geographical Science, Beijing Normal University, Beijing 100875, China
* Correspondence: lisd@cdut.edu.cn (S.L.); ywn@cdut.edu.cn (W.Y.)

Abstract: Forests are facing various threats, such as drought, in the context of global climate change. Canopy water content (CWC) is a crucial indicator of forest water stress, mortality, and fire monitoring. However, previous studies on CWC have not adequately simulated forests with heterogeneous and discontinuous canopy structures. At the same time, there is a lack of field validation. This study retrieved the forest CWC across the contiguous U.S. (CONUS) with coupled radiative transfer models (RTMs) and the random forest (RF) algorithm. A Gaussian copula and prior knowledge were used for model parameterization. The results indicated that more accurate simulations of leaf trait dependencies and canopy structure characteristics lead to better CWC inversion. In addition, GeoSail, coupled with PROSPECT-5B, showed good performance (R^2 = 0.68, $RMSE$ = 0.15 kg m^{-2}, MAE = 0.12 kg m^{-2}, $rRMSE$ = 12.78%, $Bias$ = −0.036 kg m^{-2}) for forest CWC retrieval. Large variation existed in forest CWC, spatiotemporally, and evergreen needle forest (ENF) showed strong CWC capacity. This study underscores the suitability of 3D RTMs for inversing the parameters of forest canopies.

Keywords: canopy water content; MODIS; random forest; copula; Google Earth Engine

1. Introduction

Canopy water content (CWC), the amount of water stored in the vegetation canopy, is typically determined by multiplying the leaf water content by the canopy leaf area index (LAI). This calculation incorporates information about the leaf water condition and the canopy structure [1]. CWC is a critical parameter for assessing vegetation growth and monitoring drought stress. It is influenced by soil water supply and atmospheric demand [2]. Because of global climate change, drought events are becoming more frequent and severe, and future projections suggest they may become even more extreme [3,4]. In this context, the forest, as a critical component of terrestrial ecosystems, is facing severe threats from drought, leading to increased mortality rates [5–7]. Therefore, monitoring forest CWC using remote sensing technology can provide valuable insights into forest growth, water stress status, and drought response [8].

Optical and microwave sensors can effectively detect forest CWC [9,10]. Vegetation optical depth (VOD), retrieved from microwave emissions, is favored by many researchers for large-scale studies related to water or biomass, as it is unaffected by weather conditions and offers a short revisit period [11,12]. However, VOD provides vegetation information on both water and biomass with low spatial resolution [2]. In contrast, optical sensors offer a higher spatial resolution. In optical remote sensing, water absorbs radiation across the spectrum [13]. Additionally, the estimation of CWC generally relies on water-sensitive bands, such as 970 nm, 1200 nm, 1450 nm, 1600 nm, 1940 nm, and 2500 nm [1,14].

Approaches to CWC derived from remote sensing can be categorized into the empirical statistical method, physical model inversion, and hybrid methods [15]. The empirical statistical method estimates CWC by establishing regression relationships between field observations and band reflectance or vegetation indices [16]. This method is easy to understand and convenient to operate, but it relies on observational data and lacks generalization [1]. The physical model determines the response of CWC to canopy reflectance by simplifying the simulation of solar radiation and the transfer process in the soil–leaf–canopy–atmosphere [17]. Therefore, the accuracy of the physical process description and parameterization directly determines the result [18]. The radiative transfer model (RTM) is a commonly used physical model for CWC inversion. Both SAIL (Scattering by Arbitrarily Inclined Leaves) and GeoSail (a combination of SAIL and the Jasinski geometric model) belong to the RTM, and they generally simulate canopy spectrum characteristics by coupling with the leaf PROSPECT model [13,19–21]. In nature, remote sensing inversion is typically ill-posed. The ill-posed problem is characterized by using limited observations to estimate the current state of a complex surface system, and the amount of data information is less than the number of model parameters [22]. Therefore, challenges remain in the inversion of RTMs given the complex physical process description, the large number of parameters to be acquired, and the inherent ill-posed problem [23–25]. Hybrid-based approaches retrieve CWC by combining statistical methods with RTM inversion [26]. Practical inversion algorithms include iterative optimization [23], machine learning regression [26,27], and look-up tables [28].

In CWC inversion, previous studies commonly employed PROSAIL (leaf PROSPEPCT coupled with canopy SAIL) with unconstrained inputs, which may exacerbate the ill-posed problem [29]. Additionally, validation is typically conducted within the crop, grass, or shrub-covered area [1,15,30]. The development of plant trait datasets, such as Leaf Optical Properties Experiment 93 (LOPEX93) [31], the ANGERS Leaf Optical Properties Database (ANGERS) [32], the TRY database [33], and the National Ecological Observatory Network (NEON) [34], enables the further alleviation of ill-posed problems by integrating field observations with copula or ecological rule-based approaches. However, knowledge gaps persist in forest CWC inversion because of the complex canopy structure and the scarcity of field observations [21].

In recent years, Google Earth Engine (GEE) has experienced rapid growth and has been widely utilized for vegetation, agricultural, urban, hydrological, and land cover applications [35]. GEE encompasses remote sensing datasets, such as the MODerate-resolution Imaging Spectroradiometer (MODIS), Landsat, and Sentinel series, as well as meteorological and geophysical datasets [36,37]. Given its powerful computing performance and its purpose in geospatial data analysis, GEE serves as an excellent tool for monitoring long-term dynamic changes at various scales, from local to global [26,38].

In this study, our focus was on (1) parameterizing the model inputs with a Gaussian copula and prior knowledge, (2) building a CWC-retrieving model via simulations and RF regression, (3) evaluating the accuracy of CWC inversion based on field observations, and (4) capturing spatiotemporal variation using the GEE and MODIS products. We strive to provide a more efficient strategy for forest CWC inversion.

2. Materials

2.1. Study Area

We chose the contiguous U.S. (CONUS) as the study area because of the abundant field data collected by NEON. As shown in Figure 1, the eastern region of the CONUS is dominated by deciduous broad forests (DBFs), while the western region is characterized by evergreen needle forests (ENFs). The total ENF, EBF, and DBF areas across the CONUS are 6.67×10^5 km^2, 1.68×10^5 km^2, and 2.11×10^6 km^2, respectively. In evergreen forests, the dominant species include *Abies balsamea*, *Abies lasiocarpa*, *Juniperus virginiana*, *Picea glauca*, *Picea mariana*, *Pinus contorta*, *Pseudotsuga menziesii*, *Tsuga canadensis*, etc. On the other hand, *Acer rubrum*, *Betula alleghaniensis*, *Carya tomentosa*, *Fraxinus americana*, *Liquidambar styraciflua*, *Liriodendron tulipifera*, *Populus tremuloides*, *Quercus alba*, and *Ulmus americana*, among others, dominate deciduous forests [39].

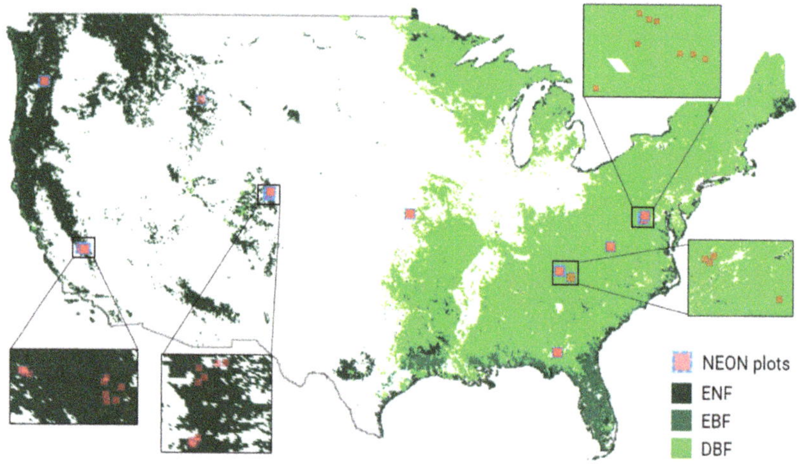

Figure 1. Forest distribution across the CONUS based on plant functional type classifications from the MODIS MCD12Q1 product. ENF: evergreen needle forest, EBF: evergreen broad forest, DBF: deciduous broad forest.

2.2. Field Observation Datasets

2.2.1. LOPEX93 Dataset

In 1993, a field campaign called the Leaf Optical Properties Experiment was implemented by the Joint Research Centre (JRC) of the European Commission in Ispra, Italy [31]. During the experiment, participants collected approximately 70 leaf samples representing more than 50 species from plants, crops, and trees at the JRC, and physical measurements and biochemical analyses were conducted [31]. During the experiment, various biochemical constituents such as leaf water, chlorophyll, lignin, cellulose, starch, and proteins (nitrogen) were measured. Directional–hemispherical reflectance and transmittance were recorded in the 400–2500 nm region [31]. The LOPEX93 database, created using this experiment's data, includes 330 records of biochemical constituents and leaf optics from 45 different species. It covers a wide range of leaf traits [40]. The LOPEX93 database provides variables such as leaf mass per area (LMA), equivalent water thickness (EWT), chlorophyll (Cab) concentration, and carotenoid (Car) concentration. As a result, it has been extensively utilized for parameterizing RTMs at the leaf scale [28,41–43].

2.2.2. ANGERS Dataset

ANGERS is a leaf optics dataset that connects leaf elements' visible/infrared spectral characteristics with biochemical analyses and physical measurements. It is part of an experiment conducted by the National Institute for Agricultural Research in Angers, France [32]. A total of 276 leaf samples, representing 39 different species, were collected for this database. Physical measurements and biochemical analyses were performed on these samples [32]. The ANGERS dataset contains directional–hemispherical reflectance and transmittance (400–2450 nm) obtained using an ASD FieldSpec spectrometer. It also includes measurements of biochemical constituents (water, carotenoid, chlorophyll, anthocyanin, etc.) obtained through biochemical analyses and physical measurements [32,43].

The ANGERS dataset, often used with LOPEX93, is utilized to parameterize leaf-level RTMs because both datasets provide the variables required by the models [28,41–43].

2.2.3. National Ecological Observatory Network (NEON)

The National Ecological Observatory Network (NEON), an observation facility funded by the National Science Foundation, aims to collect long-term ecological data to enhance understanding and forecast the effects of climate change and human activities on ecosystems across the United States [34]. The NEON established 47 terrestrial sites, ranging from deserts to tropical forests, equipped with the Terrestrial Instrument System, the Terrestrial Observation System, the Aquatic Observation System, the Aquatic Instrument System, and the Airborne Observation Platform [44].

The Terrestrial Observatory System determines the sampling frame at the site level based on the plot type and taxonomic group. Spatial sampling uses three plot types: tower, distributed, and gradient plots. Plots are selected using randomization, stratification, and sample number optimization, guided by the National Land Cover Database (NLCD) [45]. In forest ecosystems, tower base plots measure 40 m × 40 m and are divided into four 20 m × 20 m subplots. Vegetation structure, leaf area index (LAI), herbaceous clip harvest, plant diversity, and canopy foliar sampling were conducted in the core area. In contrast, soil sampling was implemented in the high-impact area [46]. Physical, chemical, and stable isotope data for plant foliage are measured through canopy sampling and analyses. Data such as LMA, EWT, Cab, and Car are provided through the Plant Foliar Traits (DP1.10026.001) data product [39].

The NEON provides Digital Hemispheric Photos (DHPs) for the Plot Vegetation data product (DP1.10017.001) to estimate the leaf area index (LAI) or the plant area index at the plot scale. The data product includes 180-degree images and metadata [47]. In the designated LAI sampling plots, 12 DHP photo points are positioned on a square grid with a spacing of four meters. The LAI of the plot was calculated as the average of the 12 points. DHPs are collected from the leaf-out-to-senescence period with a five-year interval for distributed plots and a two-week interval for towers [48].

A total of 1152 leaf samples from 195 evergreen, 134 deciduous, 24 mixed forest plots, and 12 grass plots were used for RTM parameterization. In addition, calculated LAIs from 28 evergreen and 20 deciduous forest plots were used for validation, combined with EWT.

2.3. MODIS Products

Five MODIS products, namely, MCD43A4, MCD43A2, MOD10A1, MYD10A1, and MCD12Q1, in version 6, available in GEE, were used for CWC retrieval in this study.

The MCD43A4 product contains seven bands (Table 1) and provides a 500 m daily nadir bidirectional reflectance distribution function (BRDF) adjusted reflectance (NBAR). It is produced based on 16-day composites [49]. With the BRDF, the reflectance is simulated as it is obtained from a nadir view in the local afternoon. Therefore, view angle effects are removed. In this study, all bands of the MCD43A4 were utilized as predictors for global CWC retrieval.

Table 1. Spectral specifications of the MCD43A4 bands.

Band	Wavelength
Nadir_Reflectance_Band1	620–670 nm
Nadir_Reflectance_Band2	841–876 nm
Nadir_Reflectance_Band3	459–479 nm
Nadir_Reflectance_Band4	545–565 nm
Nadir_Reflectance_Band5	1230–1250 nm
Nadir_Reflectance_Band6	1628–1652 nm
Nadir_Reflectance_Band7	2105–2155 nm

The MCD43A2 is a quality dataset for BRDF and albedo. At the same time, it contains information on land/water types and solar zenith angles for the MCD43A4 [50]. Here, we used the "BRDF_Albedo_LandWaterType" band to mask the water area and employed "BRDF_Albedo_LocalSolarNoon" to provide a solar zenith angle for pixel-based correction.

The MOD10A1 and MYD10A1 provide daily snow cover and snow albedo and related quality assessments from Terra and Aqua, respectively. The 500 m snow cover data were produced based on a normalized difference snow index mapping algorithm [51]. In this work, the snow cover band was utilized to mask the snow-covered area.

The MCD12Q1 product is a MODIS land cover type product that provides six 500 m annual land cover maps based on different classification schemes. It is produced using MODIS Terra and Aqua reflectance and supervised classifications [52]. In this study, the Annual Plant Functional Types classification was utilized, and the forest was divided into evergreen needle, evergreen broad, deciduous needle, and deciduous broad.

3. Methods

3.1. RTMs Selection and Parameterization

As per the flowchart shown in Figure 2, the spectra of the forest canopy were simulated using three leaf–canopy coupled radiative transfer models (RTMs): PROSPECT-5B + 4SAIL (PRO4SAIL), PROSPECT-5B + 4SAIL2 (PRO4SAIL2), and PROSPECT-5B + GeoSail (PROGeoSail). These coupled RTMs were parameterized based on the integration of observations and previous studies.

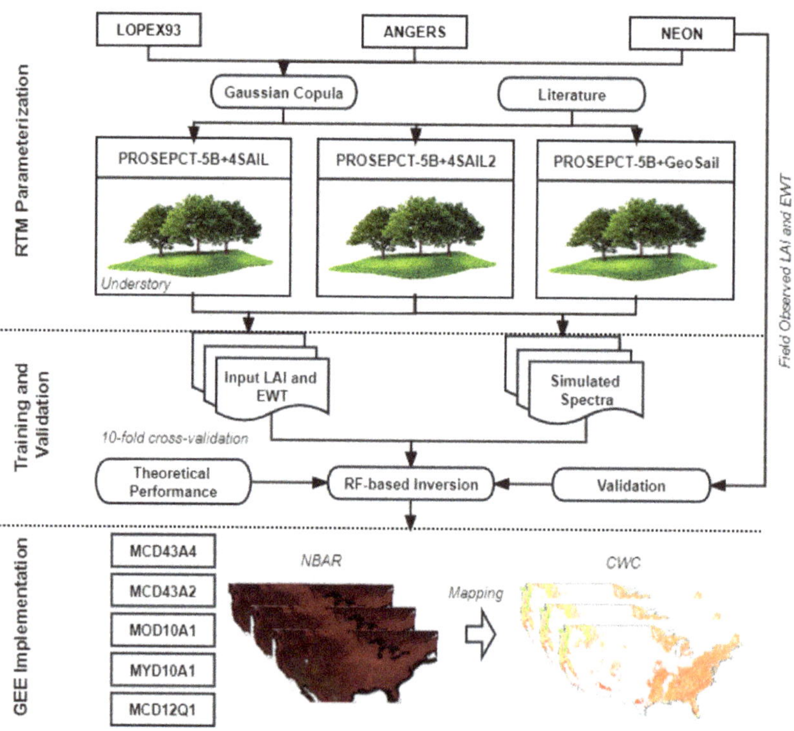

Figure 2. Flowchart of the forest CWC inversion.

The widely used PROSPECT-5 leaf optics model, developed based on the plate model, simulates reflectance and transmittance in the 400–2500 nm spectrum range [19]. In this case, the PROSPECT-5B version provided inputs for canopy RTMs, differentiating chlorophylls and carotenoids. The input variables include the leaf structure parameter and biochemical content (LMA, EWT, Cab, Cbp, and Car). The 4SAIL (an optimized version of SAIL) canopy bidirectional reflectance model, assuming horizontal uniformity, simulates the bidirectional reflectance factor based on the scattering and absorption of four upward/downward radiative fluxes [53].

Building upon the SAIL model, GeoSail incorporates the Jasinski geometric model to calculate a scene's illuminated and shadowed components separately [28]. GeoSail utilizes the optical properties of canopy components to simulate scene reflectance for heterogeneous and discontinuous forests. Within GeoSail, the reflectance and fraction of the illuminated canopy, as well as the illuminated and shadowed background, are used to calculate the forest canopy reflectance [20]

$$\rho_t = C\rho_c + S\rho_s + B\rho_b \quad (1)$$

where ρ_t is the total forest canopy reflectance; C is the canopy fraction; S is the illuminated background fraction; B is the shadowed background fraction; and ρ_c, ρ_s, and ρ_b are the reflectance of the canopy, illuminated, and shadowed background, respectively.

Specifically, the shadowed canopy can be calculated when the solar zenith angle exceeds the aspect angle and the crown shape is conical [20]:

$$\rho_t = (1 - C_s)C\rho_c + C_s\rho_c\tau_s + S\rho_s + B\rho_b \quad (2)$$

where C_s is the fraction of the shadowed canopy, and τ_s is the canopy transmittance.

4SAIL2 is a 3D, hybrid, two-layer canopy radiative transfer model that adheres to the four-stream concept. It incorporates the vertical leaf color gradient description, crown clumping, and numerical robustness from GeoSAIL, FLIM, and SAIL++ [54], respectively. Furthermore, extending a non-Lambertian soil BRDF model eliminated the limitation of 4SAIL2 to a Lambertian soil background. Thus, 4SAIL2 allows for modeling horizontal and vertical heterogeneities, resulting in more realistic simulations of forest canopies [55].

In GeoSail, the representation of the forest understory utilized the simulated canopy reflectance of grass instead of bare soil, as proposed by Quan et al. [28]. This replacement enhanced the realism of the understory reflectance simulation. The bi-hemispherical reflectance was employed, replacing the original Lambertian soil background, assuming that the canopy predominantly intercepts direct radiation.

The parameterization of RTMs relied on prior knowledge derived from field observations and the literature, and more details are shown in Table S1.

3.2. Gaussian Copula

A Gaussian copula was utilized to model the dependencies between leaf traits and constrain the parameterization of PROSPECT-5B. According to Sklar's theorem [56], there is a unique copula that holds for a multivariate absolutely continuous distribution [57]:

$$F\ (x_1, \ldots, x_p) = C\ (F_1(x_1), \ldots, F_p(x_p)) \quad (3)$$

where F is a multivariate distribution function, C is a copula, and $F_1(x_1), \ldots, F_p(x_p)$ are univariate distribution functions (marginal distributions).

The Gaussian copula takes the form [58]

$$C_{Gaussian}\ (u) = \Phi_R\left(\Phi^{-1}(u_1), \ldots, \Phi^{-1}(u_p)\right) \quad (4)$$

where $C_{Gaussian}$ is a Gaussian copula, Φ_R is the joint standard normal cumulative distribution function with positive definite covariance matrix R and zero mean, and Φ^{-1} is the standard normal quantile function. All marginal distributions and positive definite covariance matrixes allow for Gaussian copulas, and only pairwise dependencies were considered [57].

Gaussian copula modeling involves three steps: (1) calculating the univariate distributions, (2) obtaining the covariance matrix, and (3) modeling the relationships between variables using the copula function. The Gaussian copula was implemented using *openturns* in Python 3.9.5 in this study. Additionally, the univariate distributions were estimated using the Gaussian Kernel Density Estimation based on the data, and the covariance matrix was calculated using the Pearson method. Figures 3 and S1 show the distribution of Gaussian copula-based samples for trees and grass.

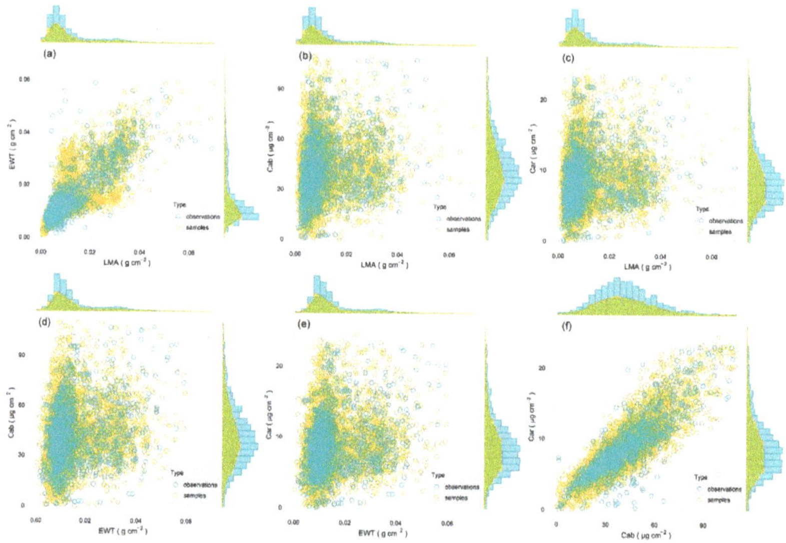

Figure 3. Constrained random samples of tree LMA, EWT, Cab, and Car based on field observations using Gaussian copula: (**a**) LMA vs. EWT; (**b**) LMA vs. Cab; (**c**) LMA vs. Car; (**d**) EWT vs. Cab; (**e**) EWT vs. Car; (**f**) Cab vs. Car.

3.3. RF-Based RTM Inversion

Random forest (RF) regression, a machine learning method, was employed for the inversion of RTMs. RF can overcome the autocorrelation of variables and overfitting, and it exhibits better prediction ability than a decision tree when dealing with many variables and observations because of its hierarchical structures and insensitivity to outliers [59,60]. RF is a combination of bagging algorithms and decision trees. It employs Bootstrap to extract n samples and constructs a decision tree for each sample [59].

Seven MODIS bands, simulated by the RTMs, were chosen as features for training an RF model. A 10-fold cross-validation was employed to evaluate the performance of the RF-based inversion model, using model efficiency (R^2), root-mean-square error (*RMSE*), and mean absolute error (*MAE*) as criteria for assessing accuracy [61]. Lastly, the forest CWC of the CONUS from 2017 to 2021 was generated in GEE using the RF model and MODIS products.

3.4. Solar Zenith Angle Correction

The geometric position of illumination and observation directly influence the reflectance of the ground target [62]. Geometric and radiometric corrections are essential for accurately retrieving land surface and vegetation parameters, particularly in time-series studies [63]. Regarding the MCD43A4, the bands were adjusted to the nadir view and the solar zenith angle at local noon. Thus, we employed the general method to correct the solar zenith angle [64]:

$$NBAR_{\beta} = \frac{NBAR_{\alpha} \times \cos\beta}{\cos\alpha} \tag{5}$$

where $NBAR_\alpha$ and $NBAR_\beta$ are the reflectance before and after correction, while α and β are the original and target solar zenith angles. In this work, the solar zenith angle is unified to 30°.

3.5. LAI Calculation from NEON DHPs

The GBOV (Ground-Based Observations for Validation) algorithm was employed to calculate field LAI based on the DHPs provided by the NEON [65]. To calculate LAIs from DHPs, the following steps were included: (1) image classification, (2) gap fraction calculation, and (3) LAI estimation.

For upward-facing DHPs, the Otsu automatic thresholding method was applied to the blue channel to classify the image cells as sky or vegetation canopy. Meanwhile, the image cells' background and vegetation canopy classification algorithm in downward-facing DHPs followed Meyer and Neto [66]. The method extracts vegetation canopy cells using two spectral indices: the excess green index (ExG, Equation (6)) and the excess red index (ExR, Equation (7) [67]). An image cell is classified as a vegetation canopy when the ExG is greater than the ExR. Otherwise, it is considered background. The ExG is calculated as follows:

$$ExG = 2DN_{\text{green}} - DN_{\text{red}} - DN_{\text{blue}} \tag{6}$$

and the ExR takes the form of

$$ExR = 1.4DN_{\text{red}} - DN_{\text{green}} \tag{7}$$

where DN_{red} is the red channel, DN_{green} is the green channel, and DN_{blue} is the blue channel.

Before gap fraction calculation, each image cell's zenith and azimuth angles need to be provided. The NEON DHPs were collected using a full-frame fisheye with a 180° field of view (FOV). Thus, assuming the zenith angle (α) of the image center is 0°, other cells can be calculated as follows [67]:

$$\alpha = \sqrt{(x - x_c)^2 + (y - y_c)^2}\frac{180}{\sqrt{w^2 + d^2}} \tag{8}$$

where x and y are the coordinates of a cell with an unknown zenith angle, x_c and y_c are the coordinates of the center cell, w is the image's width, and d is the height of the image. All these are expressed in pixels. The azimuth angle of a cell is calculated clockwise from the image center.

The classified images were divided into 36 patches with a 10° interval of azimuth angles to calculate the gap fraction. Only cells with zenith angles ranging from 52.5° to 62.5° were used to calculate the gap fraction of each patch. Therefore, the gap fraction, g_f, was calculated based on the patch as follows [67]:

$$g_f = \frac{n_{background}}{n_{total}} \tag{9}$$

where $n_{background}$ and n_{total} are the sky (or soil for downward images) cell amount and the total cell amount within a patch, respectively.

Then, the plant area index (PAI) was calculated as

$$PAI = \frac{-\overline{\ln g_f}}{0.93} \tag{10}$$

following Lang and Yueqin [68], where $\overline{\ln g_f}$ is the mean of the natural logarithm of the 36 patches' g_f values. Finally, PAI was equal to LAI for the non-woody area, while the LAI of the woody area was estimated according to

$$LAI = PAI_{up}(1 - \beta) + PAI_{\text{down}} \tag{11}$$

where PAI_{up} was calculated from the upward DHPs, PAI_{down} was calculated from the downward DHPs, and β is the area woody-to-total ratio. The means of β are 0.24, 0.16, and 0.18 for the deciduous forest, evergreen forest, and all [65], respectively.

3.6. Validation

The accuracy of CWC inversion was evaluated using NEON field observations. The NLCD land cover data were used to filter the representative NEON plots because of the difference in scale between the field plots and MODIS images. The criterion for screening the plots is that, within a 500 m buffer, one land cover type should occupy more than 90% of the area.

4. Results

4.1. Theoretical Performance of Coupled RTMs Inversion

The scatter plots between the predicted and model-set CWC values from the 10-fold cross-validation showed the theoretical performance of the three RF-based CWC inversion models. High model efficiency (R^2), low error ($RMSE$, MAE, and $rRMSE$), and weak bias ($Bias$) indicated that PRO4SAIL ($R^2 = 0.9$, $RMSE = 0.17$ kg m^{-2}, $MAE = 0.1$ kg m^{-2}, $rRMSE = 3.78\%$, $Bias = 0.002$ kg m^{-2}), PRO4SAIL2 ($R^2 = 0.88$, $RMSE = 0.17$ kg m^{-2}, $MAE = 0.1$ kg m^{-2}, $rRMSE = 3.73\%$, $Bias = 0.001$ kg m^{-2}), and PROGeoSail ($R^2 = 0.87$, $RMSE = 0.18$ kg m^{-2}, $MAE = 0.1$ kg m^{-2}, $rRMSE = 4.08\%$, $Bias = 0$ kg m^{-2}) had the theoretical ability to capture variations in CWC between different forest canopies well (Figure 4).

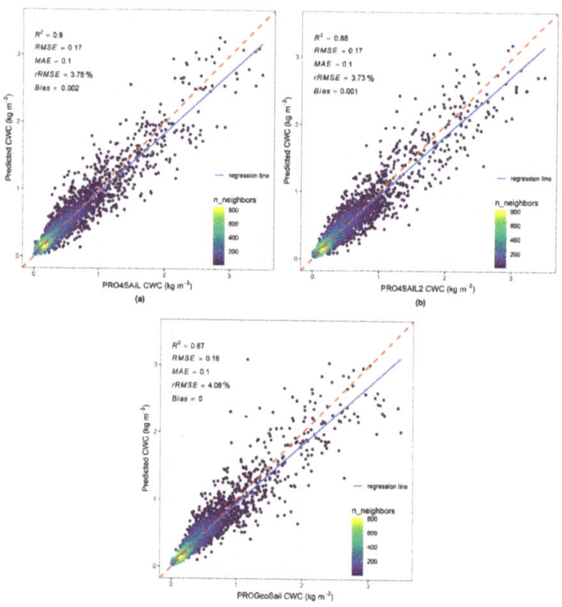

Figure 4. Theoretical performance of the RF-based inversion model over (**a**) PRO4SAIL, (**b**) PRO4SAIL2, and (**c**) PROGeoSail simulations. The 1:1 line is shown in dashed red, and the regression line is in solid blue.

4.2. Forest CWC Validation

The results of CWC validation, based on 48 NEON field observations collected at different sites during the leaf-out-to-senescence period from 2017 to 2021, are displayed in Figure 5. PRO4SAIL2 ($R^2 = 0.72$, $RMSE = 0.15$ kg m^{-2}, $MAE = 0.12$ kg m^{-2}, $rRMSE = 12.04\%$)

and PROGeoSail (R^2 = 0.68, *RMSE* = 0.15 kg m^{-2}, *MAE* = 0.12 kg m^{-2}, *rRMSE* = 12.78%) had higher coefficients of determination and lower errors than PRO4SAIL (R^2 = 0.51, *RMSE* = 0.19 kg m^{-2}, *MAE* = 0.14 kg m^{-2}, *rRMSE* = 15.86%). PROGeoSail exhibited a negative bias of −0.036 kg m^{-2}, and PRO4SAIL2 was −0.081 kg m^{-2}, while the bias of PRO4SAIL was positive: 0.071 kg m^{-2}. Weaker bias, a lower error, and a higher coefficient of determination indicated that PROGeoSail had better accuracy.

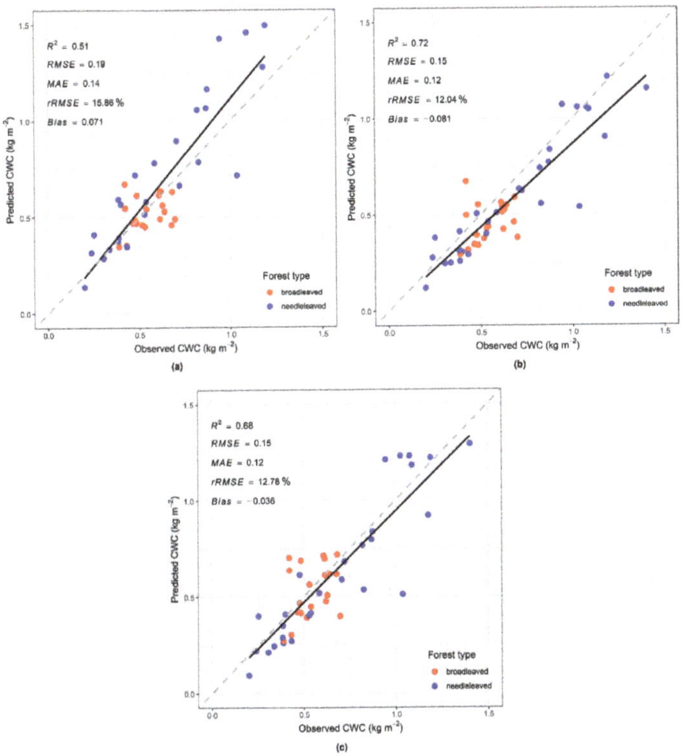

Figure 5. Prediction vs. field observations between (**a**) PRO4SAIL, (**b**) PRO4SAIL2, and (**c**) PROGeoSail-based hybrid inversion. The 1:1 line is shown in dashed gray, and the regression line is in solid black.

4.3. Spatiotemporal Variations in Forest CWC

Retrieval models were implemented in GEE using MODIS products to monitor spatiotemporal variations in forest CWC across the CONUS. The monthly mean forest CWC values from 2017 to 2021 were then generated (Figures 6, S2 and S3).

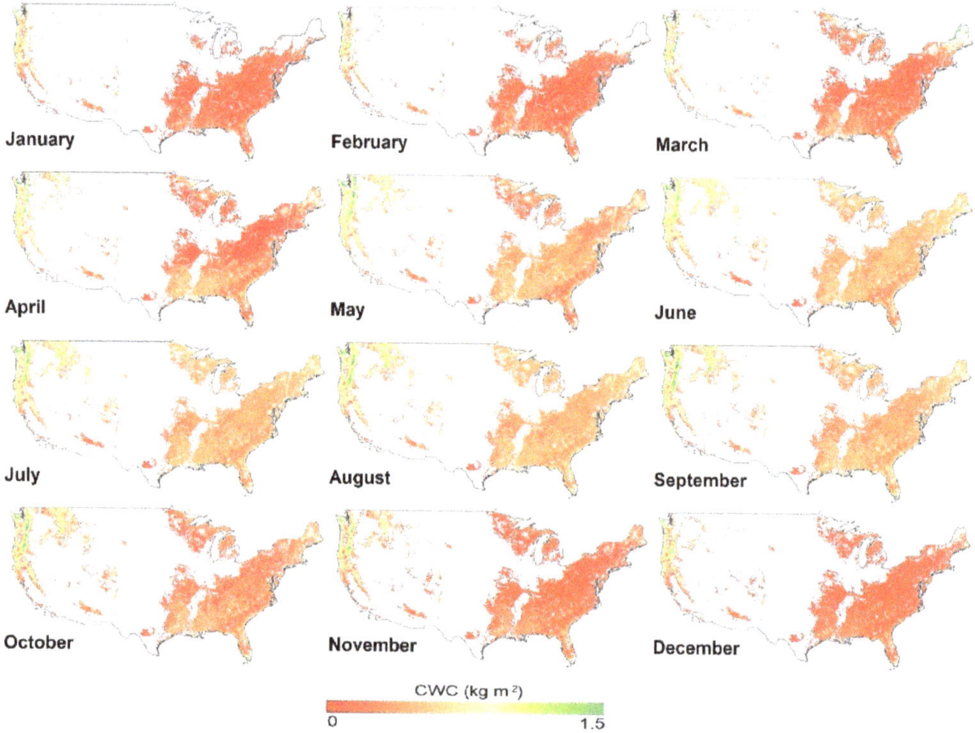

Figure 6. Monthly mean CWC spatial distribution of the CONUS at 500 m resolution in 2017–2021 based on PROGeoSail inversion.

Spatially, CWC exhibited variations between different forest types. Lower CWC values were observed in the DBF located in the eastern part of the CONUS, whereas higher CWC values were generally found in the ENF in the western region. Additionally, CWC dynamics varied with the month. As Figure 7 shows, throughout the year, forest CWC showed an increasing trend from January to August, followed by a decreasing trend until December. The DBF exhibited more significant seasonal variation compared with the ENF and the EBF.

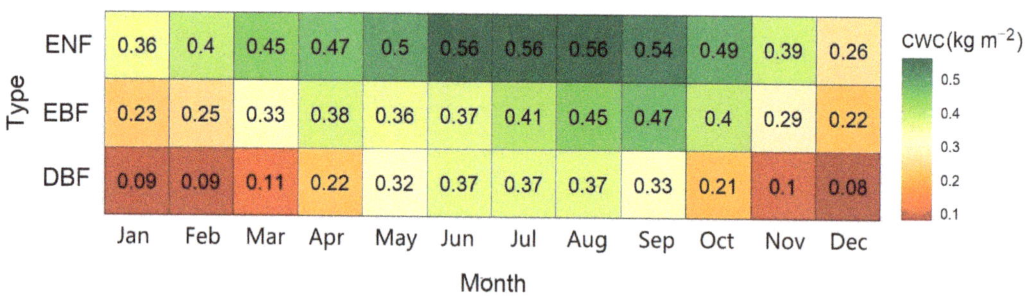

Figure 7. Monthly mean CWC of different forest types in 2017–2021 based on PROGeoSail inversion.

Furthermore, the three RTMs exhibited consistent spatial and temporal patterns. However, the retrieved CWC from PRO4SAIL was higher than that of PRO4SAIL2 and PROGeoSail, particularly in the ENF.

5. Discussion

Leaf traits are correlated, and forest canopy structures are heterogeneous and discontinuous. The accurate simulation of leaf and canopy characteristics can improve the accuracy of CWC inversion using coupled RTMs. Here, the 3D RTMs produced more accurate CWC retrieval than 1D RTMs. Our results demonstrated better theoretical performance than Campos-Taberner et al. [17] and superior validation accuracy compared with Trombetti et al. [29] and García-Haro et al. [1].

The successful inversion of forest CWC can be attributed to two key improvements. Firstly, in contrast to previous studies [21,29,69], we employed a Gaussian copula to constrain the leaf traits in the leaf-level RTM. This approach was adopted to alleviate the well-known ill-posed inversion problem in remote sensing. Prior studies [23–25,70] have highlighted the importance of constraining the fully independent variables using prior information to mitigate this issue. Quan et al. [25] and Reyes-Muñoz et al. [27] employed different methods to constrain the inputs of the RTMs for vegetation parameter inversion, resulting in favorable outcomes. As demonstrated in Figures 3 and S1, the leaf traits exhibit natural correlations with each other.

Secondly, a more accurate simulation of the forest canopy structure is needed. When retrieving canopy parameters, the widely used SAIL (or 4SAIL) model is generally applied to describe the canopy structure of grass, crops, shrubs, and forests [1,17,21,26,29,71] without considering their differences. However, the forest canopy exhibits heterogeneity and discontinuity, and modeling it solely with the turbid-based PROSAIL model may introduce bias. 4SAIL2 incorporates horizontal and vertical heterogeneities to achieve more realistic modeling by combining crown clumping and leaf color gradients [54]. GeoSail models crown clumping using Jasinski's parameterization [20]. Although including additional parameters may exacerbate the ill-posed problem, 3D canopy RTMs constrained by prior information, such as 4SAIL2 and GeoSail, are more suitable for forest modeling [28,55].

The forest CWC undergoes dynamic changes, particularly in the DBF. This variation was also observed by Trombetti et al. [29]. According to this definition, the CWC is determined by its EWT and LAI, both of which are influenced by plant phenology, climate, soil moisture, and other factors [2,72,73].

Our results also revealed a spatial pattern of higher CWC in the ENF of the CONUS and lower CWC in the DBF, which is consistent with the findings of Campos-Taberner et al. [17] and Reyes-Muñoz et al. [27]. This pattern can be attributed to the significant difference in leaf EWT, as indicated by field observations (Figures 5 and S4). However, the difference in CWC between ENF and DBF is more pronounced than in previous studies [1,17,26,29]. This difference could be explained by the variations in the selection and parameterization of RTMs [17], the inversion method [29,74], and remote sensing images [1,26]. Given the limitations of the data, more research on this difference is needed in the future.

In our study, a few limitations and uncertainties remain in the following aspects: Firstly, the field observations are limited and unevenly distributed in time and space. There are 170, 226, and 989 leaf samples from trees from LOPEX93, ANGERS, and the NEON, respectively, and the majority of these samples were obtained from deciduous trees. Limited and uneven leaf samples may lead to bias in EWT distribution and eventually affect the inversion of CWC. In addition, the NEON field observations of leaf traits and DHPs were primarily collected from leaf-out to senescence [48]. In this study, only 48 field CWC measurements (28 evergreen and 20 deciduous) were available for accuracy assessment, and there is a lack of observations for the winter period. Therefore, future research should prioritize additional efforts in forest field validation.

Secondly, this study focused on simulating the canopy while neglecting the leaf. PROSPECT-5, based on the plate model, was developed to simulate the reflectance and transmittance of broad leaves [19]. Although numerous studies have demonstrated the suitability of PROSPECT-5 for retrieving physiological parameters in coniferous species [13,28,75], the use of LIBERTY to simulate the reflectance and transmittance of coniferous species would provide greater robustness [76].

Thirdly, the simulated reflectance of the grass canopy was used instead of bare soil. This improvement can make the simulation of the forest more realistic [28]. However, the ground cover of CONUS forests is usually dominated by herbaceous and shrub layers [77]. Further study on the influence of forest underlayer on the inversion of CWC is needed.

Finally, compared with 4SAIL, 4SAIL2 and GeoSail require more inputs, which exacerbates the ill-posed problem. In this study, we employed prior information to constrain the inputs of the RTMs, including leaf-level parameters such as LAM, EWT, Cab, and Car, as well as the canopy cover fraction at the canopy level. However, there remain several independent parameters for the RTMs' inputs. Thus, future research should aim to impose further constraints on these free parameters based on additional prior information.

In this study, we estimated the forest CWC at a coarse spatial resolution. To study the species-specific characterization of CWC, hyperspectral, LiDAR, terahertz technology, and high-resolution images from unmanned aerial vehicles (cm to m level), Sentinel-2 (10 m), and Landsat 8–9 (30 m) could be used [61,78].

6. Conclusions

In this study, we aimed to compare the performance of different RTMs and provide an efficient strategy for forest CWC monitoring. A Gaussian copula and prior-knowledge-constrained parameters were input into coupled RTMs, and RF regression was used to retrieve forest CWC values across the CONUS. We found that accurately simulating the characteristics of leaf traits and canopy structures can improve the accuracy of CWC inversion. Additionally, 3D RTMs demonstrated better performance in retrieving forest CWC. We found that forest CWC showed a dynamically changed spatiotemporal pattern, and ENF had a strong CWC capacity. However, uncertainties still exist because of data limitations and the ill-posed problem. Future research should focus more on simulating forests with heterogeneous and discontinuous canopy structures and conducting field validation.

Supplementary Materials: The following supporting information can be downloaded at https://www.mdpi.com/article/10.3390/f14071418/s1: Figure S1. Constrained random samples of herbaceous LMA, EWT, Cab, and Car based on field observations using a Gaussian copula for (a) LMA vs. EWT; (b) LMA vs. Cab; (c) LMA vs. Car; (d) EWT vs. Cab; (e) EWT vs. Car; (f) Cab vs. Car. Figure S2. Monthly mean CWC spatial distribution of CONUS at 500 m resolution in 2017–2021 based on PRO4SAIL inversion; Figure S3. Monthly mean CWC spatial distribution of CONUS at 500 m resolution in 2017–2021 based on PRO4SAIL2 inversion; Figure S4. Box plot for the EWT of NEON field observations comparing deciduous and evergreen forests. The statistical significance is shown by lowercase letters (a and b) at the level of $p < 0.01$ using one-way analysis of variance (ANOVA); Table S1. Parameterization of RTM inputs for forest CWC inversion.

Author Contributions: Conceptualization, S.L. and L.L.; methodology, L.L.; software, L.L.; validation, L.L., X.W., and X.L.; resources, S.L. and W.Y.; data curation, L.L. and P.R; writing—original draft preparation, L.L.; writing—review and editing, L.L., S.L., W.Y., X.W., X.L., P.R., and H.Z.; supervision, S.L. and W.Y.; funding acquisition, W.Y. All authors have read and agreed to the published version of the manuscript.

Funding: This research was funded by the National Natural Science Foundation of China, grant number 41671432.

Data Availability Statement: The MODIS products (MCD43A4, MCD43A2, MOD10A1, MYD10A1, and MCD12Q1) used here are available on GEE. The LOPEX93 and ANGERS datasets can be downloaded from http://opticleaf.ipgp.fr/index.php?page=database (accessed on 10 May 2023). The Plant Foliar Traits (DP1.10026.001) and Digital Hemispheric Photos (DHP) of the Plot Vegetation data products (DP1.10017.001) are accessible from https://data.neonscience.org (accessed on 10 May 2023). All the scripts are available upon request.

Acknowledgments: The National Ecological Observatory Network is a program sponsored by the National Science Foundation and operated under a cooperative agreement by Battelle. The Plant Foliar Traits (DP1.10026.001) and Digital Hemispheric Photos of Plot Vegetation (DP1.10017.001) data products are based in part upon work supported by the National Science Foundation through the NEON Program.

Conflicts of Interest: The authors declare no conflict of interest.

References

1. García-Haro, F.J.; Campos-Taberner, M.; Moreno, A.; Torbern Tagesson, H.; Camacho, F.; Martinez, B.; Sanchez, S.; Piles, M.; Camps-Valls, G.; Yebra, M.; et al. A global canopy water content product from AVHRR/Metop. *ISPRS J. Photogramm. Remote Sens.* **2020**, *162*, 77–93. [CrossRef]
2. Lyons, D.S.; Dobrowski, S.Z.; Holden, Z.A.; Maneta, M.P.; Sala, A. Soil moisture variation drives canopy water content dynamics across the western US. *Remote Sens. Environ.* **2021**, *253*, 112233. [CrossRef]
3. Zhou, S.; Zhang, Y.; Williams, A.P.; Gentine, P. Projected increases in intensity, frequency, and terrestrial carbon costs of compound drought and aridity events. *Sci. Adv.* **2019**, *5*, eaau5740. [CrossRef] [PubMed]
4. Bevacqua, E.; Zappa, G.; Lehner, F.; Zscheischler, J. Precipitation trends determine future occurrences of compound hot-dry events. *Nat. Clim. Change* **2022**, *12*, 350. [CrossRef]
5. Senf, C.; Pflugmacher, D.; Yang, Z.Q.; Sebald, J.; Knorn, J.; Neumann, M.; Hostert, P.; Seidl, R. Canopy mortality has doubled in Europe's temperate forests over the last three decades. *Nat. Commun.* **2018**, *9*, 4978. [CrossRef]
6. Jiao, W.Z.; Wang, L.X.; Smith, W.K.; Chang, Q.; Wang, H.L.; D'Odorico, P. Observed increasing water constraint on vegetation growth over the last three decades. *Nat. Commun.* **2021**, *12*, 3777. [CrossRef]
7. Hartmann, H.; Bastos, A.; Das, A.J.; Esquivel-Muelbert, A.; Hammond, W.M.; Martinez-Vilalta, J.; McDowell, N.G.; Powers, J.S.; Pugh, T.A.M.; Ruthrof, K.X.; et al. Climate Change Risks to Global Forest Health: Emergence of Unexpected Events of Elevated Tree Mortality Worldwide. *Annu. Rev. Plant Biol.* **2022**, *73*, 673–702. [CrossRef]
8. Brodrick, P.G.; Anderegg, L.D.L.; Asner, G.P. Forest Drought Resistance at Large Geographic Scales. *Geophys. Res. Lett.* **2019**, *46*, 2752–2760. [CrossRef]
9. Konings, A.G.; Yu, Y.; Xu, L.; Yang, Y.; Schimel, D.S.; Saatchi, S.S. Active microwave observations of diurnal and seasonal variations of canopy water content across the humid African tropical forests. *Geophys. Res. Lett.* **2017**, *44*, 2290–2299. [CrossRef]
10. Quemada, C.; Perez-Escudero, J.M.; Gonzalo, R.; Ederra, I.; Santesteban, L.G.; Torres, N.; Iriarte, J.C. Remote Sensing for Plant Water Content Monitoring: A Review. *Remote Sens.* **2021**, *13*, 2088. [CrossRef]
11. Holtzman, N.M.; Anderegg, L.D.L.; Kraatz, S.; Mavrovic, A.; Sonnentag, O.; Pappas, C.; Cosh, M.H.; Langlois, A.; Lakhankar, T.; Tesser, D.; et al. L-band vegetation optical depth as an indicator of plant water potential in a temperate deciduous forest stand. *Biogeosciences* **2021**, *18*, 739–753. [CrossRef]
12. Frappart, F.; Wigneron, J.P.; Li, X.J.; Liu, X.Z.; Al-Yaari, A.; Fan, L.; Wang, M.J.; Moisy, C.; Le Masson, E.; Lafkih, Z.A.; et al. Global Monitoring of the Vegetation Dynamics from the Vegetation Optical Depth (VOD): A Review. *Remote Sens.* **2020**, *12*, 2915. [CrossRef]
13. Jacquemoud, S.; Verhoef, W.; Baret, F.; Bacour, C.; Zarco-Tejada, P.J.; Asner, G.P.; François, C.; Ustin, S.L. PROSPECT+SAIL models: A review of use for vegetation characterization. *Remote Sens. Environ.* **2009**, *113*, S56–S66. [CrossRef]
14. Ceccato, P.; Flasse, S.; Tarantola, S.; Jacquemoud, S.; Gregoire, J.M. Detecting vegetation leaf water content using reflectance in the optical domain. *Remote Sens. Environ.* **2001**, *77*, 22–33. [CrossRef]
15. Casas, A.; Riano, D.; Ustin, S.L.; Dennison, P.; Salas, J. Estimation of water-related biochemical and biophysical vegetation properties using multitemporal airborne hyperspectral data and its comparison to MODIS spectral response. *Remote Sens. Environ.* **2014**, *148*, 28–41. [CrossRef]
16. Ceccato, P.; Flasse, S.; Gregoire, J.M. Designing a spectral index to estimate vegetation water content from remote sensing data—Part 2. Validation and applications. *Remote Sens. Environ.* **2002**, *82*, 198–207. [CrossRef]
17. Campos-Taberner, M.; Moreno-Martinez, A.; Javier Garcia-Haro, F.; Camps-Valls, G.; Robinson, N.P.; Kattge, J.; Running, S.W. Global Estimation of Biophysical Variables from Google Earth Engine Platform. *Remote Sens.* **2018**, *10*, 1167. [CrossRef]
18. Verrelst, J.; Malenovsky, Z.; Van der Tol, C.; Camps-Valls, G.; Gastellu-Etchegorry, J.P.; Lewis, P.; North, P.; Moreno, J. Quantifying Vegetation Biophysical Variables from Imaging Spectroscopy Data: A Review on Retrieval Methods. *Surv. Geophys.* **2019**, *40*, 589–629. [CrossRef]
19. Feret, J.-B.; François, C.; Asner, G.P.; Gitelson, A.A.; Martin, R.E.; Bidel, L.P.R.; Ustin, S.L.; le Maire, G.; Jacquemoud, S. PROSPECT-4 and 5: Advances in the leaf optical properties model separating photosynthetic pigments. *Remote Sens. Environ.* **2008**, *112*, 3030–3043. [CrossRef]
20. Huemmrich, K.F. The GeoSail model: A simple addition to the SAIL model to describe discontinuous canopy reflectance. *Remote Sens. Environ.* **2001**, *75*, 423–431. [CrossRef]
21. García-Haro, F.J.; Campos-Taberner, M.; Muñoz-Marí, J.; Laparra, V.; Camacho, F.; Sánchez-Zapero, J.; Camps-Valls, G. Derivation of global vegetation biophysical parameters from EUMETSAT Polar System. *ISPRS J. Photogramm. Remote Sens.* **2018**, *139*, 57–74. [CrossRef]
22. Wang, Y. Quantitative Remote Sensing Inversion in Earth Science: Theory and Numerical Treatment. In *Handbook of Geomathematics*; Freeden, W., Nashed, M.Z., Sonar, T., Eds.; Springer: Berlin/Heidelberg, Germany, 2010; pp. 785–812.

23. Combal, B.; Baret, F.; Weiss, M.; Trubuil, A.; Mace, D.; Pragnere, A.; Myneni, R.; Knyazikhin, Y.; Wang, L. Retrieval of canopy biophysical variables from bidirectional reflectance—Using prior information to solve the ill-posed inverse problem. *Remote Sens. Environ.* **2003**, *84*, 1–15. [CrossRef]
24. Yebra, M.; Chuvieco, E. Linking ecological information and radiative transfer models to estimate fuel moisture content in the Mediterranean region of Spain: Solving the ill-posed inverse problem. *Remote Sens. Environ.* **2009**, *113*, 2403–2411. [CrossRef]
25. Quan, X.; He, B.; Li, X. A Bayesian Network-Based Method to Alleviate the Ill-Posed Inverse Problem: A Case Study on Leaf Area Index and Canopy Water Content Retrieval. *Ieee Trans. Geosci. Remote Sens.* **2015**, *53*, 6507–6517. [CrossRef]
26. Martínez-Ferrer, L.; Moreno-Martínez, Á.; Campos-Taberner, M.; García-Haro, F.J.; Muñoz-Marí, J.; Running, S.W.; Kimball, J.; Clinton, N.; Camps-Valls, G. Quantifying uncertainty in high resolution biophysical variable retrieval with machine learning. *Remote Sens. Environ.* **2022**, *280*, 113199. [CrossRef]
27. Reyes-Muñoz, P.; Pipia, L.; Salinero-Delgado, M.; Belda, S.; Berger, K.; Estévez, J.; Morata, M.; Rivera-Caicedo, J.P.; Verrelst, J. Quantifying Fundamental Vegetation Traits over Europe Using the Sentinel-3 OLCI Catalogue in Google Earth Engine. *Remote Sens.* **2022**, *14*, 1347. [CrossRef]
28. Quan, X.; Yebra, M.; Riano, D.; He, B.; Lai, G.; Liu, X. Global fuel moisture content mapping from MODIS. *Int. J. Appl. Earth Obs. Geoinf.* **2021**, *101*, 102354. [CrossRef]
29. Trombetti, M.; Riaño, D.; Rubio, M.A.; Cheng, Y.B.; Ustin, S.L. Multi-temporal vegetation canopy water content retrieval and interpretation using artificial neural networks for the continental USA. *Remote Sens. Environ.* **2008**, *112*, 203–215. [CrossRef]
30. Djamai, N.; Zhong, D.T.; Fernandes, R.; Zhou, F.Q. Evaluation of Vegetation Biophysical Variables Time Series Derived from Synthetic Sentinel-2 Images. *Remote Sens.* **2019**, *11*, 1547. [CrossRef]
31. Hosgood, B.; Jacquemoud, S.; Andreoli, G.; Verdebout, J.; Pedrini, G.; Schmuck, G. *Leaf Optical Properties Experiment 93 (LOPEX93)*; Report EUR 16095 EN.; Joint Research Centre, European Commission: Ispra, Italy, 1994.
32. Jacquemound, S.; Bidel, L.; Francois, C.; Pavan, G. ANGERS Leaf Optical Properties Database. Available online: http://opticleaf.ipgp.fr/index.php?page=database (accessed on 10 May 2023).
33. Kattge, J.; Bonisch, G.; Diaz, S.; Lavorel, S.; Prentice, I.C.; Leadley, P.; Tautenhahn, S.; Werner, G.D.A.; Aakala, T.; Abedi, M.; et al. TRY plant trait database—enhanced coverage and open access. *Glob. Change Biol.* **2020**, *26*, 119–188. [CrossRef]
34. Kao, R.H.; Gibson, C.M.; Gallery, R.E.; Meier, C.L.; Barnett, D.T.; Docherty, K.M.; Blevins, K.K.; Travers, P.D.; Azuaje, E.; Springer, Y.P.; et al. NEON terrestrial field observations: Designing continental-scale, standardized sampling. *Ecosphere* **2012**, *3*, 1–17. [CrossRef]
35. Amani, M.; Ghorbanian, A.; Ahmadi, S.A.; Kakooei, M.; Moghimi, A.; Mirmazloumi, S.M.; Moghaddam, S.H.A.; Mahdavi, S.; Ghahremanloo, M.; Parsian, S.; et al. Google Earth Engine Cloud Computing Platform for Remote Sensing Big Data Applications: A Comprehensive Review. *IEEE J. Sel. Top. Appl. Earth Obs. Remote Sens.* **2020**, *13*, 5326–5350. [CrossRef]
36. Hu, Q.; Yang, J.Y.; Xu, B.D.; Huang, J.X.; Memon, M.S.; Yin, G.F.; Zeng, Y.L.; Zhao, J.; Liu, K. Evaluation of Global Decametric-Resolution LAI, FAPAR and FVC Estimates Derived from Sentinel-2 Imagery. *Remote Sens.* **2020**, *12*, 912. [CrossRef]
37. Gorelick, N.; Hancher, M.; Dixon, M.; Ilyushchenko, S.; Thau, D.; Moore, R. Google Earth Engine: Planetary-scale geospatial analysis for everyone. *Remote Sens. Environ.* **2017**, *202*, 18–27. [CrossRef]
38. Sanchez-Ruiz, S.; Moreno-Martinez, A.; Izquierdo-Verdiguier, E.; Chiesi, M.; Maselli, F.; Gilabert, M.A. Growing stock volume from multi-temporal landsat imagery through google earth engine. *Int. J. Appl. Earth Obs. Geoinf.* **2019**, *83*, 101913. [CrossRef]
39. Weintraub, S. NEON User Guide to Plant Foliar Traits (DP1.10026.001). Available online: https://data.neonscience.org/documents/10179/2237401/NEON_CFC_userGuide_vD/3f1404e9-1a6b-9c13-7807-742893f88ef8 (accessed on 10 May 2023).
40. Hosgood, B.; Jacquemound, S.; Andreeoli, G.; Verdebout, J.; Pedrini, G.; Schmuck, G. Leaf Optical Properties Experiment Database (LOPEX93). Available online: http://opticleaf.ipgp.fr/index.php?page=database (accessed on 10 May 2023).
41. Majasalmi, T.; Bright, R.M. Evaluation of leaf-level optical properties employed in land surface models. *Geosci. Model Dev.* **2019**, *12*, 3923–3938. [CrossRef]
42. Yang, B.; He, Y.; Chen, W. A simple method for estimation of leaf dry matter content in fresh leaves using leaf scattering albedo. *Glob. Ecol. Conserv.* **2020**, *23*, e01201. [CrossRef]
43. Liu, L.; Song, B.; Zhang, S.; Liu, X. A Novel Principal Component Analysis Method for the Reconstruction of Leaf Reflectance Spectra and Retrieval of Leaf Biochemical Contents. *Remote Sens.* **2017**, *9*, 1113. [CrossRef]
44. Schimel, D. NEON Observatory Design. Available online: https://data.neonscience.org/documents/10179/11206/NEON.DOC.000001vD/af0626a5-7e95-4b0e-aaf1-c330d1b6a554 (accessed on 10 May 2023).
45. Barnett, D. TOS Science Design for Spatial Sampling. Available online: https://data.neonscience.org/documents/10179/1723439/NEON.DOC.000913vC/92b5da9f-8c35-f441-4413-c743ea4d01f7 (accessed on 10 May 2023).
46. Meier, C. NEON Terrestrial Observation System: Spatial and Temporal Sampling Strategy. Available online: https://data.neonscience.org/documents/10179/1723439/NEON.DOC.005108vB/1eda32ae-84f0-d263-0665-ab08c7d4eaca (accessed on 10 May 2023).
47. Meier, C. NEON User Guide to Digital Hemispheric Photos of Plot Vegetation (NEON.DP1.10017). Available online: https://data.neonscience.org/documents/10179/2237401/NEON_dhp_userGuide_vE/a7d3ae35-5595-cd1d-f1ad-aeed03bdb62e (accessed on 10 May 2023).
48. Meier, C. TOS Protocol and Procedure: DHP—Measurement of Leaf Area Index. Available online: https://data.neonscience.org/documents/10179/1883155/NEON.DOC.014039vM/94cd2218-598c-935f-829e-5003d7b25950 (accessed on 10 May 2023).

49. Schaaf, C.; Wang, Z. MCD43A4 MODIS/Terra+Aqua BRDF/Albedo Nadir BRDF Adjusted Ref Daily L3 Global—500m V006. Available online: https://lpdaac.usgs.gov/products/mcd43a4v006 (accessed on 20 June 2022).
50. Schaaf, C.; Wang, Z. MCD43A2 MODIS/Terra+Aqua BRDF/Albedo Quality Daily L3 Global—500m V006. Available online: https://lpdaac.usgs.gov/products/mcd43a2v006 (accessed on 20 June 2022).
51. Hall, D.K.; Riggs, G.A. MODIS/Terra Snow Cover Daily L3 Global 500m SIN Grid, Version 6. Available online: https://doi.org/10.5067/MODIS/MOD10A1.006 (accessed on 10 May 2023).
52. Friedl, M.; Sulla-Menashe, D. MCD12Q1 MODIS/Terra+Aqua Land Cover Type Yearly L3 Global 500m SIN Grid V006. Available online: https://lpdaac.usgs.gov/products/mcd12q1v006 (accessed on 10 May 2023).
53. Verhoef, W.; Jia, L.; Xiao, Q.; Su, Z. Unified optical-thermal four-stream radiative transfer theory for homogeneous vegetation canopies. *Ieee Trans. Geosci. Remote Sens.* **2007**, *45*, 1808–1822. [CrossRef]
54. Verhoef, W.; Bach, H. Coupled soil-leaf-canopy and atmosphere radiative transfier modeling to simulate hyperspectral multi-angular surface reflectance and TOA radiance data. *Remote Sens. Environ.* **2007**, *109*, 166–182. [CrossRef]
55. le Maire, G.; Marsden, C.; Verhoef, W.; Ponzoni, F.J.; Lo Seen, D.; Begue, A.; Stape, J.-L.; Nouvellon, Y. Leaf area index estimation with MODIS reflectance time series and model inversion during full rotations of Eucalyptus plantations. *Remote Sens. Environ.* **2011**, *115*, 586–599. [CrossRef]
56. Sklar, A. *Fonctions de Répartition à n Dimensions et Leurs Marges*; Publications de L'institut de Statistique de L'université de Paris: Paris, France, 1959; Volume 8.
57. Žežula, I. On multivariate Gaussian copulas. *J. Stat. Plan. Inference* **2009**, *139*, 3942–3946. [CrossRef]
58. Van de Vyver, H.; Van den Bergh, J. The Gaussian copula model for the joint deficit index for droughts. *J. Hydrol.* **2018**, *561*, 987–999. [CrossRef]
59. Breiman, L. Random forests. *Mach. Learn.* **2001**, *45*, 5–32. [CrossRef]
60. Tang, X.; Fan, S.; Zhang, W.; Gao, S.; Chen, G.; Shi, L. Global variability in belowground autotrophic respiration in terrestrial ecosystems. *Earth Syst. Sci. Data* **2019**, *11*, 1839–1852. [CrossRef]
61. de Almeida, C.T.; Galvao, L.S.; de Oliveira Cruz e Aragao, L.E.; Henry Balbaud Ometto, J.P.; Jacon, A.D.; de Souza Pereira, F.R.; Sato, L.Y.; Lopes, A.P.; Lima de Alencastro Graca, P.M.; Silva, C.V.D.J.; et al. Combining LiDAR and hyperspectral data for aboveground biomass modeling in the Brazilian Amazon using different regression algorithms. *Remote Sens. Environ.* **2019**, *232*, 111323. [CrossRef]
62. Widlowski, J.L.; Taberner, M.; Pinty, B.; Bruniquel-Pinel, V.; Disney, M.; Fernandes, R.; Gastellu-Etchegorry, J.P.; Gobron, N.; Kuusk, A.; Lavergne, T.; et al. Third Radiation Transfer Model Intercomparison (RAMI) exercise: Documenting progress in canopy reflectance models. *J. Geophys. Res. Atmos.* **2007**, *112*, D09111. [CrossRef]
63. Breunig, F.M.; Galvao, L.S.; dos Santos, J.R.; Gitelson, A.A.; de Moura, Y.M.; Teles, T.S.; Gaida, W. Spectral anisotropy of subtropical deciduous forest using MISR and MODIS data acquired under large seasonal variation in solar zenith angle. *Int. J. Appl. Earth Obs. Geoinf.* **2015**, *35*, 294–304. [CrossRef]
64. Shahi, A.P.; Rai, P.K.; Rabi ul, I.; Mishra, V.N. Chapter 5—Remote sensing data extraction and inversion techniques: A review. In *Atmospheric Remote Sensing*; Kumar Singh, A., Tiwari, S., Eds.; Elsevier: Amsterdam, The Netherlands, 2023; pp. 85–104.
65. Brown, L.A.; Fernandes, R.; Djamai, N.; Meier, C.; Gobron, N.; Morris, H.; Canisius, F.; Bai, G.; Lerebourg, C.; Lanconelli, C.; et al. Validation of baseline and modified Sentinel-2 Level 2 Prototype Processor leaf area index retrievals over the United States. *ISPRS J. Photogramm. Remote Sens.* **2021**, *175*, 71–87. [CrossRef]
66. Meyer, G.E.; Neto, J.C. Verification of color vegetation indices for automated crop imaging applications. *Comput. Electron. Agric.* **2008**, *63*, 282–293. [CrossRef]
67. Dash, L.B.H.M.J. *Algorithm Theoretical Basis Document—Vegetation Products RM4 (Tcanopy), RM6 (FIPAR) and RM7 (LAI) and FCOVER*; University of Southampton: Southampton, UK, 2021.
68. Lang, A.R.G.; Yueqin, X. Estimation of leaf area index from transmission of direct sunlight in discontinuous canopies. *Agric. For. Meteorol.* **1986**, *37*, 229–243. [CrossRef]
69. Boren, E.J.; Boschetti, L. Landsat-8 and Sentinel-2 Canopy Water Content Estimation in Croplands through Radiative Transfer Model Inversion. *Remote Sens.* **2020**, *12*, 2803. [CrossRef]
70. de Sá, N.C.; Baratchi, M.; Hauser, L.T.; van Bodegom, P. Exploring the Impact of Noise on Hybrid Inversion of PROSAIL RTM on Sentinel-2 Data. *Remote Sens.* **2021**, *13*, 648. [CrossRef]
71. Estevez, J.; Salinero-Delgado, M.; Berger, K.; Pipia, L.; Rivera-Caicedo, J.P.; Wocher, M.; Reyes-Munoz, P.; Tagliabue, G.; Boschetti, M.; Verrelst, J. Gaussian processes retrieval of crop traits in Google Earth Engine based on Sentinel-2 top-of-atmosphere data. *Remote Sens. Environ.* **2022**, *273*, 112958. [CrossRef] [PubMed]
72. Joswig, J.S.; Wirth, C.; Schuman, M.C.; Kattge, J.; Reu, B.; Wright, I.J.; Sippel, S.D.; Ruger, N.; Richter, R.; Schaepman, M.E.; et al. Climatic and soil factors explain the two-dimensional spectrum of global plant trait variation. *Nat. Ecol. Evol.* **2022**, *6*, 36–50. [CrossRef]
73. Savoy, P.; Mackay, D.S. Modeling the seasonal dynamics of leaf area index based on environmental constraints to canopy development. *Agric. For. Meteorol.* **2015**, *200*, 46–56. [CrossRef]
74. Asner, G.P.; Brodrick, P.G.; Anderson, C.B.; Vaughn, N.; Knapp, D.E.; Martin, R.E. Progressive forest canopy water loss during the 2012–2015 California drought. *Proc. Natl. Acad. Sci. USA* **2016**, *113*, E249–E255. [CrossRef]

75. Cheng, Y.-B.; Zarco-Tejada, P.J.; Riano, D.; Rueda, C.A.; Ustin, S.L. Estimating vegetation water content with hyperspectral data for different canopy scenarios: Relationships between AVIRIS and MODIS indexes. *Remote Sens. Environ.* **2006**, *105*, 354–366. [CrossRef]
76. Dawson, T.P.; Curran, P.J.; Plummer, S.E. LIBERTY—Modeling the effects of leaf biochemical concentration on reflectance spectra. *Remote Sens. Environ.* **1998**, *65*, 50–60. [CrossRef]
77. Krebs, M.A.; Reeves, M.C.; Baggett, L.S. Predicting understory vegetation structure in selected western forests of the United States using FIA inventory data. *For. Ecol. Manag.* **2019**, *448*, 509–527. [CrossRef]
78. Gente, R.; Koch, M. Monitoring leaf water content with THz and sub-THz waves. *Plant Methods* **2015**, *11*, 15. [CrossRef] [PubMed]

forests

MDPI

Article

Automatic Extraction of the Spatial Distribution of *Picea schrenkiana* in the Tianshan Mountains Based on Google Earth Engine and the Jeffries–Matusita Distance

Fujin Xu [1], Zhonglin Xu [2,3,4,*], Changchun Xu [1,3,*] and Tingting Yu [1]

1 College of Geography and Remote Sensing Sciences, Xinjiang University, Urumqi 830017, China; xfj_1027@126.com (F.X.); 107552101053@stu.xju.edu.cn (T.Y.)
2 College of Ecology and Environment, Xinjiang University, Urumqi 830017, China
3 Key Laboratory of Oasis Ecology, Xinjiang University, Urumqi 830017, China
4 Xinjiang Jinghe Observation and Research Station of Temperate Desert Ecosystem, Ministry of Education, Urumqi 830017, China
* Correspondence: zlxu@xju.edu.cn (Z.X.); xucc@xju.edu.cn (C.X.)

Abstract: As a distinct species in the Tianshan Mountains (TS) of Central Asia (CA), *Picea schrenkiana* plays a significant role in water purification, soil and water conservation, and climate regulation. In the context of climate change, rapidly and accurately obtaining its spatial distribution has critical decision-making significance for maintaining ecological security in the arid area of CA and the sustainable development of the "Silk Road Economic Belt". However, conventional methods are extremely challenging to accomplish the high-resolution mapping of *Picea schrenkiana* in the TS, which is characterized by a wide range (9.97×10^5 km^2) and complex terrain. The approach of geo-big data and cloud computing provides new opportunities to address this issue. Therefore, the purpose of this study is to propose an automatic extraction procedure for the spatial distribution of *Picea schrenkiana* based on Google Earth Engine and the Jeffries–Matusita (JM) distance, which considered three aspects: sample points, remote-sensing images, and classification features. The results showed that (1) after removing abnormal samples and selecting the summer image, the producer accuracy (PA) of *Picea schrenkiana* was improved by 2.95% and 0.24%–2.10%, respectively. (2) Both the separation obtained by the JM distance and the analysis results of eight schemes showed that spectral features and texture features played a key role in the mapping of *Picea schrenkiana*. (3) The JM distance can seize the classification features that are most conducive to the mapping of *Picea schrenkiana*, and effectively improve the classification accuracy. The PA and user accuracy of *Picea schrenkiana* were 96.74% and 96.96%, respectively. The overall accuracy was 91.93%, while the Kappa coefficient was 0.89. (4) The results show that *Picea schrenkiana* is concentrated in the middle TS and scattered in the remaining areas. In total, 85.7%, 66.4%, and 85.9% of *Picea schrenkiana* were distributed in the range of 1500–2700 m, 20–40°, and on shady slope and semi-shady slope, respectively. The automatic procedure adopted in this study provides a basis for the rapid and accurate mapping of the spatial distribution of coniferous forests in the complex terrain.

Keywords: Google Earth Engine; Sentinel-1; Sentinel-2; JM distance; Tianshan mountains; *Picea schrenkiana*

1. Introduction

As one of the most valuable terrestrial ecosystems, the forest ecosystem plays a vital role in maintaining global carbon balance and slowing down the rise in the greenhouse gas concentration [1]. The carbon stock of forest ecosystems accounts for about 33%–46% of terrestrial ecosystems [2]. Under the background of "carbon neutrality", global changes caused by both natural and human factors directly or indirectly affect forest ecosystems, which has aroused widespread concern in the whole society [3]. The arid area of Central Asia (CA), which is connected to China by mountains and rivers, is a key hub zone in the construction of "the Silk Road Economic Belt" and a hot area concerned with the international community [4]. The Tianshan Mountains (TS) is one of the most important parts in the arid area of CA, in which the forest ecosystem has critical ecological values and functions such as carbon sink [5], water resource conservation, flood regulation, water purification, and soil and water conservation. The TS is highly essential for maintaining the ecological security of the downstream areas and the sustainable development of the whole arid zone [6]. The arid areas of CA, far from the sea and interspersed with mountains and basins, have light precipitation but intense evaporation, severe water shortages, and fragile ecosystems. Therefore, this region is a dramatic response region to global changes [7,8]. In the past half-century, although the trend of "warming and humidification" has occurred in the arid area of CA [9], it has not slowed down the ecological degradation in the region, because the positive ecological effect of increasing precipitation is not sufficient to offset the negative effect of rising temperatures [10]. In addition, the normalized difference vegetation index (NDVI) of the natural vegetation growing season in CA has declined at a rate of 0.0003 per year due to a sudden increase in the temperature since 1998 [10]. All of the above studies indicate that the natural vegetation in this area shows browning as a whole. *Picea schrenkiana* is a coniferous forest species with an altitude range of 1400–2800 m on the northern slopes of the TS, and is particularly sensitive to climate change [11]. Timely and accurate acquisition of its spatial distribution information is of great significance to ensure the sustainable development of the arid area of CA and "the Silk Road Economic Belt".

Remote sensing is the most time-efficient and economical method for obtaining large spatial-scale land-cover information [12,13]. With the improvement in remote-sensing image resolution, remote-sensing technology plays a vital role in forest-cover extraction and forest-change monitoring [14]. In Vietnam, researchers utilized Landsat images from 1973 to 2020 to identify changes in the spatial distribution of mangroves in Thanh Hoa and Nghe An provinces, and analyzed the reasons for these changes [15]. The researchers first determined the range of cultivated land in the Guanzhong region based on the land-use map, and then distinguished between the grain/non-grain crops based on MODIS vegetation index data. Finally, the non-grain production of cultivated land was extracted [16]. The above studies on land-cover extraction and forest-change monitoring mainly used Landsat data with high spatial resolution and MODIS data with high temporal resolution. Landsat data are limited by the temporal resolution and rainy weather, and MODIS data are limited by the spatial resolution and mixed pixels, which when used separately do not meet the requirements for fine forest-cover extraction and forest-change monitoring. To solve this problem, some scholars [17,18] implement the method for integrating different remote-sensing data to make comprehensive use of the advantages of different sensors in time and space. However, the traditional data acquisition and data processing and analysis methods based on stand-alone off-line remote-sensing analysis software (ENVI and ERDAS) have great limitations [19], which cannot meet the needs of wide-area and high spatial resolution fine forest-cover extraction and forest-change monitoring.

The rapid development of remote-sensing cloud-computing platforms has become a new way to solve geographical and spatial-related problems. Researchers can directly invoke the huge amount of data on the platform, in conjunction with the algorithms provided by it, to implement online analysis and processing, as well as visualization operations, which considerably increase the efficiency of data analysis and processing. So far, Google Earth Engine (GEE) has been used in forestry [20], agriculture [21,22], wetlands [23], disaster monitoring [24,25], and land-use change [26,27]. In terms of vegetation information extraction, Liu et al. [28] selected Landsat images and used decision-tree classification to extract the spatial distribution information of mangroves in Yuxian County in Vietnam, and monitored the inter-annual dynamic changes in mangroves in the region from 1993 to 2017. Li et al. [29] selected Landsat OLI and MODIS NDVI data and fused spectral and phonological characteristics to extract distribution information on rubber forests in Thailand, Malaysia, and Indonesia. Xu et al. [30] used the GEE platform and multi-temporal Landsat images to extract information from citrus orchards in southern Jiangxi based on the random forest (RF) method, and the average OA reached 93.15%. The above results show that with the help of GEE, there are three main features for the extraction of vegetation information by domestic and foreign scholars. First, this vegetation is the most important cash crop in the region, such as rubber and citrus, which plays an essential role in the prosperity of farmers and the economic development of the region. Second, it has a major ecological value and ecological function. Third, it grows mainly in tropical and subtropical regions, such as Southeast Asia and Yunnan, Guangxi, Guangdong, and Fujian provinces.

At present, the research on *Picea schrenkiana* mainly includes nitrogen addition [31], individual genomics [32], stoichiometric characteristics [33], population dynamics [34], and biomass estimation [35,36]. However, the spatial distribution pattern of *Picea schrenkiana* in the TS is not clear, which has become the bottleneck for the scientific management and utilization of the plant resources. Therefore, we proposed an automatic extraction procedure for the spatial distribution of *Picea schrenkiana* based on GEE and the Jeffries–Matusita (JM) distance. The key research contents are as follows: (1) Can eliminating abnormal sample data improve the accuracy of land-cover classification in the TS of CA? (2) The difference in land-cover classification accuracy in the TS of CA under different seasonal images. (3) Feasibility analysis of using the JM distance to determine the importance of classification features. (4) Eight different scenarios were designed to determine the effects of different types of classification features and the best scenario was to extract the spatial distribution of *Picea schrenkiana* with a resolution of 10 m in the TS.

2. Materials and Methods

2.1. Study Area Description

This study was conducted in the TS (69°–95° E, 39°–46° N), with a total area of about 9.97×10^5 km^2 (Figure 1). Due to its vast area and distance from the ocean, the TS is the farthest mountain system from the ocean in the world [37] and also the largest mountain system in the arid area of the world. The main topography of the TS consists of mountains, intermountain basins, and piedmont plains. The mean altitude of the ridge is about 4000 m and the highest peak has an altitude of about 7443.8 m [38]. Located in the northern temperate zone and far from the sea, the area is characterized by a temperate continental climate. The annual average temperature is about 5.0 °C [39], and the annual average precipitation ranges from 250 to 300 mm. The vegetation distribution of the TS has a typical vertical zone. From low altitude to high altitude, there are mountain steppe, meadow steppe, mountain coniferous forest, alpine steppe, alpine cushion plants, and snow-covered glaciers [33]. *Picea schrenkiana* is the dominant species in mountain conifer forests. According to the field survey data and existing literature (Table 1), the elevation range of *Picea schrenkiana* on the northern slope of the TS in CA is from 1400 to 2800 m. To ensure an accurate extraction of *Picea schrenkiana*, a 1300 m contour was drawn from ArcGIS software and SRTM data with a spatial resolution of 90 m. The field enclosed by the contour at 1300 m is defined as the region studied in this paper.

Table 1. The main elevation range of *Picea schrenkiana* on the northern slope of the Tianshan Mountains.

Elevation	Reference
1500–2700 m	Wang, Ren et al. [34]
1600–2800 m	Li, Chang et al. [33]
1600–2800 m	Lan, Xiao et al. [40]
1500–2800 m	Luo, Xu et al. [35]
1500–2800 m	Li, Luo et al. [41]
1400–2800 m	Jiang, Zhu et al. [42]

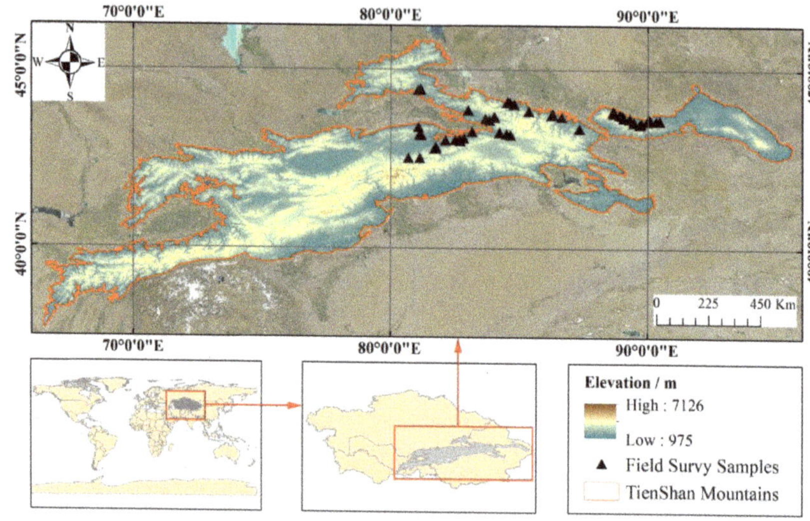

Figure 1. Location of the study area.

2.2. Remote-Sensing Data

Sentinel-1 satellites are equipped with a dual-polarization C-band synthetic aperture radar (SAR) instrument, which provides SAR images with high temporal and spatial resolution. Sentinel-1 data are interferometric wide swath mode (IW) images, which is the standard mode over land [43] and corresponds to *COPERNICUS/S1_GRD* in GEE. The Sentinel-1 Toolbox has been used to preprocess the dataset, including thermal noise removal, radiometric calibration, and terrain correction [44]. Here, we exclusively utilized the vertical transmit/horizontal receive (VH) and vertical transmit/vertical receive (VV) bands. We also used Sentinel-2 data from the GEE archive. Sentinel-2 consists of two satellites (Sentinel-2A and Sentinel-2B), each carrying a multi-spectral imager (MSI) to acquire data in 13 spectral bands along a 290 km sun-synchronous orbit path every 5 days. Since Sentinel-2 MSI, Level-2A data were not provided fully on GEE's data pool, Level-1C data were employed in this work, which corresponds to *COPERNICUS/S2* in GEE. The datasets are products of the top of the atmosphere (TOA) after radiometric calibration and terrain correction.

Due to the wide coverage of the study area, it is necessary to combine images taken at different times to obtain high-quality image data covering the study area. At the same time, considering the factors of vegetation growth seasons, the images are divided into the four seasons of spring, summer, autumn, and winter (Table 2). The *ee. Filter.calendarRange* function was used to obtain images in the time range of the four seasons. The shapefile for the study area was imported into the GEE platform and the *ee. filterBounds* filter was employed to obtain images of the region of interest. For Sentinel-2 images, *CLOUDY_PIXEL_PERCENTAGE* of the four seasons was set to different values to remove all Sentinel-2 images of poor quality and obtain the number of pixel-by-pixel high-quality observation images in the study area (Table 2). The results showed that during the study period, the number of high-quality Sentinel-2 images from the four seasons in the study area ranged from 1 to 288 (Figure 2). The *ee. Reducer.median* function was then utilized to create a single image by calculating the median of all values on each pixel across the stack of all images. Finally, the Sentinel-1 composite image and high-quality Sentinel-2 composite image covering the study area were achieved for subsequent studies.

Table 2. Sentinel-1/2 image selection schedule.

Seasons	Years	Months	Cloudy _Pixel_Percentage
Spring	2019–2020	4–6	30
Summer	2019–2020	7–9	20
Autumn	2019–2020	10–12	35
Winter	2019–2020	1–3	60

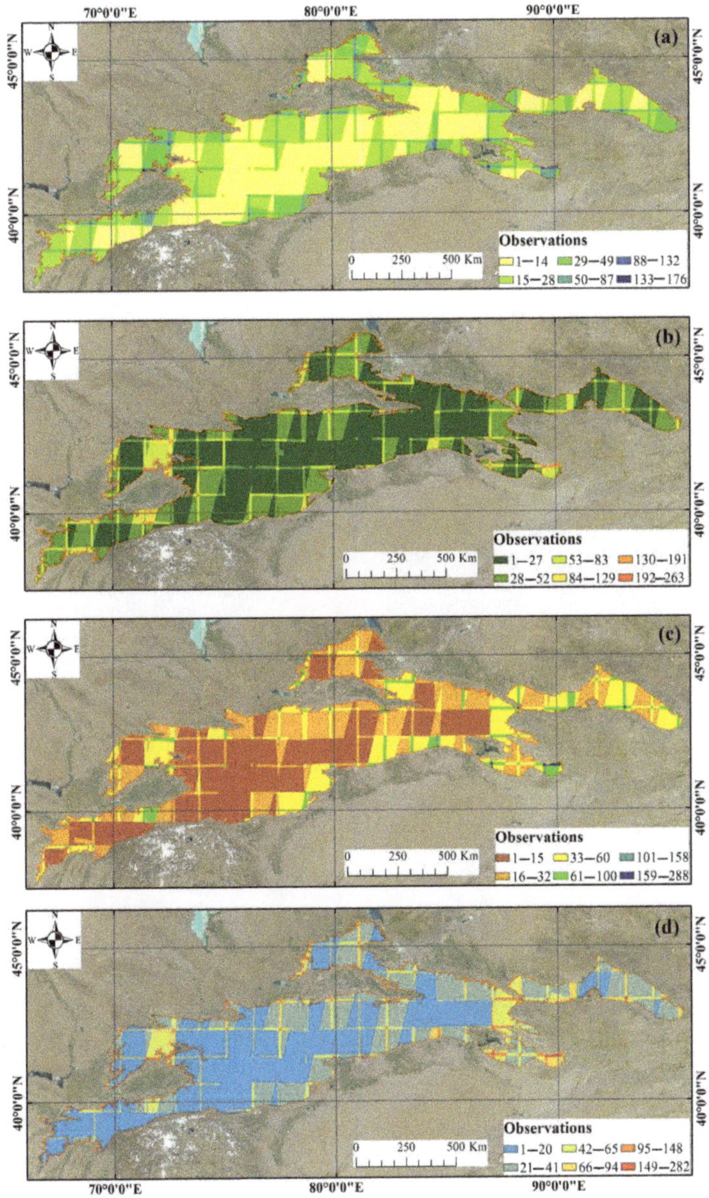

Figure 2. Statistics of Sentinel-2 valid remote-sensing data from 2019 to 2020. (**a**) Spring; (**b**) Summer; (**c**) Autumn; (**d**) Winter.

2.3. Training and Validation Sample Points

The accuracy of classification results is intimately related to the representativeness and quantity of samples [45]. Under realistic conditions for the TS and available land-cover data, the land-cover categories in this region were divided into seven types to separate *Picea schrenkiana* from the rest of the land uses. These include *Picea schrenkiana*, grassland, cropland, built, bare land, snow and ice, and waterbody. Due to the large extent of the study area, visual interpretation of a large number of sample points not only requires a lot of work, but also fails to properly distinguish between land-cover classes. To address this issue, we combined visual interpretation using Google Earth VHRI with semi-automatic extraction of sample points. On the one hand, the *Picea schrenkiana* samples were collected from VHRI on Google Earth. On the other hand, the non-*Picea schrenkiana* data (e.g., grassland, cropland, built, bare land, snow and ice, and waterbody) were collected by the semi-automatic extraction of sample points method. Specifically, semi-automatic extraction of sample points refers to the use of two tools (*Create Random Points* tool and *Extract Multi Values to Points* tool) in ArcGIS software and ESA World Cover data to obtain a large number of samples more accurately and quickly. Furthermore, the coordinates of 99 field survey *Picea schrenkiana* samples (Figure 1) were collected through GPS in the summer of 2019. Finally, a total of 8159 samples (Table 3) were collected, and 70% of these (5866) were used for model training and 30% (2293) for validation.

Table 3. The situation of Sample Collection.

Type	*Picea schrenkiana*	Grassland	Cropland	Built	Bare Land	Snow and Ice	Waterbody
Numbers	1524	3036	533	517	1704	275	570

2.4. Classification Feature Input

Seven categories of features are applied for classification in this study (Table 4). These include radar features, spectral features, red-edge features, spectral indices, texture features, and terrain features. Multiple lines of evidence showed that texture features, especially texture features based on SAR data, can avoid the phenomenon of " same spectrum foreign objects" and " same object foreign spectrum", to improve the accuracy of remote-sensing image classification [46]. A gray-level co-occurrence matrix (GLCM), proposed by Haralick in 1973 [47], is the most common statistical analysis method for calculating texture features at present. Therefore, this study used the algorithm glcmTexture provided by GEE to compute the texture metrics with a window size of 3×3 and an offset of 1×1 from the GLCM around each pixel of Sentinel-1 bands (VV and VH) [48]. Considering that *Picea schrenkiana* is a single group tree species on the north slope of the TS at an altitude of 1400–2800 m, the ee. Algorithms. Terrain and ee. Image ("USGS/SRTMGL1_003") function provided by the GEE platform was used to calculate terrain features.

Table 4. Summary of classification features applied in this research.

Categories	Features	Central Wavelength/Indices Formula	Data Source
Radar features	VV	vertical transmit/vertical receive	Sentinel-1
	VH	vertical transmit/horizontal receive	Sentinel-1
Spectral features	AEROS	443 nm	Sentinel-2
	BLUE	490 nm	Sentinel-2
	GREEN	560 nm	Sentinel-2
	RED	665 nm	Sentinel-2
	RDED1	705 nm	Sentinel-2
	RDED2	740 nm	Sentinel-2
	RDED3	783 nm	Sentinel-2
	NIR	842 nm	Sentinel-2
	RDED4	865 nm	Sentinel-2
	VAPOR	940 nm	Sentinel-2

Table 4. *Cont.*

Categories	Features	Central Wavelength/Indices Formula	Data Source
	CIRRU	1375 nm	Sentinel-2
	SWIR1	1610 nm	Sentinel-2
	SWIR2	2190 nm	Sentinel-2
Red-edge features	NDVIre1	(RDED4 − RDED1)/(RDED4 + RDED1)	Sentinel-2
	NDVIre2	(RDED4 − RDED2)/(RDED4 + RDED2)	Sentinel-2
	NDVIre3	(RDED4 − RDED3)/(RDED4 + RDED3)	Sentinel-2
	NDre1	(RDED2 − RDED1)/(RDED2 + RDED1)	Sentinel-2
	NDre2	(RDED3 − RDED1)/(RDED3 + RDED1)	Sentinel-2
Spectral indices	NDVI	(NIR − RED)/(NIR + RED)	Sentinel-2
	EVI	2.5 × (NIR − RED)/(NIR + 6.0 × RED −7.5 × BLUE + 1)	Sentinel-2
	MNDVI	(RED − GREEN)/(RED + GREEN)	Sentinel-2
	NDWI	(GREEN − NIR)/(GREEN + NIR)	Sentinel-2
	LSWI	(NIR − SWIR1)/(NIR + SWIR1)	Sentinel-2
	MNDWI	(GREEN − SWIR1)/(GREEN + SWIR1)	Sentinel-2
	NDTI	(SWIR1 − SWIR2)/(SWIR1 + SWIR2)	Sentinel-2
	NDI45	(RDED1 − RED)/(RDED1 + RED)	Sentinel-2
Texture features	VV ASM	VV Angular Second Moment	Sentinel-1
	VV CON	VV Contrast	Sentinel-1
	VV CORR	VV Correlation	Sentinel-1
	VV SVAR	VV Sum Variance	Sentinel-1
	VV ENT	VV Entropy	Sentinel-1
	VH ASM	VH Angular Second Moment	Sentinel-1
	VHCON	VH Contrast	Sentinel-1
	VH CORR	VH Correlation	Sentinel-1
	VH SVAR	VH Sum Variance	Sentinel-1
	VH ENT	VH Entropy	Sentinel-1
Terrain features	ELE	ELEVATION	SRTM
	SLO	SLOPE	SRTM
	ASP	ASPECT	SRTM
	SHA	HILL SHADE	SRTM

2.5. The Method for Eliminating Abnormal Samples

The Box-plot developed by American statisticians in 1977 [49] is a common method to check whether there are outliers in a group of data. Compared with the Grubbs method, 3σ method, and Z-score method, the Box-plot has a wider application because it does not require the data to obey normal distribution to determine outliers. The 1st quartile (Q1), 2nd quartile (Q2), and 3rd quartile (Q3) are the numbers placed at 25%, 50%, and 75% of a set of data in ascending order, respectively [50] (Figure 3). The difference between Q1 and Q3 is called the interquartile range (IQR). The criteria for identifying outliers in the Box-plot are whether the data are greater than Q3 + 1.5IQR or less than Q1 − 1.5IQR in a group of data [51]. Minimum and maximum are the minimum and maximum values except for outliers. In the process of visual interpretation, Google Earth VHRI and semi-automatic extraction of sample points were used to determine land-cover categories. It is impossible to ensure that all sample points are pure and typical, which leads to the existence of outliers in the value of the extracted classification feature. For a specific sample, if the number of outliers in the 13 extraction values based on spectral features is greater than or equal to 2, the sample is considered to be an abnormal sample. Therefore, the outliers of the extracted values are identified based on the Box-plot and then the abnormal samples are judged and eliminated. By comparing the Kappa score and OA before and after eliminating abnormal samples, the influence of abnormal samples on classification accuracy was analyzed.

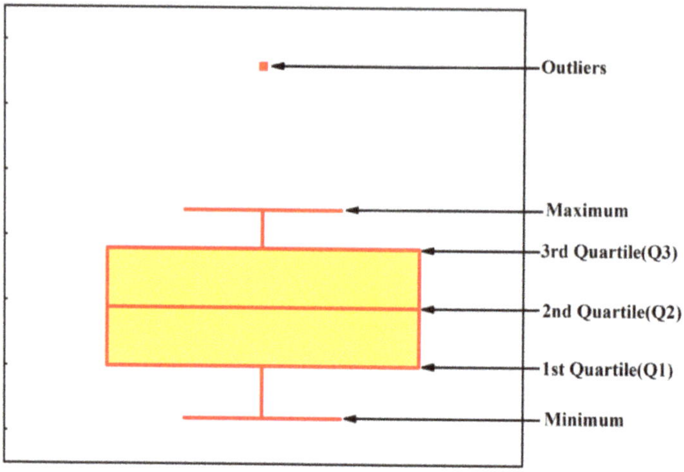

Figure 3. The structure of the Box-plot.

2.6. *The Method for Classification Feature Optimization*

Considering that some features selected in this study could not satisfy the requirement to distinguish *Picea schrenkiana* from other land-cover categories, redundant information existed in classification features [52], which affected the accuracy of mapping the spatial distribution of *Picea schrenkiana*. For this problem, the purpose of separability analysis is to evaluate the ability of each feature to distinguish *Picea schrenkiana* from other land-cover categories. There are three common methods to determine the separability between different categories: the separability index (SI) [53,54], the transformed divergence (TD) [55], and the JM distance [56,57]. Compared with the SI and TD, the JM distance can quantitatively evaluate the separability between different categories. Thus, the JM distance is used to evaluate the ability of each feature to distinguish *Picea schrenkiana* from other categories, to determine the best feature combination to extract *Picea schrenkiana*. The JM value is between 0 and 2, and its size represents the degree of separation of the seven land-cover categories under the selected features. When JM = 2, it indicates that seven types of land-cover categories are entirely separated under the selected classification features [58,59]. The expression for the JM distance is as follows:

$$\mathrm{JM} = 2(1 - \mathrm{e}^{-\mathrm{B}}) \tag{1}$$

where B represents the Bhattacharyya distance on a feature dimension. On the premise that normal distribution is satisfied, the Bhattacharyya distance between samples of different land-cover types is:

$$\mathrm{B} = \frac{1}{8}(\mathrm{e}_1 - \mathrm{e}_2)^2 \frac{2}{\delta_1^2 + \delta_2^2} + \frac{1}{2} ln\left[\frac{\delta_1^2 + \delta_2^2}{2\delta_1 \cdot \delta_2}\right] \tag{2}$$

where e_k represents the mean value of a certain type of feature; δ_k^2 represents the variance of a certain type of feature, where (k = 1, 2).

2.7. *Random Forest-Based Classifier*

RF algorithm is a non-parametric machine-learning algorithm of multiple decision-tree combinations proposed by Breiman Leo and Adele Cutler in 2001 [60]. In contrast to machine-learning algorithms such as decision trees and support vector machines, the RF algorithm is stable and effective in processing data multi-collinearity and dimensionality [61,62] without an over-fitting phenomenon [63]. Correspondingly, the RF algorithm was chosen and used in this study to map the spatial distribution of *Picea schrenkiana* in the TS of the CA.

The dominant construction procedure of the RF algorithm is as follows: based on the bootstrap sampling method, 2/3 of the samples are randomly chosen from the original sample set to construct the decision-tree model. Assuming that the features of each sample are M dimensions, m feature subsets (m << M) are randomly extracted from them, and the optimal feature is selected from these m features to split the node of the decision tree. The above procedure is iterated n times to obtain a random forest composed of n decision trees, and the optimal result is chosen from the n decision trees by voting for the classification result for each classification sample. In this study, a decision tree of size 50 was selected as optimal.

2.8. Description of Different Scenarios

Eight scenarios (Table 5) were designed to illustrate the effects of different types of classification features and the optimal classification features conducive to the extraction of *Picea schrenkiana*. We selected the best scenario and optimized the parameters of the classifier to extract the spatial distribution of *Picea schrenkiana* with a resolution of 10 m in the TS.

Table 5. Eight scenarios for mapping the spatial distribution of *Picea schrenkiana*.

Scenarios	Radar Features	Spectral Features	Red-Edge Features	Spectral Indices	Texture Features	Terrain Features	Features Optimization
S1		√					
S2		√			√		
S3		√		√			
S4		√	√				
S5	√	√					
S6		√				√	
S7							√
S8	√	√	√	√	√	√	

2.9. The Method for Accuracy Assessment

Accuracy evaluation is performed by constructing a confusion matrix based on the 2293 validation sample points and classification results from this study. The purpose of constructing the confusion matrix is to obtain accuracy assessment indexes, such as producer accuracy (PA), user accuracy (UA), the OA, and the Kappa score. The PA is the ratio of the number of pixels correctly assigned by the classifier to the actual total number of references in the category, which is used to measure the omission error of each category. The UA represents the ratio between the number of pixels correctly assigned to this category and the total number of pixels classified into this category by the classifier, which is used to measure the commission error of this category. The OA and Kappa scores are used to evaluate the overall classification accuracy of all categories. The four evaluation indices are calculated as follows:

$$\text{PA} = \frac{N_{ij}}{N_{*j}} \times 100\% \tag{3}$$

$$\text{UA} = \frac{N_{ij}}{N_{i*}} \times 100\% \tag{4}$$

$$\text{OA} = \frac{\sum_{i=1}^{q} N_{ii}}{N} \times 100\% \tag{5}$$

$$\text{Kappa} = \frac{[N \sum_{i=1}^{q} N_{ii} - \sum_{i=1}^{q} (N_{i*} * N_{*i})]}{[N^2 - \sum_{i=1}^{q} (N_{i*} * N_{*i})]} \tag{6}$$

N_{ij} is the value of row i and column j in the confusion matrix, N_{*j} is the sum of the j column of the confusion matrix, N_{i*} is the sum of row i of the confusion matrix, N is the total number of validation sample points, and q is the number of rows and columns of the confusion matrix.

3. Results

3.1. The Result of Eliminating Abnormal Samples

A total of 656 anomalous samples were selected by using the method described in Section 2.5. These include 117 *Picea schrenkiana*, 235 grassland, 56 cropland, 56 built, 139 bare lands, 24 snow and ice, and 29 waterbody. In this study, *Picea schrenkiana* was taken as an example to draw scatter plots of the extraction values of *Picea schrenkiana* samples in the spectral features (BLUE, GREEN, and RED) before and after the removal of abnormal samples. It can be seen from Figure 4 that after the removal of abnormal samples, the dispersion degree of the extraction values of *Picea schrenkiana* samples in the spectral features (BLUE, GREEN, and RED) decreased (Figure 4). In addition, the accuracy assessment index also changed. After the removal of abnormal samples, the OA and Kappa scores increased by 1.43% and 0.02, respectively, and the PA of *Picea schrenkiana* increased by 2.95%, but its UA decreased by 1.68% (Table 6).

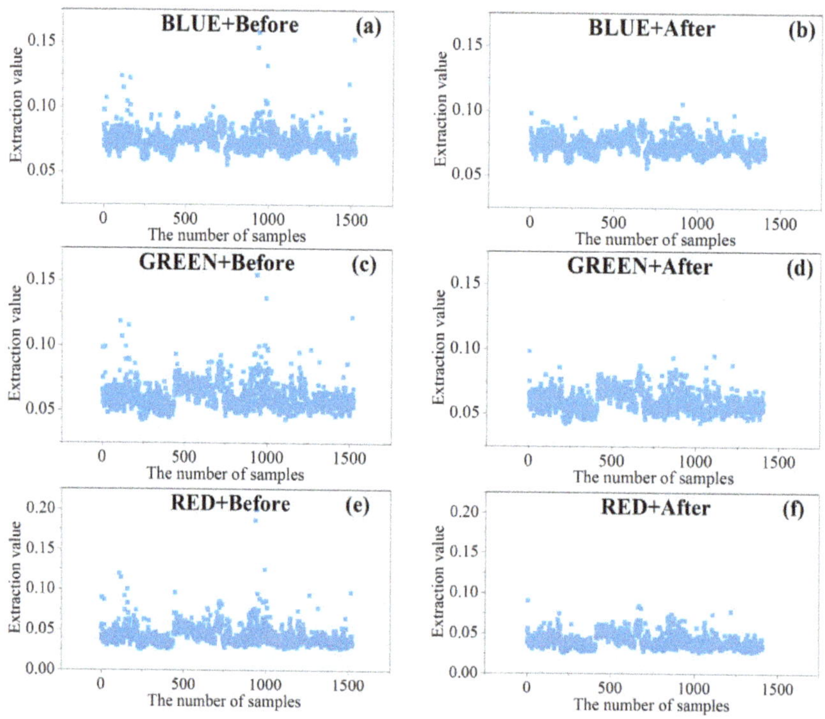

Figure 4. Scatter plot of extraction values of *Picea schrenkiana* samples in Sentinel-2 bands (Blue, Green, Red) before and after eliminating abnormal samples. (The horizontal axis represents the number of samples, the vertical axis represents the band extraction value; (**a**,**c**,**e**) represent the scatter plots of extracted values of *Picea schrenkiana* in bands BLUE, GREEN, and RED before eliminating outliers, respectively; (**b**,**d**,**f**) represent the scatter plots of extracted values of *Picea schrenkiana* in bands BLUE, GREEN, and RED after eliminating outliers, respectively).

Table 6. Classification accuracy of abnormal samples before and after removal.

Type		*Picea schrenkiana*	Grassland	Cropland	Built	Bare Land	Snow and Ice	Waterbody	OA	Kappa
Before	PA/%	93.93	89.93	66.88	81.40	85.89	70.00	91.52	87.10%	0.83
	UA/%	95.43	84.52	79.85	90.32	83.47	78.87	96.79		
After	PA/%	96.88	89.89	62.94	80.69	86.94	83.13	94.80	88.53%	0.85
	UA/%	93.75	87.65	81.08	81.82	85.78	90.79	95.35		

Note: Darker red indicates lower values, while darker green indicates higher values.

3.2. Classification Results of Single Seasonal Images

The classification accuracy of different seasonal images is different; the OA and Kappa score was higher in spring and summer than in autumn and winter (Table 7). The difference in the classification accuracy between different seasonal images is mainly caused by three land-cover categories: cropland, snow and ice, and waterbody. The main reason is that compared with other land-cover categories (*Picea schrenkiana*, grassland, built, and bare land), the seasonal variation in the three land-cover categories (cropland, snow and ice, and waterbody) is more obvious. The PA of *Picea schrenkiana* was the highest in summer and the lowest in winter (96.04% and 95.34, respectively), while the UA was the highest in spring and the lowest in winter (96.71% and 96.24%, respectively). Although the summer UA of Picea is lower than the spring UA, the summer PA of *Picea schrenkiana* increases by 0.24%–2.10% compared to the other three seasons. In addition, the OA and Kappa scores increased by 0.48%–5.28% and 0.01–0.07, respectively, in the summer compared to the other three seasons. As a result, the summer was used as the optimal sextant for mapping the spatial distribution of *Picea schrenkiana*. This conclusion is in accordance with the actual situation, as a great variety of plants are in a period of vigorous growth during the summer, and the physical properties and chlorophyll of the different plants are significantly different.

Table 7. Classification accuracy of single seasonal images.

Type		*Picea schrenkiana*	Grassland	Cropland	Built	Bare Land	Snow and Ice	Waterbody	OA	Kappa
Spring	PA/%	95.80	93.85	79.61	88.10	88.24	73.56	96.89	91.19%	0.88
	UA/%	96.71	89.91	90.98	93.28	87.88	81.01	97.50		
Summer	PA/%	96.04	94.91	74.34	89.68	87.63	85.06	96.89	91.67%	0.89
	UA/%	96.26	88.52	91.87	91.13	91.53	90.24	98.73		
Autumn	PA/%	93.94	91.48	75.00	92.06	84.79	70.11	87.58	88.36%	0.85
	UA/%	95.95	85.32	90.48	85.93	85.48	82.43	98.60		
Winter	PA/%	95.34	93.14	69.74	87.30	78.09	62.07	80.75	86.39%	0.82
	UA/%	96.24	80.97	88.33	88.71	84.06	87.10	98.48		

Note: Darker red indicates lower values, while darker green indicates higher values.

3.3. The Results of Feature Optimization

The JM distance of different classification features is considerably different (Table 8). The feature with the largest JM distance is RED, while the smallest feature is VH CORR and VV CORR. The maximum distance is approximately 50 times the minimum distance. Based on the JM distances of the 42 categorical features in order from highest to lowest, the variation in the Kappa score was observed and the best combination of categorical features for mapping the spatial distribution of *Picea schrenkiana* was determined accordingly. With the increase in the number of features involved in classification, the Kappa score in the early stage (1–7 classification features) increased considerably, rapidly increasing from 0.410 to 0.821 (Table 8). In the middle period (8–17 classification features), the increased range of the Kappa score decreased significantly, showing a steady upward trend. In the later period (18–42 classification features), the Kappa score fluctuated slightly, fluctuating around 0.88. When the number of classification features used was 32 and 42, the Kappa score was 0.894 and 0.898, respectively. Considering that the difference between them is only 0.004, the first 32 classification features are selected as the final results of the feature optimization in this study.

Table 8. The JM distance of features and the accuracy of classification results of different numbers of features.

Feature	JM Distance	Kappa	Numbers
RED	1.01	0.41	1
SWIR2	0.94	0.608	2
GREEN	0.9	0.659	3
BLUE	0.89	0.73	4
RDED1	0.87	0.764	5
CIRRU	0.82	0.793	6
SWIR1	0.81	0.821	7
MNDVI	0.72	0.821	8
AEROS	0.7	0.828	9
VH ASM	0.68	0.829	10
VH ENT	0.6	0.836	11
LSWI	0.6	0.837	12
MNDWI	0.59	0.848	13
EVI	0.58	0.856	14
VAPOR	0.58	0.858	15
RDED2	0.56	0.864	16
NDTI	0.54	0.869	17
NIR	0.52	0.869	18
VV ASM	0.52	0.862	19
RDED4	0.51	0.868	20
RDED3	0.51	0.871	21
NDre2	0.49	0.873	22
NDre1	0.49	0.869	23
NDVIre1	0.47	0.87	24
NDVI	0.47	0.866	25
VV ENT	0.46	0.863	26
VH	0.39	0.873	27
NDVIre3	0.3	0.87	28
NDI45	0.3	0.869	29
SLOPE	0.28	0.886	30
VV	0.26	0.89	31
ELEVATION	0.24	0.894	32
VV CON	0.23	0.889	33
VV SVAR	0.22	0.89	34
NDVIre2	0.18	0.884	35
ASPECT	0.13	0.894	36
VH CON	0.11	0.892	37
VH SVAR	0.1	0.894	38
NDWI	0.09	0.89	39
HILL SHADE	0.07	0.894	40
VH CORR	0.02	0.89	41
VV CORR	0.02	0.898	42

Note: Darker red indicates lower values, while darker green indicates higher values.

The JM distances of the first 32 features were classified and analyzed (Table 9). The spectral features and texture features show high separation, while the radar features and terrain features show low separation. Among the spectral features, RED, GREEN, and BLUE have a high separation. The spectral features play an important role in forest mapping and are more prominent than other features in distinguishing *Picea schrenkiana* from other land-cover categories. In addition, the texture features play a key role in mapping the spatial distribution of *Picea schrenkiana*, as *Picea schrenkiana* interleaves with other land-cover categories such as grassland and bare land in patches.

Table 9. Jeffries–Matusita distance between different types of features.

Type of Feature	The Feature with the Highest Degree of Separation (Ranking)	The Feature with the Lowest Degree of Separation (Ranking)	The Average JM Distance
Spectral features	RED (1)	RDED3 (21)	0.74
Texture features	VH ASM (10)	VV ENT (26)	0.57
Spectral indices	MNDVI (8)	NDI45 (29)	0.54
Red-edge features	NDre2 (22)	NDVIre3 (28)	0.44
Radar features	VH (27)	VV (31)	0.32
Terrain features	SLOPE (30)	ELEVATION (32)	0.26

Note: Darker red indicates lower values, while darker green indicates higher values.

3.4. Accuracy Analysis under Different Scenarios

Among the six scenarios (scenario 1- scenario 6), the PA and UA of mapping the spatial distribution of *Picea schrenkiana* were between 95.57%–97.44% and 95.40%–96.27%, respectively (Table 10). The OA of scenario 1 based on Sentinel-2 bands was 88.75%, and the Kappa score was 0.85. Based on the spectral features, the texture features (scenario 2), spectral indices (scenario 3), red-edge indices (scenario 4), radar features (scenario 5), and terrain features (scenario 6) were added, respectively, and the OA and Kappa scores were slightly increased. The OA scores increased by 0.39%, 1.09%, 0.7%, 0.61%, and 2.35%, while the Kappa scores increased by 0.01, 0.02, 0.01, 0.01, and 0.03, respectively. These results indicated that the spectral features had the best identification ability for *Picea schrenkiana*, while the other features had no obvious influence. However, the texture and terrain features still played a positive role in the land-use classification of the TS.

Table 10. Classification accuracy statistics under different scenarios.

Type		*Picea schrenkiana*	Grassland	Cropland	Built	Bare Land	Snow and Ice	Waterbody	OA	Kappa
Scenario 1	PA/%	96.27	92.90	61.84	75.4	86.00	81.61	95.03	88.75%	0.85
	UA/%	96.27	85.33	83.19	86.36	88.15	87.65	96.23		
Scenario 2	PA/%	95.57	92.9	62.5	76.19	88.03	82.76	94.41	89.14%	0.86
	UA/%	95.79	86.36	85.59	84.96	87.50	91.14	96.82		
Scenario 3	PA/%	96.50	94.08	65.79	75.40	87.63	81.61	95.03	89.84%	0.87
	UA/%	95.83	87.17	85.47	84.82	88.89	93.42	96.84		
Scenario 4	PA/%	96.74	93.73	63.82	79.37	85.80	81.61	95.03	89.45%	0.86
	UA/%	95.62	86.46	88.18	82.64	89.62	87.65	96.23		
Scenario 5	PA/%	96.74	92.54	66.45	78.57	87.02	80.46	95.03	89.36%	0.86
	UA/%	95.40	86.70	86.32	86.84	87.73	89.74	96.84		
Scenario 6	PA/%	97.44	93.73	74.34	84.92	88.24	79.31	96.27	91.10%	0.88
	UA/%	95.87	89.19	90.40	89.17	88.41	94.52	97.48		
Scenario 7	PA/%	96.74	95.50	73.68	91.27	87.63	83.91	95.65	91.93%	0.89
	UA/%	96.96	89.17	94.12	89.15	91.72	86.9	98.09		
Scenario 8	PA/%	97.20	95.62	73.03	86.51	88.03	83.91	95.65	91.84%	0.89
	UA/%	96.53	88.79	93.28	88.62	91.75	91.25	98.72		

Note: Darker red indicates lower values, while darker green indicates higher values.

Compared with the other seven scenarios, scenario 7 obtained the best classification effect after optimizing all features by using the JM distance, and the OA and Kappa scores were 91.93% and 0.89 (Table 10), respectively. Compared to scenario 8, where all features participate in the classification, the OA and Kappa scores in scenario 7 are improved by 0.09% and 0.001, respectively, and the number of features in scenario 7 are less. Therefore, scenario 7 was selected as the final scheme for mapping the spatial distribution of *Picea schrenkiana* in the TS of CA.

3.5. Spatial Distribution of *Picea schrenkiana* in the Best Scenario

By comparing and analyzing different classification scenarios in Section 3.4, scenario 7 achieves the optimal classification effect, and its confusion matrix is shown in Table 11. *Picea schrenkiana* and open water have the best mapping results, with both the PA and UA above 95%. The PA and UA of grassland, built, and bare land are both above 85%. The mapping accuracy of cropland and snow and ice is not high and the PA of cropland is only 73.68%. The spatial distribution of *Picea schrenkiana* in the optimal classification scenario (scenario 7) is shown in the figure and six places are randomly selected for detailed display. *Picea schrenkiana* is concentrated in the middle TS and scattered in the remaining areas (Figure 5). In addition, the pattern boundary of *Picea schrenkiana* based on the method presented in this paper has a strong consistency with the optical image (Figure 6).

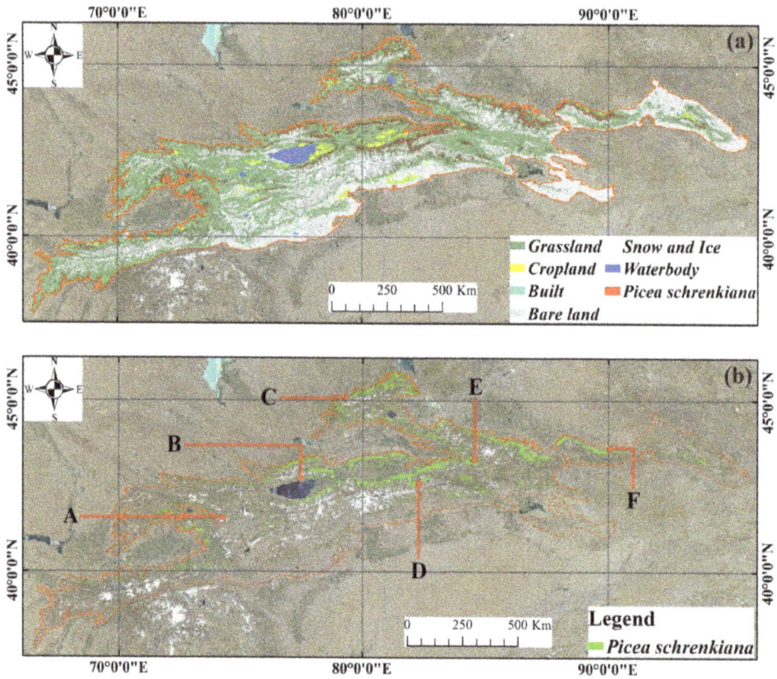

Figure 5. Land-cover classification (**a**) and spatial distribution of *Picea schrenkiana* (**b**) in the Tianshan Mountains of Central Asia. The study selected six different regions to display the extraction results of *Picea schrenkiana*, with A–F representing the positions of six different regions.

Table 11. Confusion matrix of scenario 7.

	Picea schrenkiana	Grassland	Cropland	Built	Bare Land	Snow and Ice	Waterbody
Picea schrenkiana	415	13	0	0	0	0	0
Grassland	13	807	33	4	43	0	5
Cropland	0	5	112	2	0	0	0
Built	0	2	7	115	5	0	0

Table 11. *Cont.*

	Picea schrenkiana	Grassland	Cropland	Built	Bare Land	Snow and Ice	Waterbody
Bare Land	0	18	0	5	432	14	2
Snow and Ice	0	0	0	0	11	73	0
Waterbody	1	0	0	0	2	0	154
PA/%	96.74	95.5	73.68	91.27	87.63	83.91	95.65
UA/%	96.96	89.17	94.12	89.15	91.72	86.90	98.09
	OA = 91.93%					Kappa = 0.89	

Note: Darker red indicates lower values, while darker green indicates higher values.

Figure 6. Local classification results of *Picea schrenkiana* and their corresponding Sentinel-2 images. (**A–F**) represents the extraction results of *Picea schrenkiana* in six different regions, respectively.

A total of seven intervals with a separation of 300 m were set in the elevation distribution range of *Picea schrenkiana*, and the variation characteristics of *Picea schrenkiana* with altitude were observed. The distribution area of *Picea schrenkiana* first increases and then decreases with altitude (Figure 7, Elevation). *Picea schrenkiana* occupies the largest distribution area in the altitude range of 2100 to 2400 m, accounting for 29.7%. In total, 75.7% of the *Picea schrenkiana* is distributed in a range of elevation from 1800 to 2700 m. This study shows that *Picea schrenkiana* distributes in the TS of CA with a slope range of 0–69°. A total of seven slope ranges were set at intervals of 10° to observe the change in *Picea schrenkiana* distribution. In total, 33.9% of the *Picea schrenkiana* was distributed in the slope range of 30–40°, accounting for the largest proportion of all the intervals, and 66.4% of the *Picea schrenkiana* was distributed in the slope range of 20–40° (Figure 7, Slope). According to the classification method for aspects in forestry investigation, four aspects (semi-shady slope, shady slope, semi-sunny slope, and sunny slope) are set. Only 14.1% of the *Picea schrenkiana* were distributed in sunny and semi-sunny slopes, while 85.9% of the *Picea schrenkiana* were distributed in shady and semi-shady slopes (Figure 7, Aspect).

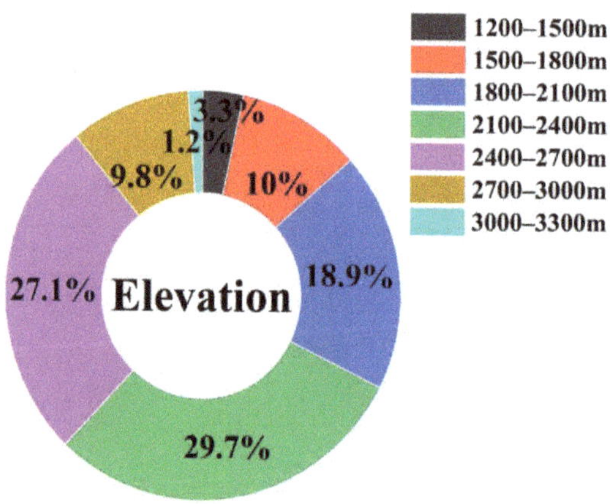

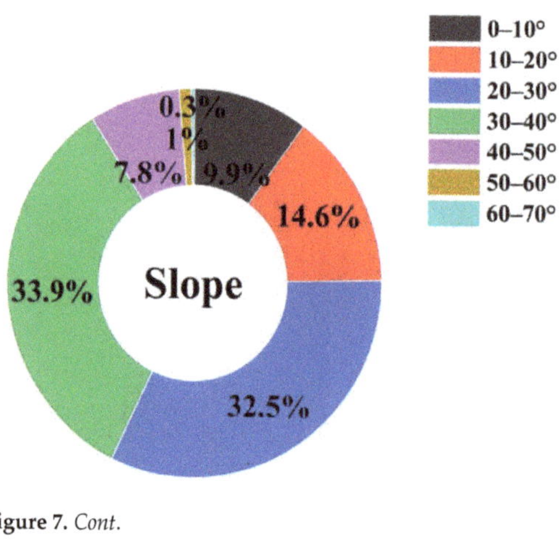

Figure 7. *Cont.*

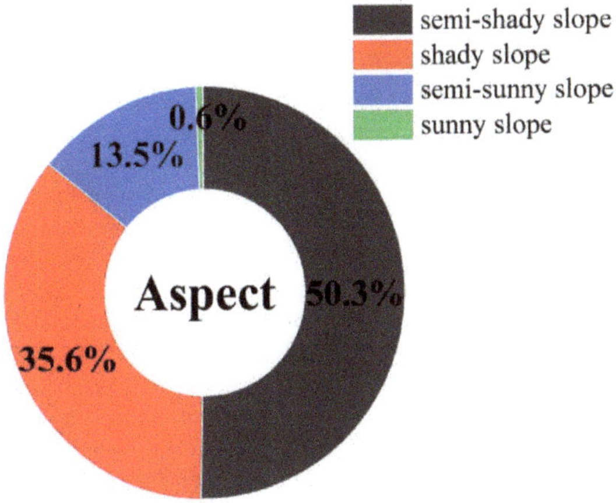

Figure 7. Distribution of *Picea schrenkiana* in different elevations, slopes, and aspect.

4. Discussion

4.1. The Influence of Feature Selection on Classification Results

In this study, the JM distance was used to explore the best classification features of land-use classification and map the spatial distribution of *Picea schrenkiana* in the TS. SWIR2, GREEN, BIUE, RDED1, CIRRU, SWIR1, MNDVI, AEROS, VH ASM, VH ENT, LSWI, MNDWI, EVI, and VAPOR were the top classification features in a sequence of importance (Table 8). Jiang et al. [42] selected an area of 10,000 km^2 in the TS for the remote-sensing identification of mountain coniferous forest and screened the ELEVATION, NDVI, spectral slope of red to near-infrared band, BLUE, RED, SWIR1. SLOPE is the most important classification feature of mountainous coniferous forest in the TS. SWIR1, RED, BLUE, SWIR1, and MNDVI (NDVI) in this study were consistent with Jiang's research conclusion. The important reasons for BLUE and SWIR1 are that the bare land of the study area is large and the two classification features are sensitive to the soil background [64]. RED and NDVI are sensitive to vegetation and play a significant role in distinguishing vegetation from other land covers. ELEVATION and SLOPE, two categorical features, are of low importance in this study, which is inconsistent with the conclusions of Jiang et al. The main reason is that this study focuses on the importance of different features for mapping the spatial distribution of *Picea schrenkiana*. Therefore, only the JM distance between *Picea schrenkiana* and other land-cover types (grassland, cropland, built, bare land, snow and ice, and waterbody) was calculated. The distance between each of the two categories was not calculated. The immediate consequence of this practice would be to obscure the topographic features of *Picea schrenkiana*'s distribution in the TS. In addition, the JM distance algorithm only evaluates the classification features based on the separation degree between different categories and does not consider the correlation between different features [65]. As a result, the classification features selected based on the JM distance selection may still have strong correlations, affecting the classification accuracy. At the same time, VH ASM and VH ENT are also of high importance in this study, because VH can obtain more vegetation structure information than VV [66].

4.2. The Spatial Distribution of Picea schrenkiana

The statistical results of this study showed that *Picea schrenkiana* was concentrated in the middle TS and scattered in other areas (Figure 5). Topographic factors such as elevation, slope, and aspect affect the spatial distribution of vegetation through the redistribution of vegetation habitat elements such as heat and water. The biomass of *Picea schrenkiana* showed a unimodal trend with the increase in altitude [35,67], and the biomass of *Picea schrenkiana* reached its peak at an altitude of 2100–2400 m, indicating that the altitude range was the most suitable for *Picea schrenkiana* to grow. The results of this study also show that the distribution area of *Picea schrenkiana* first increases and then decreases with the increase in altitude. The distribution area of *Picea schrenkiana* occupies the largest proportion in the altitude range of 2100 to 2400 m, accounting for 29.7% (Figure 7, Elevation). This is consistent with previous findings. *Picea schrenkiana* occupies the largest distribution area of 33.9% in the slope range of 30–40°, and 66.4% of the *Picea schrenkiana* is distributed in the slope range of 20–40° (Figure 7, Slope). In addition, 85.9% of the *Picea schrenkiana* is distributed in the shade slope and semi-shade slope (Figure 7, Aspect). These are consistent with the descriptions in the literature [68].

To sum up, although this paper has achieved good results in mapping the spatial distribution of *Picea schrenkiana* in the TS of CA (Figure 5), the impact of shadows was a major challenge for this work. In this area, the effect of removing shadows by terrain correction is not obvious, and the method for identifying the actual object type in the shadow according to the difference in the distribution position of shadows on the two images will have a large workload [42]. How to avoid the influence of shadows and choose the optimal feature selection algorithm [65] to more accurately identify the spatial distribution pattern of *Picea schrenkiana* in the TS is a problem worthy of attention in future studies.

5. Conclusions

The advantage of this work is being able to map the spatial distribution of *Picea schrenkiana* with 10 m resolution in the entire TS (9.97×10^5 km^2) by using GEE, the JM distance, the latest available satellite sensors, and a large number of training samples, overcoming the problem of insufficient computing capability for the large-scale monitoring of *Picea schrenkiana*. By using the Box-plot method to eliminate abnormal samples, the PA of *Picea schrenkiana* is increased by 2.95%, and the OA and Kappa scores of land-cover classification in the TS of CA are increased by 1.43% and 0.02. Compared to the other three seasons, the PA of *Picea schrenkiana* in summer increased by 0.24%–2.10%, the OA and Kappa scores of land-cover classification in the TS of CA by 0.48%–5.28% and 0.01–0.07. The separation obtained by the JM distance showed that spectral features and texture features played a key role in the mapping of *Picea schrenkiana*. The JM distance can seize the classification features that are most conducive to the mapping of *Picea schrenkiana*. The best classification features are RED, SWIR2, GREEN, BIUE, RDED1, CIRRU, SWIR1, MNDVI, AEROS, VH ASM, VH ENT, LSWI, MNDWI, EVI, VAPOR, RDED2, NDTI, NIR, VV ASM, RDED4, RDED3, NDre2, NDre1, NDVIre1, NDVI, VV ENT, VH, NDVIre3, NDI45, SLOPE, VV, and ELEVATION. *Picea schrenkiana* has a PA of 96.74%, and OA and Kappa scores of 91.93% and 0.89, respectively, for land-cover classification in the TS. *Picea schrenkiana* is concentrated in the middle TS and scattered in the remaining areas. In total, 85.7% of the *Picea schrenkiana* was distributed in the altitude range of 1500–2700 m, 66.4% of the *Picea schrenkiana* was distributed in the slope range of 20–40°, and 85.9% of the *Picea schrenkiana* was distributed in shady and semi-shady slopes.

Author Contributions: Conceptualization, Z.X. and C.X.; methodology, Z.X. and C.X.; software, F.X. and T.Y.; validation, F.X.; formal analysis, F.X. and Z.X.; investigation, Z.X., C.X., F.X. and T.Y.; resources, Z.X.; data curation, Z.X., C.X., F.X. and T.Y.; writing—original draft preparation, F.X.; writing—review and editing, Z.X., C.X. and F.X.; visualization, F.X.; supervision, Z.X.; project administration, Z.X.; funding acquisition, Z.X. and C.X. All authors have read and agreed to the published version of the manuscript.

Funding: This research was funded by the National Natural Science Foundation of China (42067062) and the Natural Science Foundation of Xinjiang Uygur Autonomous Region (2022D01B234, 2022D01C399, and 2022E01052).

Data Availability Statement: The data presented in this study are available on request from the corresponding author.

Conflicts of Interest: The authors declare no conflict of interest.

References

1. Woodwell, G.M.; Whitaker, R.H.; Reiners, W.A.; Likens, G.E.; Botkin, D.B. The Biota and the World Carbon Budget: The terrestrial biomass appears to be a net source of carbon dioxide for the atmosphere. *Science* **1978**, *199*, 141–146. [CrossRef] [PubMed]
2. Liu, W.; Wang, X.; Lu, F.; Ouyang, Z. Regional and global estimates of carbon stocks and carbon sequestration capacity in forest ecosystems: A review. *Chin. J. Appl. Ecol.* **2015**, *26*, 2881–2890.
3. Prichard, S.J.; Stevens-Rumann, C.S.; Hessburg, P.F. Tamm Review: Shifting global fire regimes: Lessons from reburns and research needs. *For. Ecol. Manag.* **2017**, *396*, 217–233. [CrossRef]
4. Pritchard, H.D. Asia's glaciers are a regionally important buffer against drought. *Nature* **2017**, *545*, 169–174. [CrossRef] [PubMed]
5. Xu, W.; Yang, L.; Chen, X.; Gao, Y.; Wang, L. Carbon storage, spatial distribution and the influence factors in Tianshan forests. *Chin. J. Plant Ecol.* **2016**, *40*, 364–373.
6. Zhang, B.; Zhou, C.; Chen, S. The Geo-info-spectrum of Montane Altitudinal Belts in China. *Acta Geogr. Sin.* **2003**, *58*, 163–171.
7. Li, C.; Zhang, C.; Luo, G.; Chen, X.; Maisupova, B.; Madaminov, A.A.; Han, Q.; Djenbaev, B.M. Carbon stock and its responses to climate change in Central Asia. *Glob. Chang. Biol.* **2015**, *21*, 1951–1967. [CrossRef]
8. Wang, T.; Ren, G.; Chen, F.; Yuan, Y. An analysis of precipitation variations in the west-central Tianshan Mountains over the last 300 years. *Quat. Int.* **2015**, *358*, 48–57. [CrossRef]
9. Li, Z.; Chen, Y.; Fang, G.; Li, Y. Multivariate assessment and attribution of droughts in Central Asia. *Sci. Rep.* **2017**, *7*, 1316. [CrossRef]
10. Li, Z.; Chen, Y.; Li, W.; Deng, H.; Fang, G. Potential impacts of climate change on vegetation dynamics in Central Asia. *J. Geophys. Res. Atmos.* **2015**, *120*, 12345–12356. [CrossRef]
11. Jiao, L.; Chen, K.; Wang, S.-J.; Liu, X.-P. Stability evaluation of radial growth of *Picea schrenkiana* in different age groups in response to climate change in the eastern Tianshan Mountains. *J. Mt. Sci.* **2020**, *17*, 1735–1748. [CrossRef]
12. Gong, P. Remote sensing of environmental change over China: A review. *Chin. Sci. Bull.* **2012**, *57*, 2793–2801. [CrossRef]
13. Weiss, M.; Jacob, F.; Duveiller, G. Remote sensing for agricultural applications: A meta-review. *Remote Sens. Environ.* **2020**, *236*, 111402. [CrossRef]
14. Potapov, P.V.; Turubanova, S.A.; Hansen, M.C.; Adusei, B.; Broich, M.; Altstatt, A.; Mane, L.; Justice, C.O. Quantifying forest cover loss in Democratic Republic of the Congo, 2000–2010, with Landsat ETM+ data. *Remote Sens. Environ.* **2012**, *122*, 106–116. [CrossRef]
15. Hao, S.; Chen, Y.; Liu, H.; Zhu, X.; Dawa, Z.; Li, W. Object-oriented Forest Classification of Linzhi County based on CART Decision Tree with Texture Information. *Remote Sens. Technol. Appl.* **2017**, *32*, 386–394.
16. Zhang, D.; Yang, W.; Kang, D.; Zhang, H. Spatial-temporal characteristics and policy implication for non-grain production of cultivated land in Guanzhong Region. *Land Use Policy* **2023**, *125*, 106466. [CrossRef]
17. Feng, G.; Masek, J.; Schwaller, M.; Hall, F. On the blending of the Landsat and MODIS surface reflectance: Predicting daily Landsat surface reflectance. *IEEE Trans. Geosci. Remote Sens.* **2006**, *44*, 2207–2218. [CrossRef]
18. Walker, J.J.; de Beurs, K.M.; Wynne, R.H.; Gao, F. Evaluation of Landsat and MODIS data fusion products for analysis of dryland forest phenology. *Remote Sens. Environ.* **2012**, *117*, 381–393. [CrossRef]
19. Baumann, P.; Mazzetti, P.; Ungar, J.; Barbera, R.; Wagner, S. Big Data Analytics for Earth Sciences: The EarthServer approach. *Int. J. Digit. Earth* **2015**, *9*, 3–29. [CrossRef]
20. Liu, C.; Xiong, T.; Gong, P.; Qi, S. Improving large-scale moso bamboo mapping based on dense Landsat time series and auxiliary data: A case study in Fujian Province, China. *Remote Sens. Lett.* **2017**, *9*, 1–10. [CrossRef]
21. Wang, S.; Di Tommaso, S.; Deines, J.M.; Lobell, D.B. Mapping twenty years of corn and soybean across the US Midwest using the Landsat archive. *Sci. Data* **2020**, *7*, 307. [CrossRef]
22. Tan, S.; Wu, B.; Zhang, X. Mapping paddy rice in the Hainan Province using both Google Earth Engine and remote sensing images. *J. Geo-Inf. Sci.* **2019**, *21*, 937–947.
23. Amani, M.; Mahdavi, S.; Afshar, M.; Brisco, B.; Huang, W.; Mohammad Javad Mirzadeh, S.; White, L.; Banks, S.; Montgomery, J.; Hopkinson, C. Canadian Wetland Inventory using Google Earth Engine: The First Map and Preliminary Results. *Remote Sens.* **2019**, *11*, 842. [CrossRef]
24. Venkatappa, M.; Sasaki, N.; Han, P.; Abe, I. Impacts of droughts and floods on croplands and crop production in Southeast Asia—An application of Google Earth Engine. *Sci. Total Environ.* **2021**, *795*, 148829. [CrossRef] [PubMed]
25. Guo, S.; Du, P.; Meng, Y.; Wang, X.; Tang, P.; Lin, C.; Xia, J. Dynamic monitoring on flooding situation in the Middle and Lower Reaches of the Yangtze River Region using Sentinel-1A time series. *Natl. Remote Sens. Bull.* **2021**, *25*, 2127–2141. [CrossRef]
26. Zhang, X.; Liu, L.; Chen, X.; Gao, Y.; Jiang, M. Automatically Monitoring Impervious Surfaces Using Spectral Generalization and Time Series Landsat Imagery from 1985 to 2020 in the Yangtze River Delta. *J. Remote Sens.* **2021**, *2021*, 9873816. [CrossRef]
27. Yang, L.; Zhang, J.; Gong, E.; Liu, M.; Ren, J.; Wang, Y. Analysis of spatio-temporal land-use patterns and the driving forces in Xi'an City using GEE and multi-source data. *Trans. Chin. Soc. Agric. Eng.* **2022**, *38*, 279–288.
28. Liu, K.; Peng, L.; LI, X.; Tan, M.; Wang, S. Monitoring the Inter-annual Change of Mangroves based on the Google Earth Engine. *J. Geo-Inf. Sci.* **2019**, *21*, 731–739.
29. Li, W.; Zhang, J.; Liu, S.; Chen, X.; Zou, H.; Bai, R.; Lyu, R. GEE-based Extraction of Rubber Forest Distribution in Main Producing Areas of Southeast Asia. *Chin. J. Trop. Crops* **2021**, *42*, 3345–3350.

30. Xu, H.; Liu, C.; Wang, J.; QI, S. Study on extraction of citrus orchard in Gannan region based on google earth engine platform. *J. Geo-Inf. Sci.* **2018**, *20*, 396–404.
31. Zhang, H.; Cong, L.; Liu, X.; Shao, K.; Li, X.-Z.; Li, R.-X. Soil Enzyme Activity in *Picea schrenkiana* and Its Relationship with Environmental Factors in the Tianshan Mountains, Xinjiang. *Huanjing Kexue* **2021**, *42*, 403–410. [PubMed]
32. Li, D.-L.; Yang, Y.; Yang, S.; Chen, Y.-K. The complete chloroplast genome sequence of *Keteleeria hainanensis* (Pinaceae). *Mitochondrial DNA Part B* **2019**, *4*, 2934–2935. [CrossRef]
33. Li, L.; Chang, Y.; Xu, Z. Stoichiometric characteristics of *Picea schrenkiana* forests with a hydrothermal gradient and their correlation with soil physicochemical factors on Tianshan Mountain. *Acta Ecol. Sin.* **2018**, *38*, 8139–8148.
34. Wang, T.; Ren, H.; Ma, K. A preliminary study on *Picea schrenkiana* population dynamics in the central Tianshan Mountains, northwestern China. *Ecol. Environ.* **2006**, *3*, 564–571.
35. Luo, Q.; Xu, Z.; Xu, Z. Estimation and spatial pattern analysis of biomass of *Picea schrenkiana* forests. *Acta Ecol. Sin.* **2020**, *40*, 5288–5297.
36. Ding, X.; Xu, Z.; Wang, Y. Application of MaxEnt Model in Biomass Estimation: An Example of Spruce Forest in the Tianshan Mountains of the Central-Western Part of Xinjiang, China. *Forests* **2023**, *14*, 953. [CrossRef]
37. Chen, Y.; Li, Z.; Fang, G.; Deng, H. Impact of climate change on water resources in the Tianshan Mountians, Central Asia. *Acta Geogr. Sin.* **2017**, *72*, 18–26.
38. Cai, X.; Li, Z.; Zhang, H.; Xu, C. Vulnerability of glacier change in Chinese Tianshan Mountains. *Acta Geogr. Sin.* **2021**, *76*, 2253–2268.
39. Liu, C.; Ding, L. A Primary Calculation of Temperature and Precipitation in Tianshan Mountains, China. *J. Glaciol. Geocryol.* **1988**, *10*, 151–159.
40. Lan, J.; Xiao, Z.; Li, J.; Zhang, Y. Biomass allocation and allometric growth of *Picea schrenkiana* in Tianshan Mountains. *J. Zhejiang AF Univ.* **2020**, *37*, 416–423.
41. Li, Z.; Luo, Q.; Xu, Z. Effects of stand density on the biomass allocation and tree height-diameter allometric growth of *Picea schrenkiana* forest on the northern slope of the western TS. *Arid Zone Res.* **2021**, *38*, 545–552.
42. Jiang, J.; Zhu, W.; Qiao, K.; Jiang, Y. An Identification Method for Mountains Coniferous in Tian- shan with Sentinel-2 Data. *Remote Sens. Technol. Appl.* **2021**, *36*, 847–856.
43. Torres, R.; Snoeij, P.; Geudtner, D.; Bibby, D.; Davidson, M.; Attema, E.; Potin, P.; Rommen, B.; Floury, N.; Brown, M.; et al. GMES Sentinel-1 mission. *Remote Sens. Environ.* **2012**, *120*, 9–24. [CrossRef]
44. Gulácsi, A.; Kovács, F. Sentinel-1-Imagery-Based High-Resolution Water Cover Detection on Wetlands, Aided by Google Earth Engine. *Remote Sens.* **2020**, *12*, 1614. [CrossRef]
45. Zhang, L.; Gong, Z.; Wang, Q.; Jin, D.; Wang, X. Wetland mapping of Yellow River Delta wetlands based on multi-feature optimization of Sentinel-2 images. *J. Remote Sens.* **2019**, *23*, 313–326.
46. Erasmi, S.; Twele, A. Regional land cover mapping in the humid tropics using combined optical and SAR satellite data—A case study from Central Sulawesi, Indonesia. *Int. J. Remote Sens.* **2009**, *30*, 2465–2478. [CrossRef]
47. Haralick, R.M. Textural features for image classification. *IEEE Trans.Syst. Man Cybern.* **1973**, *3*, 610–621. [CrossRef]
48. Stromann, O.; Nascetti, A.; Yousif, O.; Ban, Y. Dimensionality Reduction and Feature Selection for Object-Based Land Cover Classification based on Sentinel-1 and Sentinel-2 Time Series Using Google Earth Engine. *Remote Sens.* **2019**, *12*, 76. [CrossRef]
49. David, F.N.; Tukey, J.W. Exploratory Data Analysis. *Biometrics* **1977**, *33*, 768. [CrossRef]
50. Frigge, M.; Hoaglin, D.C.; Iglewicz, B. Some Implementations of the Boxplot. *Am. Stat.* **1989**, *43*, 50–54.
51. Grönroos, C. The perceived service quality concept—A mistake? *Manag. Serv. Qual.* **2001**, *11*, 150–152. [CrossRef]
52. Bellman, R. Dynamic Programming. *Science* **1966**, *153*, 34–37. [CrossRef]
53. Somers, B.; Asner, G.P. Multi-temporal hyperspectral mixture analysis and feature selection for invasive species mapping in rainforests. *Remote Sens. Environ.* **2013**, *136*, 14–27. [CrossRef]
54. Hu, Q.; Sulla-Menashe, D.; Xu, B.; Yin, H.; Tang, H.; Yang, P.; Wu, W. A phenology-based spectral and temporal feature selection method for crop mapping from satellite time series. *Int. J. Appl. Earth Obs. Geoinf.* **2019**, *80*, 218–229. [CrossRef]
55. Chen, L.; Jin, Z.; Michishita, R.; Cai, J.; Yue, T.; Chen, B.; Xu, B. Dynamic monitoring of wetland cover changes using time-series remote sensing imagery. *Ecol. Inform.* **2014**, *24*, 17–26. [CrossRef]
56. Wardlow, B.D.; Egbert, S.L.; Kastens, J.H. Analysis of time-series MODIS 250 m vegetation index data for crop classification in the U.S. Central Great Plains. *Remote Sens. Environ.* **2007**, *108*, 290–310. [CrossRef]
57. Arvor, D.; Jonathan, M.; Meirelles, M.S.P.; Dubreuil, V.; Durieux, L. Classification of MODIS EVI time series for crop mapping in the state of Mato Grosso, Brazil. *Int. J. Remote Sens.* **2011**, *32*, 7847–7871. [CrossRef]
58. Vanniel, T.; McVicar, T.; Datt, B. On the relationship between training sample size and data dimensionality: Monte Carlo analysis of broadband multi-temporal classification. *Remote Sens. Environ.* **2005**, *98*, 468–480. [CrossRef]
59. Murakami, T.; Ogawa, S.; Ishitsuka, N.; Kumagai, K.; Saito, G. Crop discrimination with multitemporal SPOT/HRV data in the Saga Plains, Japan. *Int. J. Remote Sens.* **2010**, *22*, 1335–1348. [CrossRef]
60. Breiman, L. Random Forests. *Mach. Learn.* **2001**, *45*, 5–32. [CrossRef]
61. Rodriguez-Galiano, V.F.; Ghimire, B.; Rogan, J.; Chica-Olmo, M.; Rigol-Sanchez, J.P. An assessment of the effectiveness of a random forest classifier for land-cover classification. *ISPRS J. Photogramm. Remote Sens.* **2012**, *67*, 93–104. [CrossRef]

62. Belgiu, M.; Drăguţ, L. Random forest in remote sensing: A review of applications and future directions. *ISPRS J. Photogramm. Remote Sens.* **2016**, *114*, 24–31. [CrossRef]
63. Teluguntla, P.; Thenkabail, P.S.; Oliphant, A.; Xiong, J.; Gumma, M.K.; Congalton, R.G.; Yadav, K.; Huete, A. A 30-m landsat-derived cropland extent product of Australia and China using random forest machine learning algorithm on Google Earth Engine cloud computing platform. *ISPRS J. Photogramm. Remote Sens.* **2018**, *144*, 325–340. [CrossRef]
64. Tateishi, T.P.R.; Ishiyama, T.; Honda, Y. Relationships between percent vegetation cover and vegetation indices. *Int. J. Remote Sens.* **1998**, *19*, 3519–3535.
65. Zhu, M.; Li, G.; Peng, Z. Remote sensing crop classification method based on feature selection. *Sci. Surv. Mapp.* **2022**, *47*, 122–128.
66. Ning, X.; Chang, W.; Wang, H.; Zhang, H.; Zhu, Q. Extraction of marsh wetland in Heilongjiang Basin based on GEE and multi-source remote sensing data. *Natl. Remote Sens. Bull.* **2022**, *26*, 386–396. [CrossRef]
67. Lei, Z.; Pengtao, Y.; Yanhui, W.; Shunli, W.; Xiande, L. Biomass Change of Middle Aged Forest of Qinghai Spruce along an Altitudinal Gradient on the North Slope of Qilian Mountains. *Sci. Silvae Sin.* **2015**, *51*, 4–10.
68. Wang, X. Picea schrenkiana survey on the north slope of Tianshan Mountains. *Xinjiang Agric. Sci.* **1963**, *7*, 285–289.

forests

MDPI

Article

Forest Segmentation with Spatial Pyramid Pooling Modules: A Surveillance System Based on Satellite Images

Fung Xin Ru [1], Mohd Asyraf Zulkifley [1,*], Siti Raihanah Abdani [2] and Martin Spraggon [3]

[1] Department of Electrical, Electronic and Systems Engineering, Faculty of Engineering and Built Environment, Universiti Kebangsaan Malaysia, Bangi 43600, Selangor, Malaysia
[2] School of Computing Sciences, College of Computing, Informatics and Media, Universiti Teknologi MARA, Shah Alam 40450, Selangor, Malaysia
[3] Rabdan Academy, 65, Al Inshirah, Al Sa'adah, Abu Dhabi P.O. Box 22401, United Arab Emirates
* Correspondence: asyraf.zulkifley@ukm.edu.my

Abstract: The global deforestation rate continues to worsen each year, and will eventually lead to various negative consequences for humans and the environment. It is essential to develop an effective forest monitoring system to detect any changes in forest areas, in particular, by monitoring the progress of forest conservation efforts. In general, changes in forest status are difficult to annotate manually, whereby the boundaries can be small in size or hard to discern, especially in areas that are bordering residential areas. The previously implemented forest monitoring systems were ineffective due to their use of low-resolution satellite images and the inefficiency of drone-based data that offer a limited field of view. Most government agencies also still rely on manual annotation, which makes the monitoring process time-consuming, tedious, and expensive. Therefore, the goal of this study is to overcome these issues by developing a forest monitoring system that relies on a robust deep semantic segmentation network that is capable of discerning forest boundaries automatically, so that any changes over the years can be tracked. The backbone of this system is based on satellite imaging supplied to a modified U-Net deep architecture to incorporate multi-scale modules to deliver the semantic segmentation output. A dataset of 6048 Landsat-8 satellite sub-images that were taken from eight land parcels of forest areas was collected and annotated, and then further divided into training and testing datasets. The novelty of this system is the optimal integration of the spatial pyramid pooling (SPP) mechanism into the base model, which allows the model to effectively segment forest areas regardless of their varying sizes, patterns, and colors. To investigate the impact of SPP on the forest segmentation system, a set of experiments was conducted by integrating several variants of SPP ranging from two to four parallel paths with different combinations of pooling kernel size, placed at the bottleneck layer of the U-Net model. The results demonstrated the effectiveness of the SPP module in improving the performance of the forest segmentation system by 2.57%, 6.74%, and 7.75% in accuracy (*acc*), intersection over union (*IoU*), and F1-score ($F1_{score}$), respectively. The best SPP variant consists of four parallel paths with a combination of pooling kernel sizes of 2×2, 4×4, 6×6, and 8×8 pixels that produced the highest *acc*, *IoU*, and $F1_{score}$ of 86.71%, 75.59%, and 82.88%, respectively. As a result, the multi-scale module improved the proposed forest segmentation system, making it a highly useful system for government and private agencies in tracking any changes in forest areas.

Keywords: automated forest monitoring; machine learning; deep learning; convolutional neural network; semantic segmentation

1. Introduction

Forests are vital natural resources for sustaining life on Earth, even though they cover only 38% of the land surface [1]. Generally, forests serve as water catchment areas that reduce the likelihood of natural disasters, such as floods, landslides, and droughts. Forests absorb carbon dioxide from the atmosphere, which lessens the greenhouse effect, and they also supply oxygen through an effective process of photosynthesis. To ensure the sustainability of forests, various government and non-government organizations have implemented a variety of strategic actions to preserve and conserve them. However, the reduction rate of forests continues to be a worrying issue, mainly due to uncontrollable global deforestation. In 2020 alone, 11,088 km^2 of forest was cut down in Brazil [2]. Furthermore, the average global deforestation rate for primary forests from 2016 to 2020 reached 4.55 mega hectares, implying that a forest area the size of Switzerland is being destroyed each year [3].

Wildfires, forestry exploitation, and illegal deforestation are the three major contributors to deforestation. Normally, wildfires are considered unavoidable disasters, especially during summer and drought seasons. The exploitation of forests, on the other hand, is usually undertaken for agriculture, logging, mining, and infrastructure construction, such as power plants, dams, and roads [4]. Although deforestation activities appear to be carried out for the benefit of mankind, especially for urbanization and economic development, they have a direct negative impact on the environment. These activities also destroy vast natural habitats, contribute to the extinction of some animal species, and disrupt the ecosystem balance. The loss of the forest canopy on the Earth's surface will also raise global temperatures and cause soil erosion during rainy seasons [5]. Generally, roots from the plants and trees in the forest will trap the soil; without them, precipitation absorption will be less efficient during hydrological cycles.

However, controlled deforestation activities, including sustainable forest management that has been implemented in Sweden, have proven to alleviate this problem through an efficient reforestation policy, whereby every harvested tree must be replaced by two or more trees [6]. If deforestation activities, especially the illegal ones, are left without any restorative efforts, forest areas will surely continue to shrink. According to Koh et al. [7], 6.3 million hectares of Malaysia's forest has been legally and illegally exploited since 1957. Generally, the suspects were accused of illegally removing logs from the forest and selling them for a quick profit, which clearly shows the ineffectiveness of the deployed forest monitoring system in tracking illegal deforestation activities.

In order to address this issue, a high-accuracy and reliable monitoring system for detecting changes in forest areas must be developed. Over the last few decades, this topic has prompted numerous studies in the field of forest mapping that have focused on the use of automated detection to classify forest and non-forest areas. Drones and satellite imagery are two commonly used modalities in the development of the monitoring systems, but both have limitations [8]. The use of drones involves relatively higher operating costs with a limited field of view, which makes data acquisition a more time-consuming process compared to satellite imagery [9]. On the other hand, the resolution of satellite images and their view angles are more limited compared to drone imaging. According to the "garbage in, garbage out" (GIGO) theory of Kilkenny and Robinson [10], low-resolution and cloud-induced satellite images make monitoring less effective. For this study, we propose to develop our own datasets of forest segmentation tasks by using a high-resolution format of Landsat-8 satellite images for fixed periodical years. Some of the forest areas being considered have different color patterns and scales, making the task of automated forest segmentation more difficult. As a result, this study aims to develop a multi-scale embedded deep-learning-based forest segmentation system that automatically maps forest areas over the years. The main novelty of this study is the optimal design of a multi-scale module by using the spatial pyramid pooling approach, embedded into the U-Net model. This study aims to identify the best network configurations that include multi-scale module placement, number of parallel paths, and the best multi-scale kernel set.

In this work, U-Net is chosen to be the base architecture because of its ability to perform well in semantic segmentation tasks, even without embedding any multi-scale module. Therefore, this work aims to embed the multi-scale capability to the base network by adding a spatial pyramid pooling (SPP) module, placed at the bottleneck layer of the U-Net [11]. A set of experiments are conducted to observe the impact of several SPP variants by using a different combination of pooling kernels and the total number of parallel paths. The overall network will also be configured to produce the optimal forest segmentation system. Then, the periodical changes in forest areas will be analyzed using the time series output of the segmented maps. Hence, governments can utilize these segmented forest maps to keep track of all the reductions in forest sizes automatically. Crucially, this information can be used to detect possible illegal logging activities when the forest areas shrink even though those particular areas are gazetted as reserved forest areas. Besides that, the government can also keep track of the conservation effort by periodically mapping the targeted forest areas to measure progress. This paper is organized into five main sections. The following section discusses the related works, followed by detailed explanations of the modified U-Net with a spatial pyramid pooling mechanism. The results and discussion are explained in Section 4, while the conclusion is given in the final section.

2. Related Works

2.1. *Satellite Technology in Forest Monitoring Systems*

Landsat-8, Terra, and Sentinel-2 are the three satellites that are commonly used to develop datasets for forest monitoring systems, as used in Krasovskii et al. [12], Wyniawskyj et al. [13], and Torres et al. [14]. The spatial resolutions of these satellites are listed in Table 1, in which they have been successfully used for various other applications, including forest monitoring systems.

With respect to the equiangular grid standard, the spatial resolution of a satellite refers to the maximum land distance represented by a single pixel [15,16]. According to the information given in Table 1, the Terra dataset is unsuitable for use in forest monitoring systems due to the low spatial resolution of the moderate resolution imaging spectroradiometer (MODIS) sensor (250m), which makes it incapable of detecting small-scale changes that are important for forest deterioration because of illegal logging [17]. On the other hand, Mutanga and Kumar [18] revealed that the satellite images from Sentinel-2 performed relatively well, but the quality of the dataset was deemed to be inefficient for model training. Therefore, Singhal and Goel [19] are credited with the inspiration for using Landsat-8 to develop the forest segmentation dataset. They accessed the satellite images through the Google Earth Pro application, which produces satisfactory image quality and resolution.

Table 1. Spatial resolution of the satellite used for automatic forest monitoring systems.

Satellite	Spatial Resolution (m)	Source
Landsat-8	30	NASA, 2021 [20]
Sentinel-2	10	ESA, 2021 [21]
Terra	250	TERRA, 2021 [22]

2.2. *Review of Forest Monitoring Systems*

There are two approaches that have been employed to develop automated forest monitoring systems: conventional machine learning (ML) and deep learning (DL). An intact forest landscape (IFL) is one of the conventional machine learning methods that is used to identify forest and non-forest areas. It uses a binary scale based on two criteria: rate of change caused by human activities and anthropogenic segmentation [23]. This IFL-based system requires a short processing time, but it is only suitable for analyzing areas that are larger than 50,000 hectares and requires a geographic information system (GIS) application to detect significant changes in forest areas [24]. Since this system relies on a binary scale system, not much fine-detail information can be extracted from the satellite images alone, and hence, it needs to be supported by the GIS to further enrich the decision-making module of the monitoring system.

Souza et al. [25], Schultz et al. [26], Telkenaroglu et al. [27], and Othman et al. [28] developed their forest monitoring systems using a combination of remote sensing technology and vegetation index (VI), calculated according to the wavelength reflected from the Earth's surface to the satellite. Therefore, different surfaces of the Earth will absorb and reflect light at different rates, and thus, they can be distinguished by adjusting the VI threshold value. There are a few types of VIs that have shown promising performance in detecting and classifying the Earth's surface into various classes, including the normalized difference vegetation index (NDVI), normalized difference moisture index (NDMI), normalized difference flow index (NDFI), and normalized burn ratio (NBR). The NDFI metric outperforms the other indexes in terms of sensitivity in detecting forest canopy cover by analyzing the members of green vegetation, non-photosynthetic vegetation, and land. Although these approaches were meant for multi-class detection, they are unreliable because of the manually selected VI thresholds for each class without performing systematic analysis.

On the other hand, deep-learning-based technology has driven the study of automated forest monitoring to achieve better performance measures. The convolutional neural network (CNN) is widely used in forest monitoring studies due to its effectiveness in image recognition and processing, especially in distinguishing objects of various classes [29]. The forest monitoring system proposed by Wyniawskyj et al. [13] demonstrated the efficacy of a conventional CNN in forest area segmentation with satisfactory accuracy, but its performance is inferior when compared to other popular models such as SegNet and U-Net [30]. Nonetheless, Lee et al. [31] compared the performance of SegNet and U-Net in landscape segmentation, and found that U-Net produced higher accuracy than SegNet. The study conducted by Pashaei et al. [32] also supports Lee et al.'s findings [31] by confirming that U-Net outperformed SegNet in semantic segmentation, mainly because of SegNet's expected information loss during pooling layers, coupled with the simplified pooling index being transferred to an expansion path.

In fact, U-Net moved the entire feature map from a contraction to an expansion path, potentially improving the segmentation performance of forest and non-forest areas [29]. As a result, U-Net is popularly implemented as the base architecture for forest monitoring systems. However, there is still ample room for improvement when it comes to the development of forest monitoring systems using the U-Net architecture. Maretto et al. [33] focused on improving U-Net using the early and late fusion methods through spatio-temporal analysis. This study shows that U-Net has a high potential to produce better average accuracy but such improvements require significant modifications to deep learning model architecture.

The improvement of U-Net proposed by Abdani et al. [34] demonstrated that segmentation accuracy increment can be achieved by applying the spatial pyramid pooling (SPP) mechanism to oil palm plantation segmentation. To improve the multi-scale capability of the system, the researchers designed a minor modification by adding an SPP module just before the bottleneck layer of the U-Net model. Liu et al. [35] also showed that U-Net with a SPP mechanism (USPP) produces excellent detection accuracy performance. They revealed that the added SPP module manages to produce better performance for automated forest monitoring systems using the Landsat-8 OLI dataset. Table 2 shows the summary of the review and investigation of the forest monitoring systems developed in the previous studies.

Table 2. Summary on automatic forest monitoring systems.

No.	Source	Dataset	Development Approaches	Strengths	Weaknesses
1	Margono et al. [23]	Landsat-5, Landsat-7	Intact Forest Landscape	Fast development	Unable to detect forest area less than 50,000 hectares
2	Othman et al. [28]	MODIS	Vegetation Index (VI)	Able to detect multi-class surface on Earth	Unreliable handcrafted feature with low-spatial-resolution remote sensor
3	Maretto et al. [33]	Landsat-8	U-Net	Simple network architecture improved with early and late fusion	Requires larger set of parameters but accuracy improvement is too small
4	Lee et al. [31]	KOMPSAT-3	SegNet	Transferring only pooling indices to expansion path requires less memory space	Significant information loss during pooling layers
5	Syrris et al. [30]	Sentinel-2	Conventional CNN	Short training and prediction time	Moderate accuracy in segmentation due to its inability to capture spatially invariant features of the input images
6	Pashaei et al. [32]	UAS imagery	U-Net	Skip connection contributes to faster training and high accuracy	No improvements made to detect multi-scale forest areas

3. Methodology

The automated forest segmentation system in this study was developed in five steps: dataset development, image pre-processing, deep learning model training, system performance evaluation, and image post-processing, as shown in Figure 1. Since the deep learning model requires a set of moderate-resolution satellite images for training and testing, this system collected and annotated a dataset based on Landsat-8 satellite images acquired from the Google Earth Pro application. From the Operational Land Imager sensor of Landsat-8, only red, green, and blue (RGB) channels were utilized in this study because of the three input channels requirement of the pre-trained U-Net model. Furthermore, the chosen RGB bands allow multi-scale complex feature extraction of unique patterns that represented the forest areas compared to the other bands, which are more homogenous in patterns. The downloaded satellite images were then manually annotated by two annotators to produce ground truth maps. These annotated maps were then sliced into smaller land parcels before being fed into the deep learning model after performing the pre-processing steps. Only two class problems are considered for this study, a pixel is either labeled as forest or non-forest. The forest class encompasses various types of forests with different canopy appearances from the South-East Asia region. Therefore, this study aims to address this challenge by introducing the multi-scale capability to the deep semantic segmentation network.

Our work also employs the CNN as the base building block to automatically segment forest and non-forest areas. Generally, the CNN is a network composed of an input layer, hidden layers, and an output layer coupled with normalization and activation functions, which are trained recursively to produce the trained network. Therefore, U-Net, which is one of the CNN architectures designed specifically for semantic segmentation, was used as the base model in this study to build the forest segmentation system. To improve segmentation accuracy, multi-scale capability through an SPP mechanism was embedded at the bottleneck layer of the U-Net model. This experiment was built and trained on Google Colab platform with the help of TensorFlow and Keras libraries as used in [36].

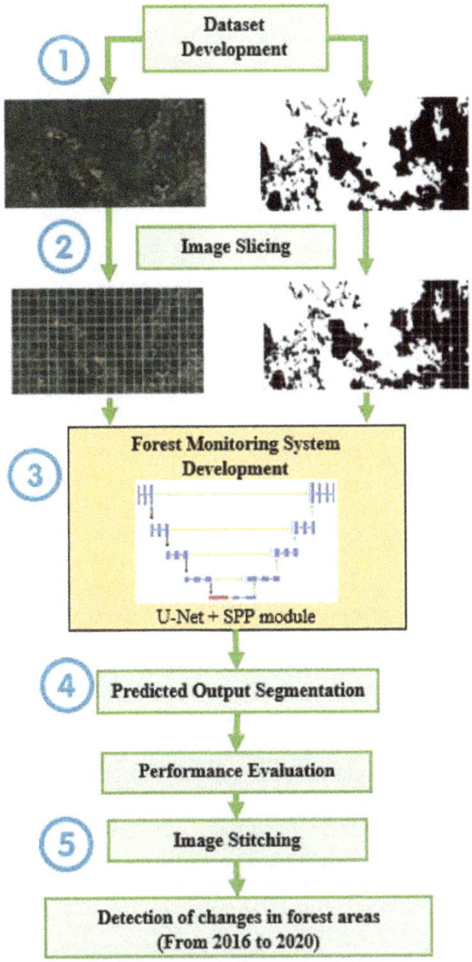

Figure 1. Workflow of the proposed automatic forest monitoring system development.

3.1. Basic Network Architecture

In view of the success of U-Net as a deep learning model in semantic segmentation, this study utilizes U-Net [37] as the base network in forest monitoring systems. Figure 2 illustrates the U-Net architecture with the SPP module integrated at its bottleneck. U-Net consists of two main network paths, namely contraction (encoder) and expansion (decoder) paths, each consisting of four down-sampling and up-sampling blocks. It is a symmetric encoder–decoder network connected via a bottleneck layer that contributes to the "U-shaped" network. The blue blocks in Figure 3 are the feature maps, with its dimension represented by $W \times H \times D$, where W, H, and D stand for width, height, and depth, respectively. In this study, the numerical values indicated above the blue blocks denote the depth of feature maps. The satellite input images that were fed into U-Net have a dimension of $224 \times 224 \times 3$ pixels in RGB format.

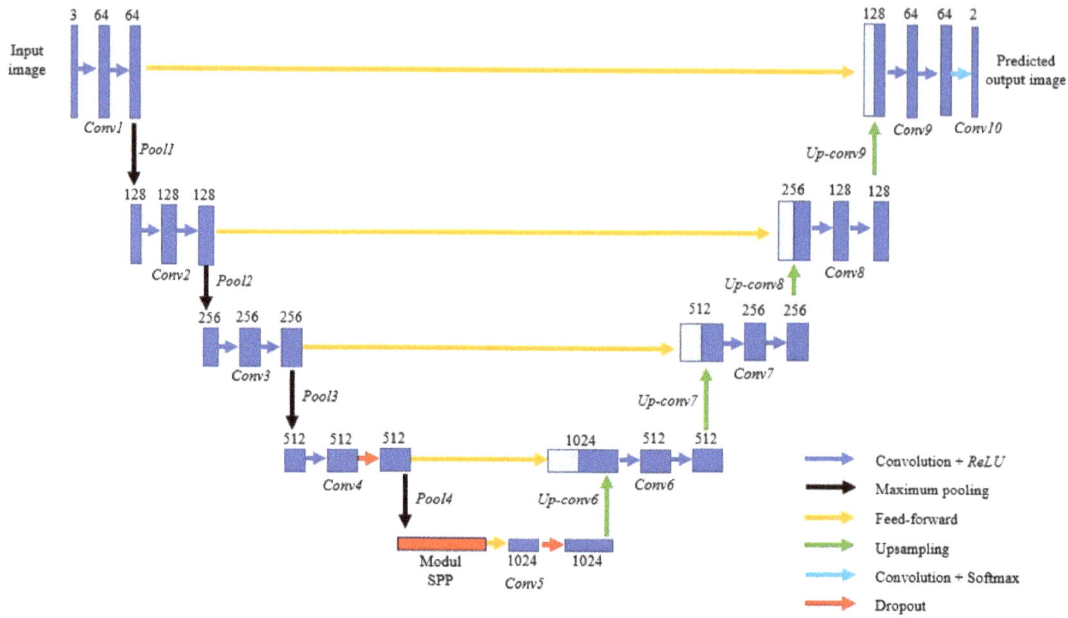

Figure 2. U-Net with a multi-scale module embedded at the bottleneck layer.

The U-Net architecture starts with the encoder with down-sampling blocks, consisting of two convolutional layers, followed by a rectified linear unit (ReLU) and a maximum pooling layer to capture the context and extract the image features for segmentation. The two-dimensional convolution uses a 3×3 convolutional kernel with a stride size of 2 to minimize the loss of spatial information. In addition, since the output feature map is usually smaller than the input feature map after the convolution, the zero-padding technique is used to ensure that the output feature map is the same size as the input feature map. The homogenous kernel is also used in pooling layers with the same height and width for the pooling kernel (2×2) to optimize the processing speed for Tensorflow. An up-sampling block in the decoder path is used to perform the precise localization by merging contextual information from the encoder path through skip connection, which allows U-Net to perform well in image segmentation. The predicted output image will have a dimension of $224 \times 224 \times 2$ at the end of the up-sampling process, whereby the RGB input image is segmented into a binary image, which is in black and white colors.

3.2. The Spatial Pyramid Pooling Mechanism

The SPP mechanism is well known for its excellent performance in multi-size feature extraction, as it effectively preserves the spatial information of an image. Figure 3 depicts an example of the architecture of an SPP mechanism with three parallel paths with pooling kernel sizes of 2×2, 4×4, and 6×6 pixels. The SPP operation started by feeding a 14×14 pixels feature map into each average pooling layer (parallel path), which was then followed by a convolution operator, a batch normalization function, and a ReLU activation function to extract the multi-scale features of forest areas of various color intensities. The output feature maps from these three paths were resized back to their original size of 14×14 pixels before being concatenated with the input feature map (feed-forward path) and fed back into the U-Net model for further processing of the decoder side. The spatial information was preserved by carrying out the feature extraction three times on the same input feature map, thereby extracting the features at different scales. As an SPP module is always composed of multiple paths, it will produce a larger model and is suited to be placed at the bottleneck layer of the U-Net model. This is because the smallest feature map size at the bottleneck layer uses the lowest memory space compared to the module placement at other layers.

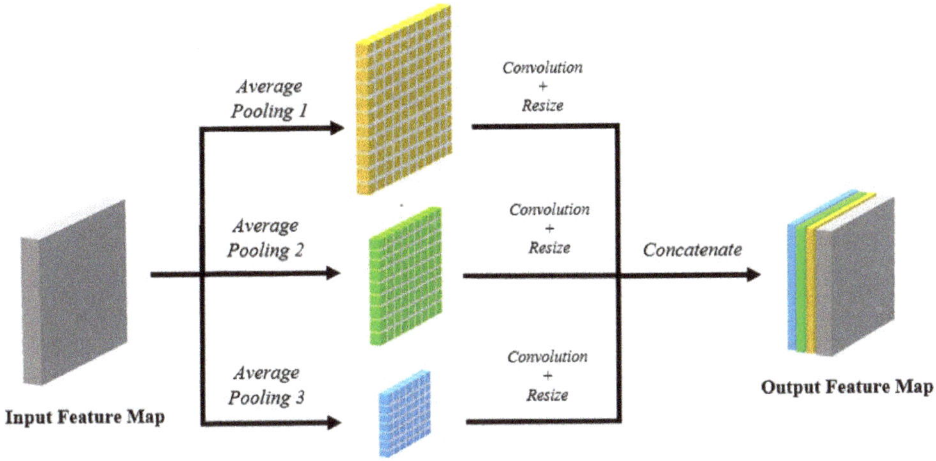

Figure 3. Network architecture of a spatial pyramid pooling mechanism.

To study the impact of optimum SPP on the forest monitoring system, an experiment was conducted using six variants of the SPP modules with the basic settings shown in Table 3. The number of parallel paths and the combination of pooling kernel sizes was adjusted so that the resultant total number of filters remained at 1536. This is because a conventional SPP is made up of three parallel paths, with each path composed of 512 filters, or 1536 filters in total. Due to the adoption of equal division methodology, this study only tested SPP modules with two, three, and four parallel paths (the minimum number of paths for an SPP module is two). The SPP network with five parallel paths was not preferred in this study because of the feature map size limitation at the bottleneck layer, whereby it becomes zero if a larger pooling kernel is used, resulting in too much information loss.

Table 3. SPP module variants.

No.	Number of Parallel Paths	Pooling Kernel Sizes
1	2	$2 \times 2, 4 \times 4$
2		$3 \times 3, 6 \times 6$
3		$4 \times 4, 8 \times 8$
4	3	$2 \times 2, 4 \times 4, 6 \times 6$
5		$3 \times 3, 6 \times 6, 9 \times 9$
6	4	$2 \times 2, 4 \times 4, 6 \times 6, 8 \times 8$

The theory of homogeneous pooling kernel was also applied to the SPP module, leading to more coherent feature maps, especially when dealing with the multi-scale issue. The output feature map from the average pooling layers must also be greater than 3×3 pixels, as the applied convolution operator right after the pooling layer has a kernel size of 3×3 pixels. Hence, the pooling kernel sizes used in this study only ranged from 2×2 to 9×9 pixels, with the kernel size selection being a multiple of each other. A pooling kernel size of 1×1 pixel is not feasible, as it carries minimal impact, while a pooling operation with a kernel size of 10×10 pixels is inappropriate because the resultant output feature map will be 5×5 pixels, which is too small to extract any meaningful features.

4. Results and Discussion

4.1. Landsat-8 Satellite Image Dataset

The evolution of satellite technology with the deep learning approach has led to a higher level of accuracy in forest mapping studies. While the Landsat-9 satellite is equipped with the best sensor specifications in the Landsat series, it is not suitable for time series studies, as it was only launched in September 2021, and lacks temporal data from previous years. In contrast, Landsat-8 is the most suitable satellite for this study, as it has the best radiometric resolution (12 bits) with sufficient temporal data over more than 10 years. The satellite imaging dataset used in this study was saved from Landsat-8 using the Google Earth Pro application. Google Earth Pro is a relatively similar application to Google Earth Engine, but it is better suited for use in this study, primarily because it is free of charge.

The selection of study locations is the first step in dataset development, and its quality has a direct impact on segmentation accuracy. The forest areas of South-East Asia were the best choice for this research because they offer a clear distinction between forest and non-forest areas. The dataset was extracted from the Landsat-8 satellite, whereby the size for each land parcel is 4800 × 2782 pixels, with a resolution of 1.34 m per pixel. This selection was made so that the satellite images had the widest field of view without any disturbances, such as cloud cover and mosaic-like images that degrade the system's performance. To track the changes in forest status, this study focused on developing the dataset with time series Landsat-8 satellite images taken from 2016, 2018, and 2020, and sampled at eight different locations throughout the South-East Asia region. Initially, this study focused on changes in forest status for every two years to ensure that small changes could still be clearly detected. If a one-year sampling period is chosen, the changes in the forest status are not too obvious, which is not informative for reporting purposes. Therefore, two years are chosen so that the changes in the forest status are clearly visible to support the intended applications. Please note that Landsat-8 was introduced in 2013, and so the early imaging quality is relatively low as a result of using the Google Earth Pro application. As a result, the sampling process starts from the year 2016 until 2020. For each sampling location, three land parcel samples were extracted, which resulted in a total of 24 Landsat-8 satellite images being extracted and annotated. The acquired satellite images for a particular location must also have the same angle, scale, and resolutions to ensure data consistency.

The ground truth satellite images were manually annotated using the Adobe Photoshop application. Since the performance of the system is heavily dependent on labeled data, the manual annotation of forest and non-forest areas must be performed meticulously and consistently. Thus, forest areas were annotated as the foreground with a white color RGB code of (255, 255, 255), while non-forest areas were annotated as the background with a black color RGB code of (0, 0, 0). Then, the pre-processing phase was completed by slicing the RGB images and their ground truth maps from 4800 × 2782 pixels to subsets of 224 × 224 pixels, which was the input format for U-Net. The satellite image of a selected land parcel was sliced into 252 images, which resulted in 6,048 smaller samples, taken from the original satellite images of 24 land parcels. The dataset was then divided into two subsets of 2268 training data and 756 testing data. Table 4 summarizes the study locations for both annotated datasets.

Table 4. Landsat-8 satellite images dataset information.

Dataset	Location	Latitude	Longitude	Type of Data
Moderate level of difficulty	A	5°38′31.56″ N	117°6′45.39″ E	Training data
	B	3°10′26.10″ N	101°58′7.99″ E	Training data
	C	1°25′32.01″ N	117°12′46.65″ E	Training data
	D	3°5′38.77″ N	98°4′11.72″ E	Testing data
High level of difficulty	E	2°7′49.52″ N	117°17′0.27″ E	Training data
	F	0°47′49.27″ S	115°48′43.03″ E	Training data
	G	2°38′28.69″ N	103°2′6.68″ E	Training data
	H	3°54′34.15″ N	102°52′5.11″ E	Testing data

To analyze the system's capability to segment the forest area at different difficulty levels, the satellite images were further divided into two sets of data based on their difficulty level to distinguish the forest and non-forest areas, as shown in Figure 4. The dataset with a moderate level of difficulty has a clear distinction between forest and non-forest areas, as well as fewer rivers, lakes, and vegetation areas. In contrast, the dataset with a high-difficulty level consists of satellite images with similar features between forest and non-forest areas, and with more rivers, lakes, and vegetation areas.

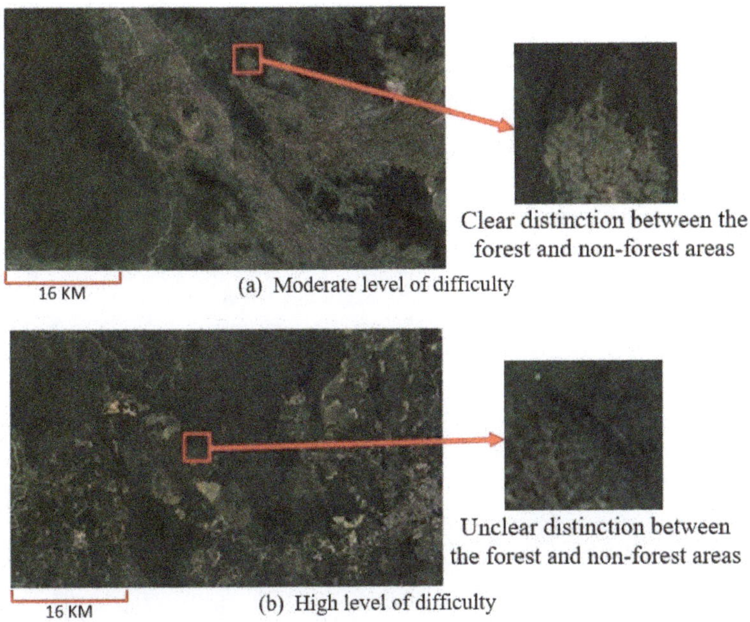

Figure 4. Satellite images from dataset of (**a**,**b**).

4.2. Performance Evaluation of the Forest Monitoring System

The performance of the forest segmentation system was measured using three metrics: average accuracy (acc), Intersection over Union (IoU), and F1-Score ($F1_{score}$). acc is used to validate system performance by calculating the correct prediction for both classes based on the confusion metrics shown in Table 5. $F1_{score}$ is a performance metric similar to acc, but with advantages for real-world classification, especially when there is an imbalanced class distribution [38]. Finally, IoU is the most commonly used metric in evaluating a segmentation model that focuses on the overlapping areas between predicated maps and ground truth maps. It represents the ratio of the area of intersection over the total area of the ground truth and predicted segmentation maps.

Table 5. Terminology of the confusion matrix.

Terminology	Description
True positive (TP)	Detect correctly forest areas as forest areas
True negative (TN)	Detect correctly non-forest areas as non-forest areas
False positive (FP)	Detect non-forest areas as forest areas
False negative (FN)	Detect forest areas as non-forest areas

$$acc = \frac{TP + TN}{TP + TN + FP + FN} \quad (1)$$

$$F1_{score} = \frac{TP}{TP + 0.5(FP + FN)} \quad (2)$$

$$IoU = \frac{TP + TN}{TP + FP + FN} \quad (3)$$

The first step in building an optimal base U-Net model using semantic segmentation methodology is finding the optimized set of hyperparameters. A set of experiments was conducted using different sets of hyperparameters to observe their respective training accuracy and loss, as well as to evaluate the segmentation accuracy of the trained model on the test data. The optimal hyperparameter configurations used in this study are listed in Table 6. The selected optimizer was stochastic gradient descent (SGD). The number of epochs required for training convergence is different for each SPP module variant, and it is also dependent on the difficulty level of the dataset and the spatial relationship between features in the satellite images. Therefore, the optimum number of epochs for a certain model is determined by observing and ensuring that the training accuracy and loss converge at the end of the training process with satisfactory testing accuracy and loss.

Table 6. Hyperparameter setting.

Hyperparameter	Value
Learning rate	0.0001
Batch size	8
Epoch	100
Optimizer Momentum	0.9

Table 7 shows the performance of the proposed forest segmentation system at moderate and high levels of difficulty with and without the SPP module for both datasets. The forest segmentation system developed by the dataset with a moderate level of difficulty has an acc, IoU, and $F1_{score}$ of 88.23%, 78.80%, and 87.02%, respectively. It can also be seen that the base U-Net performance is improved by all SPP variants, especially for the system embedded with SPP comprising four parallel paths with pooling kernel sizes of 2×2, 4×4, 6×6, and 8×8. This optimal SPP module successfully produces the highest acc, IoU, and $F1_{score}$ of 90.81%, 83.17%, and 90.72%, respectively, with a performance improvement of 2.92%, 5.55%, and 4.25%, respectively. The optimal SPP module is made up of four pooling layers, which effectively preserves the spatial information of various scales, resulting in more robust segmentation mapping. The forest segmentation system produces the best performance with an SPP module comprising four parallel paths, followed by SPP modules with three and two parallel branches. Meanwhile, the experimental results also show that a combination of smaller pooling kernel sizes is able to detect smaller features in satellite images, hence allowing the more efficacious segmentation of multi-scale forest areas. This is the reason why the system integrated with an SPP module with two parallel branches has the highest performance with pooling kernel sizes of 2×2 and 4×4, followed by 3×3 and 6×6, and finally 4×4 and 8×8. Coincidentally, the patterns are similar for the SPP module with three parallel paths. It is observable that the forest segmentation system would be more sensitive to small changes in forest areas by using a combination of smaller SPP module pooling kernel sizes.

Table 7. Performance results of the U-Net with an embedded spatial pyramid pooling module.

Dataset	Number of Parallel Paths	Pooling Kernels	Performance Metrics (%)			Improvement (%)		
			acc	IoU	$F1_{score}$	acc	IoU	$F1_{score}$
Moderate level of difficulty	Original U-Net	-	88.23	78.80	87.02	-	-	-
	2	$2 \times 2, 4 \times 4$	89.57	81.04	88.85	1.52	2.84	2.10
		$3 \times 3, 6 \times 6$	89.54	80.98	88.71	1.49	2.77	1.94
		$4 \times 4, 8 \times 8$	89.33	80.64	88.54	1.25	2.34	1.75
	3	$2 \times 2, 4 \times 4, 6 \times 6$	90.73	82.99	90.24	2.83	5.32	3.71
		$3 \times 3, 6 \times 6, 9 \times 9$	89.84	81.52	89.32	1.82	3.45	2.64
	4	$2 \times 2, 4 \times 4, 6 \times 6, 8 \times 8$	90.81	83.17	90.72	2.92	5.55	4.25
High level of difficulty	Original U-Net	-	84.54	70.84	76.92	-	-	-
	2	$2 \times 2, 4 \times 4$	86.42	74.76	81.64	2.22	5.53	6.14
		$3 \times 3, 6 \times 6$	86.10	74.42	81.62	1.85	5.05	6.11
		$4 \times 4, 8 \times 8$	83.90	71.73	80.85	−0.75	1.26	5.11
	3	$2 \times 2, 4 \times 4, 6 \times 6$	85.25	73.23	80.87	0.84	3.37	5.14
		$3 \times 3, 6 \times 6, 9 \times 9$	84.86	72.81	80.88	0.38	2.78	5.15
	4	$2 \times 2, 4 \times 4, 6 \times 6, 8 \times 8$	86.71	75.59	82.88	2.57	6.71	7.75

The forest segmentation system developed using a dataset of high-level difficulty on the base U-Net architecture has an acc, IoU, and $F1_{score}$ of 84.54%, 70.84%, and 76.92%, respectively. It performs worse than the system developed by the dataset with a moderate level of difficulty. In this case, all SPP module variants enhanced the system's performance, with the exception of the SPP module composed of two parallel paths with pooling kernel sizes of 4×4 and 8×8 pixels. This variant only achieved an acc of 83.90%, which is 0.75% lower than the system developed using the base U-Net model. This is likely caused by the large kernel sizes used in both pooling layers that neglect the smaller features in the feature maps and result in less accurate segmentation maps of forest and non-forest areas and lower the accuracy performance. As the drop in accuracy performance is very small, with observable improvements for the $F1_{score}$ and IoU, this SPP module is still considered the one that can enhance system performance in multi-scale segmentation problems.

Similar to the system developed with a moderate difficulty dataset, the optimal SPP module variant developed for a high-difficulty dataset comprises four parallel paths with pooling kernel sizes of 2×2, 4×4, 6×6, and 8×8 pixels. The performance of the forest segmentation system optimized by this variant of the SPP module produces the highest acc, IoU, and $F1_{score}$ of 86.71%, 75.59%, and 82.88%, respectively, with a performance improvement of 2.57%, 6.71%, and 7.75%, respectively. However, the performance of the forest segmentation system with two and three parallel paths is significantly different from the system developed with a moderate difficulty dataset. The SPP module with two parallel paths is better suited to the forest segmentation system for a set of high-difficulty datasets because it produces better performance for acc, IoU, and $F1_{score}$ compared to the SPP module with three parallel paths.

Further experiments were also conducted to compare the performance of the proposed multi-scale model with the other deep semantic segmentation models. The tests were carried out using a high-difficulty dataset to test the performance limit of the tested models. There are four deep semantic segmentation models that were tested with the optimized individual set up, which are FCN [39], TernausNet [40], SegNet [29], and U-Net [37]. Table 8 shows the performance comparison between the models. FCN and TernausNet return the worst performance with acc, IoU, and $F1_{score}$ of 38.39%, 19.19%, and 55.48%, respectively. Interestingly, both of these models use VGG network [41] as their encoder to extract the respective features of forest and non-forest areas. This encoder design that comprises 13 layers of CNN with ReLU activation function is not suitable for the forest segmentation task. Besides that, both SegNet and U-Net have performed relatively well compared to the FCN and TernausNet. SegNet returns the third best performance values of 84.54%, 70.84%, and 76.92% for acc, IoU, and $F1_{score}$, respectively, while U-Net returns the second best performance values of 84.54%, 70.84%, and 76.92% for *acc*, *IoU*, and $F1_{score}$, respectively. Furthermore, both SegNet and U-Net utilize a symmetric network architecture between their encoder and decoder. However, the best deep semantic segmentation model is the proposed method with the best performance values across all metrics. The proposed method also adopts symmetric design architecture because the SPP module is added at the bottleneck layer.

Table 8. Performance comparison with the other deep semantic segmentation models.

Method	Performance Metrics (%)		
	acc	*IoU*	$F1_{score}$
FCN [39]	38.39	19.19	55.48
TernausNet [40]	38.39	19.19	55.48
SegNet [29]	86.38	75.59	80.28
U-Net [37]	84.54	70.84	76.92
Proposed method	86.71	75.59	82.88

4.3. Training Performance: Accuracy and Loss

Training *acc* describes how well the system performs in achieving correct segmentation for both forest and non-forest areas on training data, while training loss refers to the segmentation error of the corresponding model, which is also used as a variable to update the weights during the learning process. Figure 5a illustrates the training accuracy and loss of a forest segmentation system developed using the moderate difficulty dataset without an SPP module. The training process converged after 80 epochs, despite there being several spikes in the loss graph at certain epochs. The spikes in loss values are normal because they are used to update the new weights by the optimizer. Thus, training accuracy drops for every single epoch that experiences a surge of loss and then increases during subsequent learning epochs after the new weights are updated with a high loss value.

The addition of the optimal SPP module to the base U-Net architecture resulted in a higher training loss compared to the system without the SPP module, which converged faster after 50 epochs, as shown in Figure 5b. The training loss for this model reached 35.2393, which is 34.8725 higher than the base system without the SPP module (0.3368). This is most likely due to the batch normalization layers, which increased the number of untrainable parameters and contributed to an increase in the training loss. This small increment is still tolerable because the batch normalization layer is important in normalizing the input to speed up the training process.

As shown in Figure 6a,b, using a high-difficulty dataset with and without an SPP module, the training accuracy and loss of the forest monitoring system converge successfully after 70 and 80 epochs, respectively. If we scrutinize the results, the training accuracy for the system without the SPP module increases very slowly after the loss spike at the ninth epoch. This is because the learning rate used for the training process is very small, only 0.0001 to ensure that the learning update can be carried out at an optimal level. Despite the fact that the training loss is higher in the system with SPP due to the batch normalization layers, the training accuracy is still successfully increased by 1.7%. Therefore, the embedded multi-scale module has managed to improve the performance of the forest segmentation algorithm, which is a very important low-level function in a forest monitoring system. The usage of a multi-scale module can be extended to any problem that encounters the issue of varying sizes of input. For example, in eye disease screening applications [42], the multi-scale capability has further improved the detection rate of the disease by better analyzing affected signals. Besides that, a multi-scale approach has also been implemented for agriculture applications that allow the system to detect leaf diseases of various sizes [43]. Hence, a multi-scale module can be optimally embedded into a base segmentation network in order to improve the model's capability to extract features of various scales.

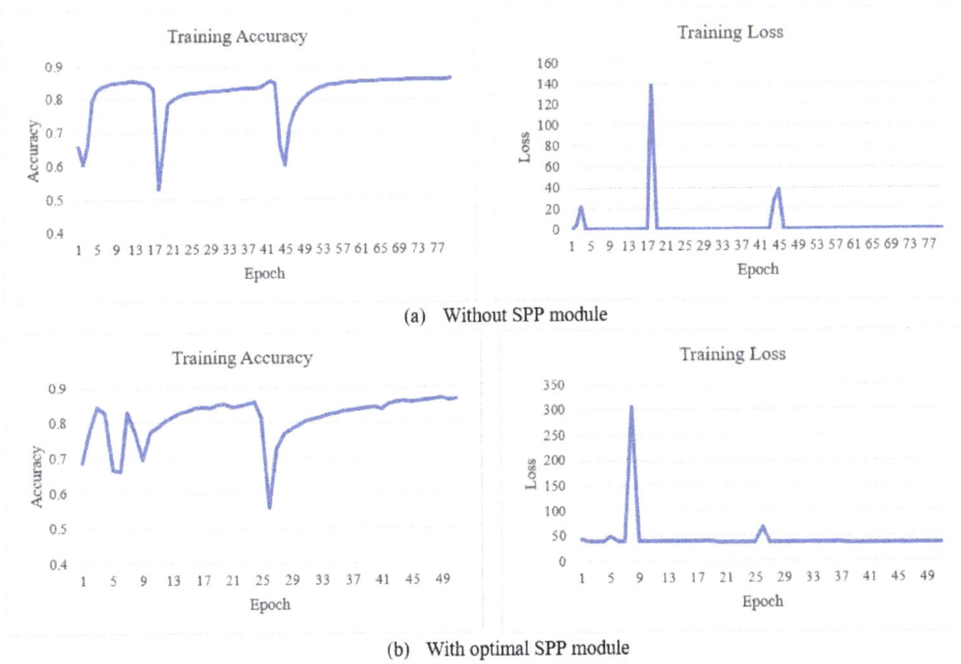

Figure 5. Training accuracy and loss of forest monitoring system using moderately difficult dataset.

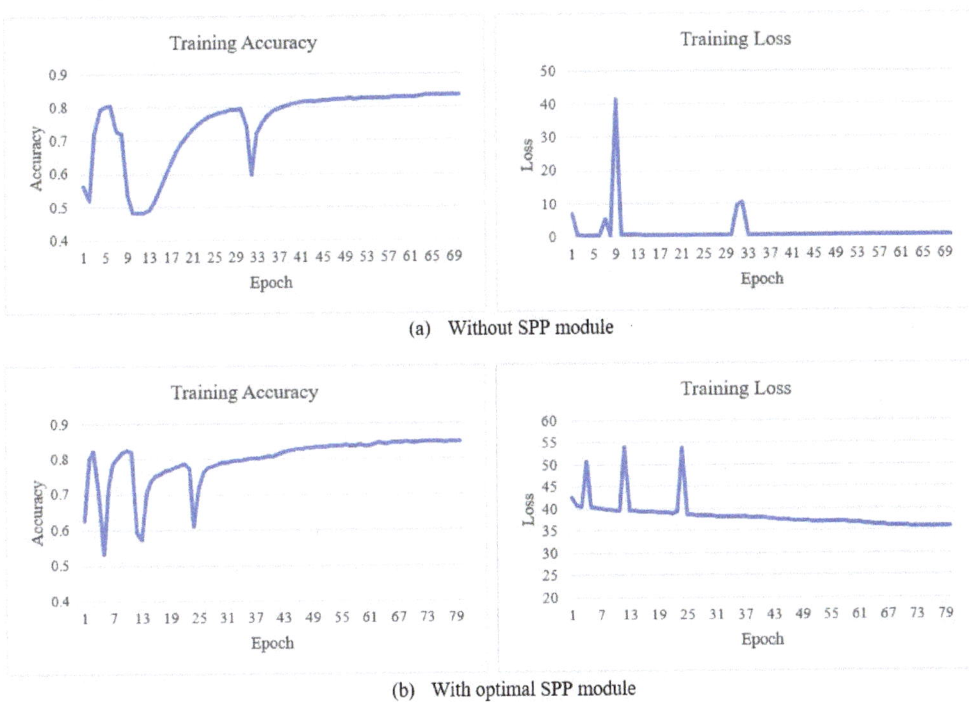

Figure 6. Training accuracy and loss of forest monitoring system using high-difficulty dataset.

4.4. Qualitative Discussion on the Segmentation Output

The impact of SPP module integration into the base U-Net model was also studied by comparing qualitatively the predicted output images to the ground truth maps. Figure 7 depicts several samples of raw satellite images, ground truth maps, and predicted output maps produced by the proposed forest segmentation system, with and without the SPP module, tested on a moderate-difficulty dataset. Generally, the base system without the integration of the SPP module has vague and unclear segmentation maps between forest and non-forest areas. This weakness is more obvious along the segmentation lines near the boundaries of both areas, which is the main cause of low segmentation accuracy. On the other hand, the output maps from the system with an optimal SPP module had successfully mitigated this weakness. The maps show more precise segmentation lines along the boundaries of both classes with minimal unwanted interference of white and black dots. In conclusion, the proposed forest segmentation system with an embedded SPP module is found to be more capable of detecting and segmenting forest and non-forest areas of varying scales.

Figure 8 illustrates the predicted segmentation output at locations D and H from 2016 to 2020. The changes in forest status were detected by observing the changes in white and black areas on the maps, which refer to the forest and non-forest areas, respectively. Figure 8a depicts the segmentation map of the satellite image at location D, taken from 2016, 2018, and 2020. The circled area in the upper right corner of the respective satellite image clearly shows the forest areas have shrunk from 2016 to 2020. In 2016, there were more forest areas than non-forest areas, as shown by the presence of more white pixels compared to black pixels. Later on, this area was exploited in 2018, and the proposed system has successfully recognized the reduction in forest areas as depicted by an increment in black-colored pixels. In 2020, it was detected that this particular area contained a larger proportion of forest areas again, mainly caused by the COVID-19 pandemic that halted deforestation activities.

4.5. Changes in the Forest Status

In order to track the changes in forest areas, post-processing of the image was carried out to stitch the smaller predicted maps from the size of 224×224 pixels to produce the original land parcel image of size 4704×2688 pixels. A total of 252 segmented maps will be stitched together to produce a single large segmentation map. In this case, the segmented output maps were taken from the optimized forest segmentation system, improved by the implementation of an optimal SPP module to enhance the segmentation performance.

Almost similar patterns can be observed in the land parcel at location H, as shown in Figure 8b. The circled area indicates that there are clear changes in the forest areas, which have decreased significantly, whereby the deteriorated situation has worsened year after year. Overall, these predicted forest maps have successfully demonstrated the performance of our forest segmentation system's effectiveness in detecting the changes from 2016 to 2020. The addition of a multi-scale module has further improved segmentation performance, which enables the system to detect deforestation activities that may be caused by illegal logging.

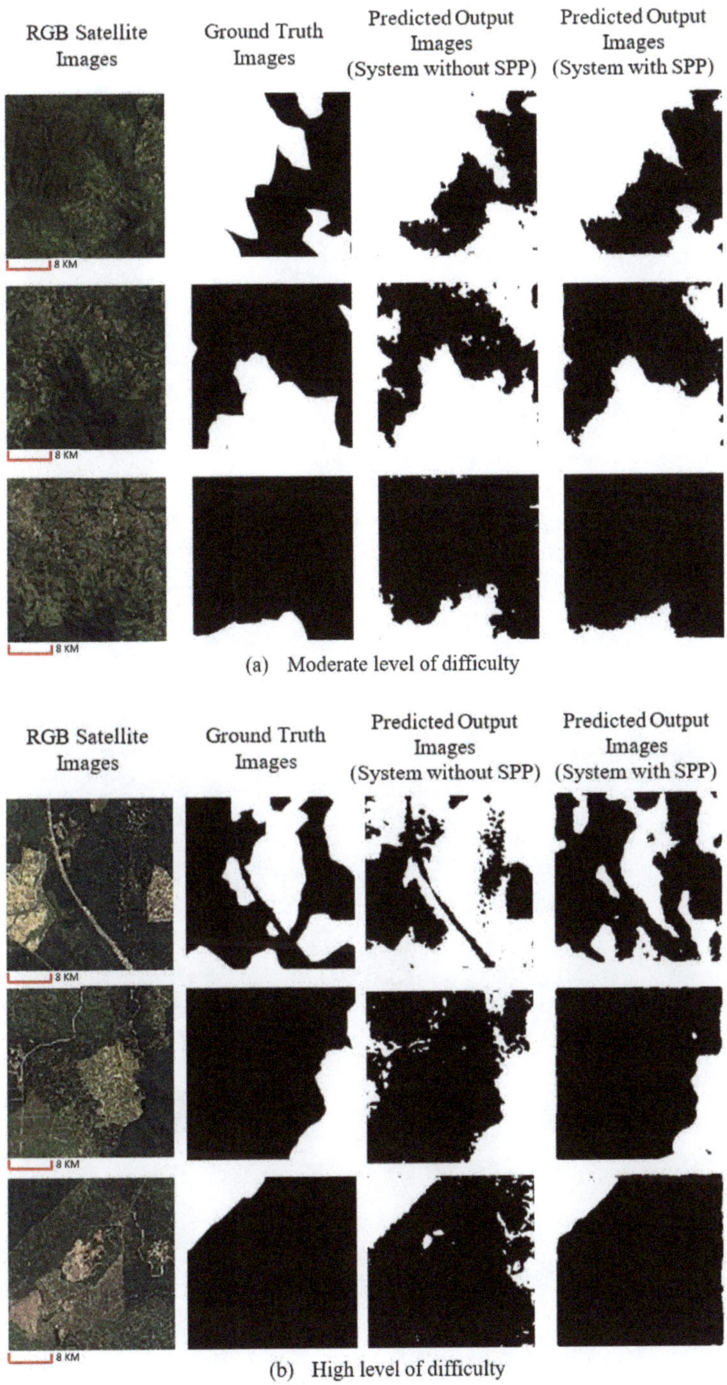

Figure 7. Predicted output images produced by forest monitoring system using dataset of (**a**,**b**).

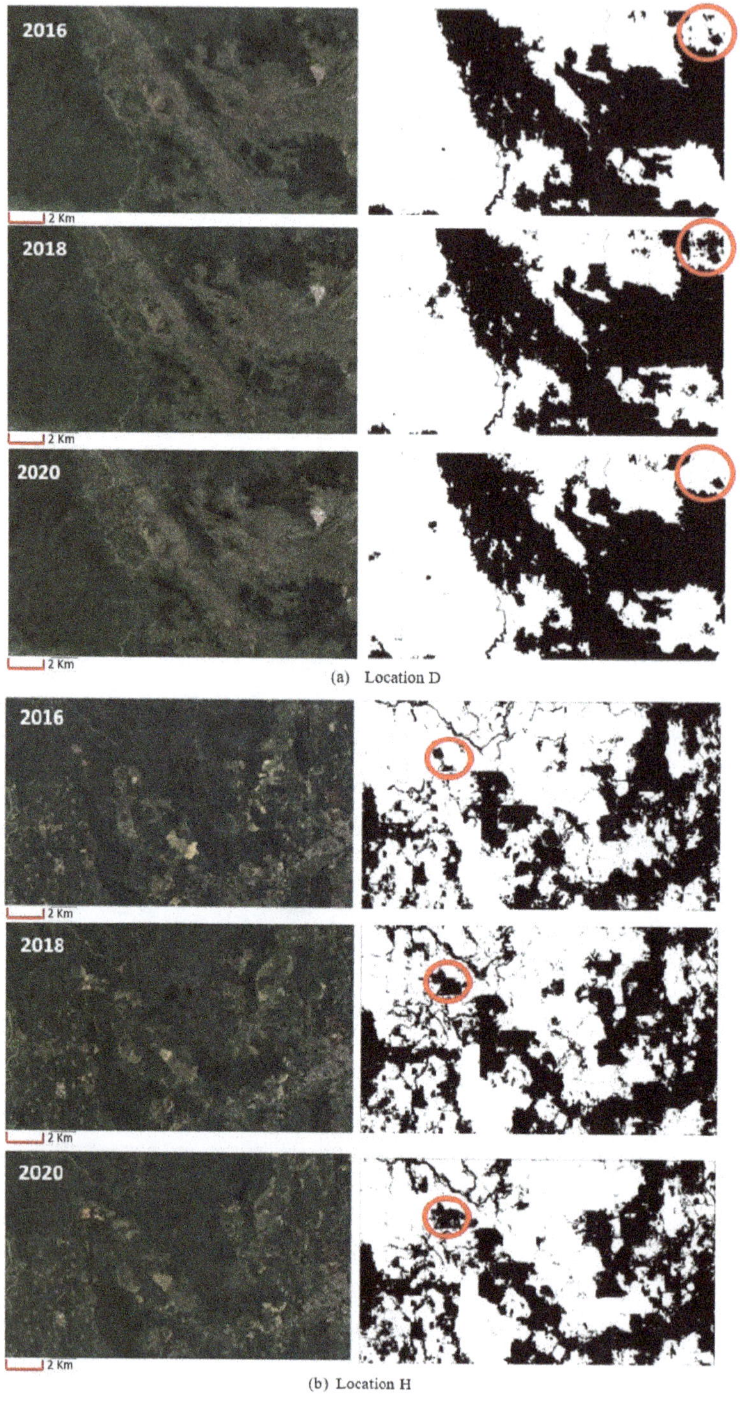

Figure 8. Predicted output images at (**a**,**b**) from 2016 to 2020.

5. Conclusions and Future Work

This study successfully achieved its initial objectives of developing a forest segmentation system based on remote sensing technology and an improved deep learning model through embedding multi-scale capability. Two sets of satellite image datasets, consisting of 6,048 samples, with satisfactory image quality and resolution were successfully developed for training and testing purposes using Landsat-8 satellite images. The system was built on the base U-Net architecture to automate the segmentation of forest areas so that any changes in status could be detected at pixel level with acc, IoU, $F1_{score}$ of 84.54%, 70.84%, and 76.92%, respectively.

The results reveal that all SPP variants managed to improve the system's multi-scale capability in detecting changes in forest areas at varying scales, especially for the SPP module made up of four parallel paths with pooling kernel sizes of 2×2, 4×4, 6×6, and 8×8 pixels. Since satellite images from high-difficulty datasets are deemed to be more reflective of real situations, the overall performance of the system was reported using this dataset, which produced the highest acc, IoU, and $F1_{score}$ of 86.71%, 75.59%, and 82.88%, respectively with an improvement of 2.57%, 6.71%, and 7.75%, respectively. The forest segmentation system developed in this study can monitor changes in forest status at the pixel level through an automated segmentation process with low operating costs.

This study encountered an issue with Google Colab due to its limited random-access memory (RAM) and graphics processing unit (GPU) runtime, making the training process difficult. There is room for improvement to develop a more robust forest segmentation system in the future. A larger training dataset is proposed for future work so that the training process can be conducted more comprehensively to learn a greater variety of features. This study could also be improved by testing different base deep learning models with other multi-scale modules such as atrous spatial pyramid pooling and a waterfall version of the SPP.

Author Contributions: Conceptualization, F.X.R., M.A.Z.; software, F.X.R., M.A.Z.; formal analysis, F.X.R.; writing—original draft preparation, F.X.R., M.A.Z., S.R.A.; writing—review and editing, F.X.R., M.A.Z., S.R.A., M.S. All authors have read and agreed to the published version of the manuscript.

Funding: This research was funded by Ministry of Higher Education Malaysia under Fundamental Research Grant Scheme with grant number FRGS/1/2022/TK07/UKM/02/4 and Asia-Pacific Telecommunity under The Extra Budgetary Contribution from the Republic of Korea Fund with grant number KK-2022-026.

Institutional Review Board Statement: Not applicable.

Informed Consent Statement: Not applicable.

Data Availability Statement: The dataset can be personally requested from the corresponding author at asyraf.zulkifley@ukm.edu.my.

Conflicts of Interest: The authors declare no conflict of interest.

Abbreviations

The following abbreviations are used in this manuscript:

CNN	Convolutional Neural Networks
SGD	Stochastic Gradient Descent
SPP	Spatial Pyramid Pooling
IoU	Intersection over Union
ReLU	Rectified Linear Unit
GIGO	Garbage In, Garbage Out
ML	Machine Learning
DL	Deep Learning
IFL	Intact Forest Landscape
GIS	Geographic Information System

VI	Vegetation Index
NDVI	Normalized Difference Vegetation Index
NDMI	Normalized Difference Moisture Index
NBR	Normalized Burn Ratio

References

1. Bohn, F.J.; Huth, A. The importance of forest structure to biodiversity–productivity relationships. *R. Soc. Open Sci.* **2017**, *4*, 160521. [CrossRef] [PubMed]
2. Silva Junior, C.H.; Pessôa, A.C.; Carvalho, N.S.; Reis, J.B.; Anderson, L.O.; Aragão, L.E. The Brazilian Amazon deforestation rate in 2020 is the greatest of the decade. *Nat. Ecol. Evol.* **2021**, *5*, 144–145. [CrossRef] [PubMed]
3. Watch, G.F. Global Primary Forest Loss. 2021. Available online: https://gfw.global/3JJS9Sm (accessed on 10 December 2021).
4. Petrenko, C.; Paltseva, J.; Searle, S. *Ecological Impacts of Palm Oil Expansion in Indonesia;* International Council on Clean Transportation: Washington, DC, USA, 2016.
5. Hartanto, H.; Prabhu, R.; Widayat, A.S.; Asdak, C. Factors affecting runoff and soil erosion: Plot-level soil loss monitoring for assessing sustainability of forest management. *For. Ecol. Manag.* **2003**, *180*, 361–374. [CrossRef]
6. Lidestav, G.; Westin, K. The impact of Swedish forest owners' values and objectives on management practices and forest policy accomplishment. *Small-Scale For.* **2023**, 1–22. [CrossRef]
7. Koh, J.; Johari, S.; Shuib, A.; Siow, M.L.; Matthew, N.K. Malaysia's forest pledges and the Bornean state of Sarawak: A policy perspective. *Sustainability* **2023**, *15*, 1385. [CrossRef]
8. Abdani, S.R.; Zulkifley, M.A.; Siham, M.N.; Abiddin, N.Z.; Aziz, N.A.A. Paddy fields segmentation using fully convolutional network with pyramid pooling module. In Proceedings of the IEEE 5th International Symposium on Telecommunication Technologies, Virtual, 9–11 November 2020 ; pp. 30–34.
9. de Araújo, M.; Andrade, E.; Machida, F. Performance analysis of machine learning-based systems for detecting deforestation. In Proceedings of the 2021 XI Brazilian Symposium on Computing Systems Engineering, Online, 22–25 November 2021; pp. 1–8.
10. Kilkenny, M.F.; Robinson, K.M. Data quality:"Garbage in–garbage out". *Health Inf. Manag. J.* **2018**, *47*, 103–105. [CrossRef]
11. Abdani, S.R.; Zulkifley, M.A. Optimal selection of pyramid pooling components for convolutional neural network classifier. In Proceedings of the International Conference on Decision Aid Sciences and Application, Sakheer, Bahrain, 8–9 November 2020; pp. 586–591.
12. Krasovskii, A.; Maus, V.; Yowargana, P.; Pietsch, S.; Rautiainen, M. Monitoring deforestation in rainforests using satellite data: A pilot study from Kalimantan, Indonesia. *Forests* **2018**, *9*, 389.
13. Wyniawskyj, N.S.; Napiorkowska, M.; Petit, D.; Podder, P.; Marti, P. Forest monitoring in Guatemala using satellite imagery and deep learning. In Proceedings of the IEEE International Geoscience and Remote Sensing Symposium, Yokohama, Japan, 28 July–2 August 2019; pp. 6598–6601.
14. Torres, D.L.; Turnes, J.N.; Soto Vega, P.J.; Feitosa, R.Q.; Silva, D.E.; Marcato Junior, J.; Almeida, C. Deforestation detection with fully convolutional networks in the Amazon forest from Landsat-8 and Sentinel-2 images. *Remote. Sens.* **2021**, *13*, 5084. [CrossRef]
15. Liang, S.; Wang, J. *Advanced Remote Sensing: Terrestrial Information Extraction and Applications*; Academic Press: Cambridge, MA, USA, 2019.
16. Vishwakarma, B.D.; Devaraju, B.; Sneeuw, N. What is the spatial resolution of GRACE satellite products for hydrology? *Remote. Sens.* **2018**, *10*, 852. [CrossRef]
17. Lunetta, R.S.; Knight, J.F.; Ediriwickrema, J.; Lyon, J.G.; Worthy, L.D. Land-cover change detection using multi-temporal MODIS NDVI data. *Remote. Sens. Environ.* **2006**, *105*, 142–154. [CrossRef]
18. Mutanga, O.; Kumar, L. Google Earth engine applications. *Remote. Sens.* **2019**, *11*, 591. [CrossRef]
19. Singhal, A.; Goel, S. Spatio-temporal analysis of open waste dumping sites using GoogleEarth: A case study of Kharagpur City, India. In *Spatial Modeling and Assessment of Environmental Contaminants*; Springer: Berlin/Heidelberg, Germany, 2021; pp. 137–151.
20. Science, N.L. Satellite Landsat Series. Available online: https://landsat.gsfc.nasa.gov/ (accessed on 20 December 2021).
21. Agency, E.S. Sentinel-Online: Sentinel-1 and Sentinel-2. Available online: https://sentinel.esa.int/web/sentinel/missions/sentinel-2 (accessed on 25 December 2021).
22. Terra, N. Terra Instruments: Moderate Resolution Imaging Spectroradiometer (MODIS). Available online: https://terra.nasa.gov/about/terra-instruments/modis (accessed on 25 December 2021).
23. Margono, B.A.; Turubanova, S.; Zhuravleva, I.; Potapov, P.; Tyukavina, A.; Baccini, A.; Goetz, S.; Hansen, M.C. Mapping and monitoring deforestation and forest degradation in Sumatra (Indonesia) using Landsat time series data sets from 1990 to 2010. *Environ. Res. Lett.* **2012**, *7*, 034010. [CrossRef]
24. Potapov, P.; Yaroshenko, A.; Turubanova, S.; Dubinin, M.; Laestadius, L.; Thies, C.; Aksenov, D.; Egorov, A.; Yesipova, Y.; Glushkov, I.; et al. Mapping the world's intact forest landscapes by remote sensing. *Ecol. Soc.* **2008**, *13*, 51. [CrossRef]
25. Souza Jr, C.M.; Siqueira, J.V.; Sales, M.H.; Fonseca, A.V.; Ribeiro, J.G.; Numata, I.; Cochrane, M.A.; Barber, C.P.; Roberts, D.A.; Barlow, J. Ten-year Landsat classification of deforestation and forest degradation in the Brazilian Amazon. *Remote. Sens.* **2013**, *5*, 5493–5513. [CrossRef]
26. Schultz, M.; Clevers, J.G.; Carter, S.; Verbesselt, J.; Avitabile, V.; Quang, H.V.; Herold, M. Performance of vegetation indices from Landsat time series in deforestation monitoring. *Int. J. Appl. Earth Obs. Geoinf.* **2016**, *52*, 318–327. [CrossRef]

27. Telkenaroglu, C.; Dikmen, M. Deforestation due to urbanization: A case study for Trabzon, Turkey. *ISPRS Ann. Photogramm. Remote. Sens. Spat. Inf. Sci.* **2017**, *4*, 379. [CrossRef]
28. Othman, M.; Ash'Aari, Z.; Aris, A.; Ramli, M. Tropical deforestation monitoring using NDVI from MODIS satellite: A case study in Pahang, Malaysia. In *Proceedings of the IOP Conference Series: Earth and Environmental Science*; IOP Publishing: Bristol, UK, 2018; Volume 169, p. 012047.
29. Badrinarayanan, V.; Kendall, A.; Cipolla, R. SegNet: A deep convolutional encoder-decoder architecture for image segmentation. *IEEE Trans. Pattern Anal. Mach. Intell.* **2017**, *39*, 2481–2495. [CrossRef]
30. Syrris, V.; Hasenohr, P.; Delipetrev, B.; Kotsev, A.; Kempeneers, P.; Soille, P. Evaluation of the potential of convolutional neural networks and random forests for multi-class segmentation of Sentinel-2 imagery. *Remote. Sens.* **2019**, *11*, 907. [CrossRef]
31. Lee, S.H.; Han, K.J.; Lee, K.; Lee, K.J.; Oh, K.Y.; Lee, M.J. Classification of landscape affected by deforestation using high-resolution remote sensing data and deep-learning techniques. *Remote. Sens.* **2020**, *12*, 3372. [CrossRef]
32. Pashaei, M.; Kamangir, H.; Starek, M.J.; Tissot, P. Review and evaluation of deep learning architectures for efficient land cover mapping with UAS hyper-spatial imagery: A case study over a wetland. *Remote. Sens.* **2020**, *12*, 959. [CrossRef]
33. Maretto, R.V.; Fonseca, L.M.; Jacobs, N.; Körting, T.S.; Bendini, H.N.; Parente, L.L. Spatio-temporal deep learning approach to map deforestation in amazon rainforest. *IEEE Geosci. Remote. Sens. Lett.* **2020**, *18*, 771–775. [CrossRef]
34. Abdani, S.R.; Zulkifley, M.A.; Mamat, M. U-Net with spatial pyramid pooling module for segmenting oil palm plantations. In Proceedings of the IEEE 2nd International Conference on Artificial Intelligence in Engineering and Technology, Kota Kinabalu, Sabah, 26–27 September 2020; pp. 1–5.
35. Liu, Y.; Gross, L.; Li, Z.; Li, X.; Fan, X.; Qi, W. Automatic building extraction on high-resolution remote sensing imagery using deep convolutional encoder-decoder with spatial pyramid pooling. *IEEE Access* **2019**, *7*, 128774–128786. [CrossRef]
36. Zulkifley, M.A.; Mohamed, N.A.; Zulkifley, N.H. Squat angle assessment through tracking body movements. *IEEE Access* **2019**, *7*, 48635–48644. [CrossRef]
37. Ronneberger, O.; Fischer, P.; Brox, T. U-Net: Convolutional networks for biomedical image segmentation. In *Proceedings of the Medical Image Computing and Computer-Assisted Intervention*; Navab, N., Hornegger, J., Wells, W.M., Frangi, A.F., Eds.; Springer International Publishing: Cham, Germany, 2015; pp. 234–241.
38. Abdani, S.R.; Zulkifley, M.A.; Moubark, A.M. Pterygium tissues segmentation using densely connected DeepLab. In Proceedings of the IEEE 10th Symposium on Computer Applications Industrial Electronics, Penang, Malaysia, 18–19 April 2020; pp. 229–232.
39. Long, J.; Shelhamer, E.; Darrell, T. Fully convolutional networks for semantic segmentation. In Proceedings of the IEEE Conference on Computer Vision and Pattern Recognition (CVPR), Boston, MA, USA, 7–12 June 2015; pp. 3431–3440. [CrossRef]
40. Iglovikov, V.; Shvets, A. Ternausnet: U-Net with VGG11 encoder pre-trained on imagenet for image segmentation. *arXiv* **2018**, arXiv:1801.05746.
41. Simonyan, K.; Zisserman, A. *Very Deep Convolutional Networks for Large-Scale Image Recognition*; Technical report; University of Oxford: Oxford, UK, 2014.
42. Araya-Arriagada, J.; Garay, S.; Rojas, C.; Duran-Aniotz, C.; Palacios, A.G.; Chacón, M.; Medina, L.E. Multiscale entropy analysis of retinal signals reveals reduced complexity in a mouse model of Alzheimer's disease. *Sci. Rep.* **2022**, *12*, 8900. [CrossRef] [PubMed]
43. Wang, D.; Cao, W.; Zhang, F.; Li, Z.; Xu, S.; Wu, X. A review of deep learning in multiscale agricultural sensing. *Remote. Sens.* **2022**, *14*, 559. [CrossRef]

forests

MDPI

Article

Using the Error-in-Variable Simultaneous Equations Approach to Construct Compatible Estimation Models of Forest Inventory Attributes Based on Airborne LiDAR

Chungan Li [1,*], Zhu Yu [1,2], Xiangbei Zhou [1,3], Mei Zhou [4] and Zhen Li [2]

[1] Forestry College, Guangxi University, 100, East Daxue Rd., Nanning 530004, China
[2] Guangxi Forest Inventory & Planning Institute, 14, Zhonghua Rd., Nanning 530011, China
[3] Guangxi Natural Resources Vocational and Technical College, 29, East Airport Ave., Fushui 532199, China
[4] School of Computer, Electronics and Information, Guangxi University, 100, East Daxue Rd., Nanning 530004, China
* Correspondence: chungan@gxu.edu.cn

Abstract: Airborne LiDAR has been extensively used for estimating and mapping forest attributes at various scales. However, most models have been developed separately and independently without considering the intrinsic mathematical relationships and correlations among the estimates, which results in the mathematical and biophysical incompatibility of the estimates. In this paper, using the measurement error model approach, the error-in-variable simultaneous equation (SEq) for airborne LiDAR-assisted estimations of four forest attributes (stand volume, V; basal area, G; mean stand height, H; and diameter at breast height, D) for four forest types (Chinese fir, pine, eucalyptus, and broad-leaved forest) is developed and compared to the independence models (IMs). The results indicated that both the SEqs and IMs performed well, and the rRMSEs of the SEqs were slightly larger than those of the IMs, while the increases in rRMSE were less than 2% for the SEqs. There were statistically significant differences ($\alpha = 0.05$) in the means of the estimates between SEqs and IMs, even though their average differences were less than ±1.0% for most attributes. There were no statistically significant differences in the mean estimates between SEqs, except for the estimates of the D and G of the eucalyptus forest. The SEqs with H and G as the endogenous variables (EVs) to estimate V performed slightly better than other SEqs in the fir, pine, and broad-leaved forests. The SEq that used D, H, and V as the EVs for estimating G was best in the eucalyptus forests. The SEq ensures the definite mathematical relationship among the estimates of forest attributes is maintained, which is consistent with forest measurement principles and therefore facilitates forest resource management applications, which is an issue that needs to be addressed for airborne LIDAR forest parameter estimation.

Keywords: independence model; measurement error model approach; mathematical relationship; area based approach (ABA)

1. Introduction

Forests play an irreplaceable role in the sustainable production of woody and non-woody products, community development, food security, biodiversity conservation, maintaining suitable conditions for life, and coping with global climate changes. Therefore, forest monitoring can help improve the sustainable management, restoration, and rehabilitation of degraded forest landscapes [1]. Light detection and ranging (LiDAR) has revolutionized the capability to accurately measure and depict the three-dimensional (3D) structures of the forest canopy. As a result, in the last 20 years, airborne LiDAR, an advanced remote sensing technology, has been widely utilized in estimating forest attributes at various scales [2–7], including the quadratic mean diameter (QMD), mean height, basal area, stand volume, stem density and/or stand density index, aboveground biomass, carbon storage [8–11], and leaf area index [12]. Since 2002, airborne LiDAR has replaced conventional field measurements and has served in operational large-scale forest resource inventories in the Scandinavian countries [13,14], and Canada [15]. In applications over large areas, airborne LiDAR forest attribute estimation and mapping were performed using an area-based approach (ABA) through a two-stage procedure [16,17]. When using LiDAR variables for forest attribute estimation, forests have generally been stratified by tree species groups, forest types (e.g., coniferous forests, mixed forests), age classes, site qualities, and topographic reliefs [13,18–23]. For all attributes of a stratum, separate estimation models were developed [24–28].

There are rigorous and exact mathematical relationships among stem volume, stem diameter (or basal area), and tree height, between basal area and QMD. In forest mensuration, these relationships are expressed in terms of an allometric equation. These mathematical relationships extend to the calculation of stand attributes when the study object is extended from a single tree to a forest stand. For example, the stand volume is equal to the multiplication of the basal area and mean height and the form factor (the ratio of stand volume to the multiplication of basal area and mean height) [29], or calculated by an allometric equation based on stand basal area and mean height [30]. The stand basal area can be determined through stem density and QMD [31]. In addition, there is an intrinsic mathematical relationship between stand carbon density and stand mean height, basal area, and timber density [11,32,33]. These mathematical relations are weakened in mixed and non-uniform multi-story forests to various degrees. In contrast, they are directly apparent in even-aged pure forests. If all forest attributes are estimated independently, parametric or nonparametric models usually lead to mathematical or biological inconsistencies among the estimates of forest attributes [34], ignoring the intrinsic mathematical relationships among forest attributes. Although this issue has long been addressed in the model development of tree and stand volume [35], stand growth and yield simulation [36–38], site index table construction [39], biomass estimation models [40–42], and carbon stock estimation [43]. However, it has not been investigated in airborne LiDAR-based forest attribute estimation until recent years [44,45].

Due to the LiDAR system configuration and plot positioning, there are random or systematic errors in the LiDAR-derived metrics. Similarly, there are random or systematic errors in the measured forest attributes of the field plots due to plot sampling, field measurements, and allometric equations. All these errors may increase the residual variance in LiDAR-based forest attribute estimation [45], which may lead to invalid statistical tests [46]. In both parametric and nonparametric models, errors are constantly inevitable in forest attribute estimation. When an estimate of a forest attribute containing errors is used as a predictor variable to estimate another forest attribute, a substantial bias occurs due to error transfers. An appropriate solution to this problem is to use error-in-variable (EIV) modeling, which takes the errors into consideration and guarantees compatibility among the estimates of forest attributes [44]. The EIV model is a model in which both the observations of the independent and dependent variables contain measurement errors [47] and can be employed in both linear and nonlinear models. Kinane et al. applied error-in-variable methods to evaluate multiple vegetation indices derived from Landsat 5 and 7 images, and then calculated errors associated with their observations and corrected them in the modeling process [48]. Yang et al. developed a compatible simultaneous equation system for DBH and height to crown base (HCB) for *Picea crassifolia* Kom. trees in northwestern China using LiDAR metrics and field measurement data of the single tree; they discovered that the effect of errors associated with the regressors on the response variables (DBH and HCB) could be illustrated by the simultaneous equation system, and guaranteed the compatibility between the DBH and HCB models at an individual level [45]. Using the same dataset, Fu et al. developed a compatible simultaneous system for individual tree DBH and aboveground biomass (AGB) using error-in-variable models. They argued that, by combining error-in-variable modeling and LiDAR data, the approach could provide the potential to expand the estimations of both DBH and AGB from individual trees to stands and improve their estimation [44]. However, there are insufficient studies regarding the compatibility estimation of the airborne LIDAR forest attributes. Therefore, additional systematic and in-depth studies are needed for different forest types, different mathematical relationships among forest attributes (different allometric equations for calculating forest attributes), and the inconsistency of independent estimation.

This study focused on even-aged pure stands of four forest types in a large subtropical study area. A system of compatible prediction models for mean height, mean diameter, basal area, and stand volume was established using the error-in-variable simultaneous equations approach, which maintained the intrinsic mathematical relationships among the estimates of forest attributes. The specific objectives are: (1) to investigate the differences in estimates between the simultaneous equation (SEq) and independence model (IM); and (2) to investigate the differences in estimates among different SEqs and to select the best SEq. We hope that this study will help to improve the consistency in the estimation of forest attributes.

2. Materials and Methods

2.1. Study Site

The study area covered the entire Guangxi Zhuang Autonomous Region (104°28′–112°04′ E, 20°54′–26°24′ N) in southern China, with an area of 237.6 × 10^3 km^2 (Figure 1). In this study, airborne LiDAR data acquisition and field plot measurements were performed over three years in three regions based on financial allocations: the Nanning region (with an area of 22.1 × 10^3 km^2), the eastern region (128.4 × 10^3 km^2), and the western region (87.1 × 10^3 km^2).

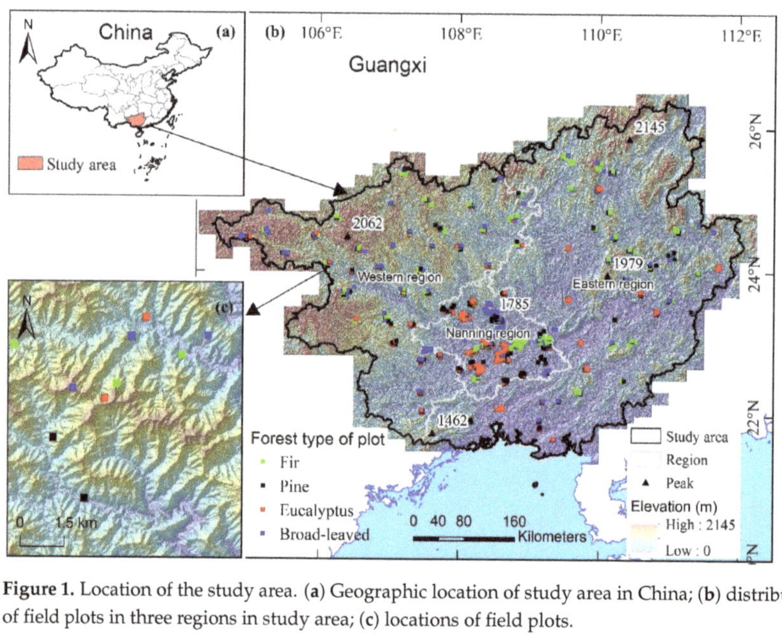

Figure 1. Location of the study area. (**a**) Geographic location of study area in China; (**b**) distribution of field plots in three regions in study area; (**c**) locations of field plots.

Various landforms, such as mountains, hills, platforms, and plains, are distributed alternately in the study area, accounting for 62.1%, 14.5%, 9.1%, and 14.3% of the total land area, respectively. The topography is high in the northwest and low in the southeast, and the study area appears to be tilted from north and northwest to south and southeast. It is surrounded by mountains and plateaus; the central and southern parts are hilly with platforms and plains, forming the shape of a basin, which is known as the "Guangxi basin". The highest elevation above sea level is 2145 m. The Tropic of Cancer crosses the central part of the study area, while the southern part borders a tropical ocean, and the study area is in the subtropical monsoon climate zone. In 2016, the mean annual temperature around the study area was 17.6–23.8 °C, and the mean annual temperature of the whole region was 21.4 °C. The total annual rainfall ranges from 723.9 to 2983.8 mm, with a region-wide mean annual rainfall of 1647.7 mm, including a region-wide average rainfall of 1234.4 mm during the flood season (April–September), accounting for 75% of the total annual rainfall.

From south to north, the study area comprises the northern tropics, southern subtropics, and central subtropics, and the representative forest vegetation is the seasonal rainforests, monsoon broad-leaved evergreen forests, and typical evergreen broad-leaved forests, respectively. The coniferous and bamboo forests are scattered in each vegetation zone. The Karst landscapes are widely distributed in the study area, with natural forests mainly including limestone (Rocky Mountain) monsoon forests, limestone mixed evergreen and deciduous broad-leaved forests, and limestone deciduous broad-leaved forests.

According to the 5th Forest Management Inventory of Guangxi (2017–2020), the forest area in the study area was 11,741.0 × 10^3 ha, accounting for 49.4% of the total land area, of which the Chinese fir (*Cunninghamia lanceolata*) planted forests, pine forests, eucalyptus plantations, and broad-leaved forests accounted for 16.5%, 17.5%, 24.8%, and 41.2% of the total forest area, respectively. The pine forests are mainly natural *Pinus massoniana* forests, with a few natural stands of *P. yunnanensis var. tenuifolia* and a small area of planted *P. massoniana* and *P. elliottii* forests. Industrial *Eucalyptus* spp. plantations are intensively managed plantations with a short rotation (5–8 years). The broad-leaved forests are multi-storied mixed forests composed of numerous tree species.

2.2. Field Plot Data

The field plots in the Nanning, eastern, and western regions were measured from October 2016 to January 2017, November 2018 to May 2019, and August 2019 to January 2020, respectively. The forests in the study area were categorized into four types: Chinese fir, pine, eucalyptus, and broad-leaved forest. The field plots were installed in the forest resources database in 2015 using ArcGIS Desktop v10.2 (ESRI, Redlands, CA, USA) based on the representativeness of the forest attributes, such as the mean height and stand volume. Approximately 100 plots were installed for each forest type in each region. The plots were distributed in clusters. The minimum spatial interval between the sample plots was 500 m (200 m in the Nanning region). There were 107 clusters of plots, each with 5–30 field plots (Figure 1). A hand-held dual-frequency differential global positioning system (GPS) was used to navigate to the stand on which the plot was located. The rectangular plot size was 30 × 20 m, all set in a north–south orientation, and was subdivided into four subplots each with a size of 15 × 10 m. The boundaries of the plot and subplots were measured and set using a compass and a handheld laser rangefinder and marked by nylon ropes. The Trimble GNSS (Global Navigation Satellite System) receiver with a real-time kinematic (RTK) positioning method was employed to position the northwestern and southeastern corners of the plot. Two RTK-GNSS instruments were used as the base stations, which were located in a nearby open field. Using the post-correction approach, the positioning accuracy was better than 1 m.

All live trees in each subplot with a DBH (1.3 m above the ground) greater than or equal to 5.0 cm were measured using a steel tape and identified by species or species groups. Three trees of average height and the tallest tree were selected to measure tree height with a Haglöf Vertex IV hypsometer. The calculation attributes of the subplots included the mean square diameter, mean height, basal area, stand volume, and stem density. The provincial species-specific allometric equations based on the basal area and mean height [49] were used to calculate the stand volumes. This study focused on the compatible models for estimating the stand attributes of pure forests. A stand is generally assumed to be a pure forest stand if the stems of the dominant species account for 80% of the total stem number. As the species (species group)-specific allometric equations were used to calculate the stand volume, we specified in this study that a forest stand was a pure forest stand only if the stand volume of the dominant species accounted for 95% of the total volume. The following methods were used to calculate the stand attributes of the field plots: The basal areas (G) of the four subplots were summed to obtain the basal area of the plots; the mean diameter (D) and mean height (H) of the plots were obtained from the weighted average of the basal areas of the subplots. The species (group)-specific allometric equations for calculating the stand volume (V) are as follows [49]:

$$V_{\text{Fir}} = G \times H \times (0.4523 + \frac{1.3133}{H+2}) \quad (1)$$

$$V_{\text{Pine}} = G \times H \times (0.3645 + \frac{1.9427}{H+2}) \quad (2)$$

$$V_{\text{Eucalyptus}} = 0.9767 \times G \times D^{-0.06843} \times H^{0.8140} \quad (3)$$

$$V_{\text{Broadleaf}} = G \times H \times (0.4049 + \frac{3.3787}{H+20}) \quad (4)$$

Since the stand volumes of the broad-leaved forests were calculated using the same allometric equation and were not divided into stand layers during the field measurement, all mixed broad-leaved stands were treated as pure forests. There were 782 pure forest field plots in the study area; the basic statistics of these plots are shown in Table 1.

Table 1. Summary statistics of the field plot data. CV is the coefficient of variation.

Forest Type	Sample Size	Stem Density (Stems ha^{-1})	Diameter at Breast Height (D)		Stand Height (H)		Basal Area (G)		Stand Volume (V)	
			Mean (cm)	CV (%)	Mean (m)	CV(%)	Mean (m^{2}ha^{-1})	CV (%)	Mean (m^{3}ha^{-1})	CV (%)
Fir	139	683–6883	11.8	26.2	10.65	27.76	33.32	30.20	205.67	46.73
Pine	170	350–3967	19.5	28.0	14.32	27.26	28.57	32.30	206.91	47.95
Eucalyptus	267	517–3350	11.2	21.4	16.10	20.63	17.14	33.87	141.14	44.58
Broad-leaved	206	233–4800	13.6	34.5	10.49	27.25	19.27	40.62	110.13	58.88

2.3. LiDAR Data

LiDAR data were collected by several contractors from October 2016 to April 2017, October 2018 to October 2019, and August 2019 to January 2020 in Nanning, eastern, and western regions. The Riegl VQ−1560 and Riegl VQ−1560i laser scanning systems (Riegl Laser Measurement Systems GmbH: Horn, Austria) were applied to collect LiDAR data in all three regions with identical standards. The pulse repetition frequency of the laser scanner was 1000 kHz, the scan frequency was 146 Hz, and the maximum scan zenith angle was ±30°. The beam divergence was 0.50 mrad with an average footprint size of approximately 0.5 m. The relative aircraft flight altitude was approximately 2500 m, the flight speed was 200–240 km h^{-1}, the swath width was 1.3–2.2 km, and the overlap between adjacent swathes was 21%–25%. The final average point density was ≥2.0 points m^{-2}. The point clouds were geo-referenced to a projection system of the China Geodetic Coordinate System 2000 (CGCS2000).

Before being delivered to the end-users, the raw data were preprocessed by the contractors using TerraScan software (TerraSolid, Ltd., Helsinki, Finland). The echo signals were classified as ground returns or nonground returns using the adaptive triangulation network (TIN) filter algorithm. A digital terrain model (DTM) with grid cells of 2 m was generated using the ground returns. Then, we eliminate the effect of topography and obtain the DTM-normalized LiDAR point clouds using the DTM.

LiDAR-derived metrics were extracted using all the echoes of the normalized point clouds, including the mean (Hmean), standard deviation (Hstdev), and coefficient of variation (Hcv) of the point cloud distribution, the 95th height percentile (hp95); canopy closure (CC), and the 50th and 75th density percentiles (dp50 and dp75). In addition, to characterize the canopy vertical heterogeneity and understory vegetation within the stand, the leaf area density (LAD) profiles and vertical foliage profile (VFP)-related metrics were extracted from the LiDAR point clouds, which include the mean (LADmean), standard deviation

(LADstdev), and coefficient of variation (LADcv) of LAD following Bouvier et al. [28], and the mean (VFPmean), standard deviation (VFPstdev), and coefficient of variation (VFPcv) of VFP following Harding et al. [50]. The LiDAR-derived metrics used in this study are shown in Table 2.

Table 2. List of LiDAR-derived metrics used in the establishment of the prediction model.

Acronym	Explanation of Metric	Structural Aspect	Predictor (P_x)
H	Mean stand height (m)	Target variable	-
D	Diameter at breast height (cm)	Target variable	-
V	Stand volume (m^3 ha^{-1})	Target variable	-
G	Basal area (m^2 ha^{-1})	Target variable	-
Hmean	Mean height of point clouds (m)	Canopy height	Ph_m
hp95	95th height percentile	Canopy height	Ph_m
Hstdev	Standard difference of point height distribution (m)	Canopy height	Ph
Hcv	Coefficient of variation of point height distribution	Canopy height	Ph
CC	Canopy cover	Canopy density	Pd_m
dp50	50th density percentile	Canopy density	Pd
dp75	75th density percentile	Canopy density	Pd
LADmean	Mean of vertical leaf area density (LAD) profile	Vertical heterogeneity	Pv
LADstdev	Standard difference of vertical LAD profile	Vertical heterogeneity	Pv
LADcv	Coefficient of variation of vertical LAD profile	Vertical heterogeneity	Pv
VFPmean	Mean of vertical foliage profile (VFP)	Vertical heterogeneity	Pv
VFPstdev	Standard difference of VFP	Vertical heterogeneity	Pv
VFPcv	Coefficient of variation of VFP	Vertical heterogeneity	Pv

2.4. Independence Model

Forest attributes are closely related to the 3D structures of their canopies. The LiDAR point clouds can accurately depict the 3D structures of forest canopies [26,36,51]. The above 13 variables can be divided into three groups: height, density, and vertical structure variable groups. Then, the independent model (IM) of forest attribute estimation can be separately established by the multiplicative power model, and the model structure formulation is as follows [52]:

$$\hat{y} = a_0 P^{a_1} H^{a_2} S^{a_3} + \varepsilon \quad (5)$$

where $\hat{y}$ is the estimate of the stand attribute (V, H, G, and D); P, H, and S are a group of LiDAR-derived variables depicting the canopy structure aspect of the density, height, and vertical structure, respectively, with each group containing 1–2 variables; and $a_0, a_1, \cdots, a_n$ are the parameters of the model.

To achieve a robust model for forest attribute estimation with good explanatory power, high accuracy, and conformity with the principles of forest mensuration and ecology, a multiplicative power model based on Equation (5) was constructed using a rule-based exhaustive combination method. One or two variables in each variable group were selected and include in Equation (5). A model formulation comprising 2–5 variables was obtained for a specific parameter of a specific forest type. The selection and combination of variables were performed according to the following rules:

(1) The variable combination scheme: 1–2 height variables + 1–2 density variables +/1 vertical structure variable.
(2) The model must comprise at least one primary height variable (Hmean or hp95) and one primary density variable (CC). When two height variables are selected, one primary and one secondary height variable (Hstdev or Hcv) can be included. When two density variables are selected, one secondary density variable is selected in addition to the CC.
(3) Each variable in the group of vertical structure variables can appear in the model separately.
(4) When the model comprises two variables, Equation (5) consists of one primary height variable and one primary density variable. When the model comprises three variables,

Equation (5) consists of one primary height variable, one primary density variable, and one vertical structure variable. When the model comprises four variables, Equation (5) consists of one primary height variable, one secondary height variable, one primary density variable, and one vertical structure variable. When the model comprises five variables, Equation (5) consists of one primary height variable, one secondary height variable, one primary density variable, one secondary density variable, and one vertical structure variable.

Eighty-six model formulations were obtained using the above rule-based exhaustive combination approach, which was applicable to all forest attribute estimations of all forest types. To achieve the best forest type-specific model formulation for specific forest attribute estimation, each model formulation was subjected to fifty iterations of model calibration and validation using the field plot dataset of that forest type. For each iteration (using the Gauss–Newton algorithm), the model was calibrated using 80% of the samples selected randomly from the dataset and validated using the remaining 20%. After all 50 repeats had been completed, the mean rRMSE and mean R^2 of that model formulation were calculated. Repeating the above procedures, we achieved the best model formulation with the smallest rRMSE and the largest R^2 after traversing all 86 models.

All plots of the same forest type were used to calibrate the best model formulation, which was validated by a 10-fold cross-validation method to obtain the best model for the independent estimation of forest attributes.

2.5. Error-in-Variable Simultaneous Equations

From Equations (1)–(4), a forest attribute can be calculated with other attributes. In other words, there is an intrinsic mathematical relationship among the forest attributes. When they are estimated independently, the mathematical relationship among the estimates may not satisfy Equations (1)–(4) because of estimation errors. LiDAR variables can be used as exogenous variables to estimate two of the three stand attributes mentioned above (for the eucalyptus forests, three of the four attributes were estimated). The two (or three) attribute estimates can be used as endogenous variables to estimate the remaining stand attribute using Equations (1)–(4). As estimation errors exist in the forest attributes, the error-in-variable method (EIV) [47,53] was used to establish the simultaneous equations (SEq) for estimating the above forest attributes. The vector form of the multiple nonlinear error-in-variable SEq is:

$$\begin{cases} f(y_i, x_i, c) = 0 \\ Y_i = y_i + e_i \qquad\qquad i = 1, 2, \cdots, n \\ E(e_i) = 0, \operatorname{cov}(e_i) = \sigma^2 \psi \end{cases} \tag{6}$$

where x_i is the measurement value of the q-dimensional error-out-variable; y_i is the measurement value of the p-dimensional error-in-variable; f is the m-dimensional function; Y_i is the unknown true value of y_i; the covariance matrix of the error is denoted as $\Phi = \sigma^2\Psi$, Ψ is the error structure matrix of e_i, and σ^2 is the estimation error.

For the fir, pine, and broad-leaved forests, three SEqs were established, and for the eucalyptus forests, four SEqs were established. As an example, SEqs of the fir forests were as follows.

$$\begin{cases} \hat{H} = a_0 \prod_{i=1}^{n} x_i^{a_i} + \varepsilon_H \\ \hat{G} = b_0 \prod_{j=1}^{m} y_j^{b_j} + \varepsilon_G \\ \hat{V} = b_0 \prod_{j=1}^{m} y_j^{b_j} \times a_0 \prod_{i=1}^{n} x_i^{a_i} \times \left(0.4523 + 1.3133/(a_0 \prod_{i=1}^{n} x_i^{a_i} + 2)\right) + \varepsilon_V \end{cases} \tag{7}$$

$$\begin{cases} \hat{V} = c_0 \prod_{k=1}^{l} z_i^{c_k} + \varepsilon_V \\ \hat{H} = a_0 \prod_{i=1}^{n} x_i^{a_i} + \varepsilon_H \\ \hat{G} = c_0 \prod_{k=1}^{l} z_i^{c_k} / a_0 \prod_{i=1}^{n} x_i^{a_i} \times \left(0.4523 + 1.3133/(a_0 \prod_{i=1}^{n} x_i^{a_i} + 2)\right) + \varepsilon_G \end{cases} \quad (8)$$

$$\begin{cases} \hat{V} = c_0 \prod_{k=1}^{l} z_i^{c_k} + \varepsilon_V \\ \hat{G} = b_0 \prod_{j=1}^{m} y_j^{b_j} + \varepsilon_G \\ \hat{H'} = c_0 \prod_{k=1}^{l} z_i^{c_k} / b_0 \prod_{j=1}^{m} y_j^{b_j} + \varepsilon_{H'} \end{cases} \quad (9)$$

Note: H and G are endogenous variables in SEquation (7); V and H are endogenous variables in SEquation (8); and V and G are endogenous variables in SEquation (9).

The terms $x_i(i = 1,2,\cdots,n)$, $y_j(j = 1,2,\cdots,m)$, and $z_k(k = 1,2,\cdots,l)$ are the LiDAR-variables used to construct the prediction models of H, G, and V, respectively; the variables are the same as those used to in Equation (5); n, m and l are the number of LiDAR-variables; $a_0, a_1, \cdots, a_n, b_0, b_1, \cdots, b_m$, and $c_0, c_1, \cdots, c_l$ are the estimates of the model parameters; $\hat{H'} = 0.4523\hat{H} \times (\hat{H} + 3.9036)/(\hat{H} + 2)$, once $\hat{H'}$ has been calculated via SEquation. (9), $\hat{H}$ can be calculated; ε_H, ε_G, ε_V and $\varepsilon_{H'}$ are the residual errors of H, G, V, and H', respectively, and they all are assumed to obey a normal distribution with a mean of 0. SEquations (7)–(9) ensure that the mathematical relationships defined by Equation (1) are satisfied among the estimates of the fir forest attributes. The SEqs for the remaining forest types can be established by referring to the above method.

The parameter estimations of the SEq were performed using the nlssystemfit (nonlinear equation system estimation) function of R software version 3.6.1 [54]. The validating methods of the equations were the same as those described in Section 2.4.

3. Result

3.1. Performance of the Independence Model

Generally, the best independence models for all forest attribute estimations of all four forest types performed well. For the fir, pine, and eucalyptus forests, the rRMSEs of the stand volume (V) and mean height (H) prediction models were approximately 20% and 10%, respectively, and the rRMSE of the basal area (G) prediction models ranged from 15.77% to 19.71%. However, for the broad-leaved forests with a complex structure, the performances of all IMs were slightly worse, with the rRMSE of the V, H, and G prediction models being 31.41%, 15.74%, and 27.22%, respectively (Table 3). Overall, compared to the reported studies [13,21,24,28,55], and considering the large area and structural heterogeneity in this study area, we considered the performances of the IMs excellent.

A paired t-test was performed on $\hat{V}_{IMi}$ and V_{Ci}, and the results indicated that there were no statistically significant differences ($\alpha = 0.05$) between the mean $\hat{V}_{IMi}$ and the mean V_{Ci} in the fir, pine, and broad-leaved forests, but there was statistically significant difference ($\alpha = 0.05$) in the eucalyptus forests.

Table 3. The parameter estimates of the best independence models and simultaneous equations of varied forest types and their goodness-of-fit statistics.

Forest Type	Model Type	Attribute	Variables and Their Parameter Estimates														Fitting Statistic		Validation Statistic	
			a0	Hmean	hp95	Hstdev	Hcv	CC	dp50	dp75	LADmean	LADstdev	LADcv	VFPmean	VFPstdev	VFPcv	R^2	RMSE(%)	R^2	RMSE(%)
Fir	IM	V	5.0365		1.2623		−0.3574	1.2661				0.03870					0.867	15.96	0.858	15.99
		H	1.4389		0.7897	0.02164		0.2902	0.08133				0.05820				0.858	10.07	0.850	10.08
		G	8.1460	0.7273		−0.1939		0.9878			0.03551						0.692	15.86	0.673	15.77
	SEq_G	V	4.1305		1.2945		−0.3923	0.9746				0.04636					0.863	16.21	0.853	16.22
		H	1.3651		0.8040	0.005625		0.18520	0.06160				0.06417				0.856	10.14	0.847	10.14
		G															0.689	15.93	0.671	15.80
	SEq_V	V															0.864	16.14	0.854	16.16
		H	1.4038		0.8016	0.01272		0.19910	0.06775				0.04897				0.857	10.12	0.848	10.13
		G	6.9021	0.7962		−0.2401		0.8014			0.04387						0.685	16.03	0.666	15.90
	SEq_H	V	4.1094		1.2942		−0.3979	0.9664				0.04423					0.863	16.22	0.853	16.23
		H															0.840	10.70	0.829	10.74
		G	7.0148	0.7866		−0.2254		0.8740			0.04492						0.686	16.00	0.667	15.87
Pine	IM	V	6.1982	1.4577				0.4028					−0.06050				0.825	19.30	0.824	19.01
		H	0.7293		1.0751			−0.01010						0.1209			0.895	8.69	0.889	8.73
		G	6.6866	0.7472		−0.04336		0.2011	0.3149							−0.1095	0.702	17.25	0.700	17.01
	SEq_G	V	5.9387	1.4838				0.6551					−0.04736				0.821	19.50	0.821	19.14
		H	0.7407		1.0697			0.007975						0.1112			0.894	8.70	0.889	8.74
		G															0.675	18.02	0.673	17.73
	SEq_V	V															0.833	18.85	0.832	18.53
		H	0.7596		1.0588			−0.01068						0.07242			0.893	8.77	0.887	8.80
		G	7.0295	0.7428		−0.03240		0.3490	0.3502							−0.1117	0.697	17.39	0.696	17.10
	SEq_H	V	6.1386	1.4617				0.6392					−0.03340				0.822	19.47	0.822	19.11
		H															0.893	8.77	0.888	8.78
		G	4.8815	0.8090		−0.08092		0.5478	0.1012							−0.01289	0.689	17.63	0.688	17.33
Eucalyptus	IM	V	5.6930		1.3300	−0.01835		0.2280		0.3609					−0.1124		0.777	20.99	0.769	20.92
		H	2.0316		0.7329	0.006645		−0.1352		0.08831	−0.01158						0.764	9.95	0.757	9.76
		G	2.9668		0.8470	−0.05055		0.2734		0.2742					−0.1500		0.657	19.83	0.650	19.71
		D	1.5173		0.7018	0.04169		−0.1269		0.07403		−0.01798					0.694	11.77	0.683	11.62
	SEq_G	V	4.6363		1.3431	0.007167		−0.003278		0.3653					−0.07611		0.773	21.20	0.765	21.12
		H	2.3845		0.6689	0.02462		−0.20860		0.1018	−0.01090						0.761	10.02	0.755	9.85
		G															0.655	19.90	0.649	19.77
		D	1.8010		0.6344	0.05877		−0.1909		0.08666		−0.01665					0.691	11.83	0.682	11.69
	SEq_V	V															0.773	21.20	0.765	21.12
		H	2.3393		0.6708	0.02836		−0.2194		0.1019	−0.001516						0.759	10.05	0.753	9.89
		G	2.4802		0.8435	−0.01378		0.1630		0.2884					−0.08184		0.654	19.91	0.648	19.78
		D	1.7722		0.6337	0.06019		−0.1975		0.08687		−0.009687					0.690	11.85	0.680	11.73
	SEq_H	V	4.1199		1.3622	0.01500		0.008849		0.3622					−0.04535		0.773	21.19	0.765	21.11
		H															0.759	10.06	0.753	9.90
		G	2.3542		0.8584	−0.01386		0.1857		0.2851					−0.07427		0.655	19.90	0.649	19.77
		D	1.7558		0.6368	0.05998		−0.2179		0.08665		−0.01231					0.689	11.86	0.680	11.73
	SEq_D	V	4.9610		1.3252	0.008399		−0.006183		0.3712					−0.08210		0.773	21.20	0.765	21.12
		H	2.3575		0.6678	0.02798		−0.2306		0.1033	−0.002148						0.759	10.06	0.753	9.91
		G	2.6718		0.8236	−0.01287		0.1580		0.2936					−0.08883		0.654	19.91	0.648	19.78
		D															0.671	12.20	0.661	12.11
Broad-leaved	IM	V	5.0340	1.2488				0.2287				0.07030					0.678	31.38	0.669	31.41
		H	2.2698	0.6531				−0.07921						0.09380			0.620	15.76	0.610	15.74
		G	3.7473	0.7143				0.4489			0.06407						0.507	27.21	0.501	27.22
	SEq_G	V	4.5078	1.2782				0.2167				0.08187					0.677	31.42	0.667	31.45

Table 3. *Cont.*

Forest Type	Model Type	Attribute	Variables and Their Parameter Estimates														Fitting Statistic		Validation Statistic	
			a0	Hmean	hp95	Hstdev	Hcv	CC	dp50	dp75	LADmean	LADstdev	LADcv	VFPmean	VFPstdev	VFPcv	R^2	RMSE(%)	R^2	RMSE(%)
		H	2.5117	0.6072				−0.08318						0.01247			0.617	15.82	0.606	15.80
		G															0.488	27.71	0.483	27.69
	SEq_V	V															0.688	30.91	0.678	30.92
		H	2.1363	0.6785				−0.07103						0.11390			0.619	15.78	0.609	15.75
		G	3.1293	0.7583				0.3296			0.11350						0.496	27.51	0.492	27.46
	SEq_H	V	4.5103	1.2779				0.2220				0.08284					0.677	31.42	0.667	31.45
		H															0.608	16.00	0.598	15.99
		G	3.2912	0.7501				0.2817			0.07504						0.503	27.32	0.498	27.30

Note: IM was the independence model; for the fir, pine, and broad-leaved forest, SEq_V was the simultaneous equation in which H and G were endogenous variables to estimate V; SEq_G was the simultaneous equation in which H and V were endogenous variables to estimate G; SEq_H was the simultaneous equation in which V and G were endogenous variables to estimate H; the naming of the simultaneous equations for the eucalyptus forest was similar to above. Assuming that $\hat{H}_{IMi}$, $\hat{G}_{IMi}$, $\hat{D}_{IMi}$, and $\hat{V}_{IMi}$ were the estimates of H, G, D, and V of plot *I* obtained by the IMs, respectively. By using Equations (1)–(4), we were able to calculate the stand volume (V_{Ci}) of plot *i*. The variations (means) of the difference ($\frac{(V_{Ci}-\hat{V}_{IMi})}{(V_{Ci}+\hat{V}_{IMi})} \times 200$) between $\hat{V}_{IMi}$ and V_{Ci} of the fir, pine, eucalyptus, and broad-leaved forests were −12.19%–4.95% (−0.03%), −14.25%–10.9% (−0.58%), −4.43%–8.86% (−0.20%), and −17.41%–8.68% (−0.27%), respectively. Although the mean differences between $\hat{V}_{IMi}$ and V_{Ci} were insignificant, the variations in the differences were large, and none of the differences were equal to 0. These results indicated that the independence estimates of H, D, G, and V were incompatible, i.e., they did not satisfy the mathematical relationship defined by Equations (1)–(4).

3.2. Performance of the Simultaneous Equations

The SEqs performed well for all four forest types. For the fir, pine, and eucalyptus forests, the variations in rRMSEs of the V, H, G, and D estimation were 16.16%–21.12%, 8.74%–10.74%, 15.80%–19.78%, and 11.69%–12.11%, respectively. For the broad-leaved forests, the rRMSEs of the V, H, and G estimations of the three SEqs were 30.92%–31.45%, 15.75%–15.99%, and 27.30%–27.369%, respectively (Table 3). The F-test results showed that the regression effects were statistically significant ($\alpha = 0.05$) for all 43 equations in all 13 SEqs.

The differences in R^2 and rRMSE were extremely small among different SEqs for the same attribute of the same forest type (Table 3). For example, in the pine forests, the differences between the largest and smallest values of R^2 among the three SEqs for the V, H, and G estimations were 3.25%, 0.62%, and 3.62%, respectively. The relative differences between the largest and smallest values of rRMSE were 1.33%, 0.19%, and 3.39%, respectively. The differences in rRMSEs were less than 2.0% in over 80% of all 43 equations, with the largest being 5.85%, indicating that all the SEqs performed essentially the same and any two out of three forest attributes (four for eucalyptus forests) could be used as the endogenous variables to estimate another attribute in the SEqs without obvious differences in estimates. In addition, the R^2 and rRMSE of the validation dataset for all attributes were relatively close to those of the calibration dataset for all SEqs, indicating that all the SEqs were quite robust. Based on the principle of the smallest rRMSE, the SEqs that used H and G as the endogenous variables to estimate V (SEq_V) in the fir, pine, and broad-leaved forests performed slightly better than the other two SEqs. In the eucalyptus forests, the SEq that used D, H, and V as the endogenous variables to estimate G (SEq_G) outperformed the three other SEqs.

The average differences in the estimates of the same attributes among the different SEqs were insignificant, not exceeding ±1.0% (Table 3). However, the variations in difference among the plots remained significant. Taking the fir forest as an example, the differences in H, G, and V between SEq_V and SEq_G varied between 2.25% and 1.09%, −9.76% and 4.17%, and −10.88% and 5.01%, respectively. Between SEq_V and SEq_H, the variations in the differences in H, G, and V were −12.69%–6.48%, −0.75%–6.68%, and −10.45%–5.15%, respectively (Figure 2).

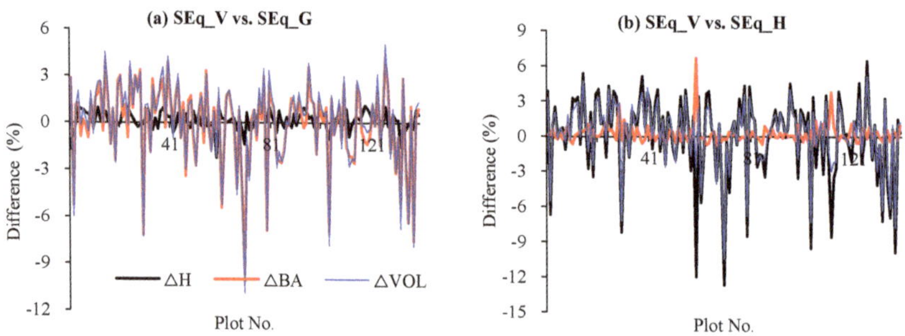

Figure 2. The plot-level differences in the estimates of H, G, and V between the simultaneous equations of the fir forest: (**a**) SEq_V vs. SEq_G, (**b**) SEq_V vs. SEq_H.

The results of the paired *t*-test showed that there were no statistically significant differences ($\alpha = 0.05$) in the means of the four forest attribute estimates between most of the SEqs in the fir, pine, and broadleaf forests. In the eucalyptus forests, there were statistically significant differences ($\alpha = 0.05$) in the means estimates of D and G between the different SEqs. However, there were no statistically significant differences ($\alpha = 0.05$) in H and V (Table 4).

Table 4. Results of paired *t*-test of the mean stand attribute estimates for various forest types between the simultaneous equations.

Forest Type	Equation/Model for Comparison	ΔH			ΔG			ΔV			ΔD		
		Mean (m)	Mean (%)	Std. (m)	Mean (m^2ha^{-1})	Mean (%)	Std. (m^2ha^{-1})	Mean (m^3ha^{-1})	Mean (%)	Std. (m^3ha^{-1})	Mean (cm)	Mean (%)	Std. (cm)
Fir	SEq_V vs. SEq_G	0.02 ***	0.16	0.05	0.02 ns	−0.05	0.73	0.54 ns	0.09	4.58			
	SEq_V vs. SEq_H	0.02 ns	0.03	0.33	0.06 ***	0.24	0.19	0.89 *	0.28	4.57			
	SEq_G vs. SEq_H	0.00 ns	−0.12	0.31	0.04 ns	0.29	0.75	0.35 ***	0.19	0.53			
Pine	SEq_V vs. SEq_D	−0.02 *	−0.36	0.15	0.08 ns	0.19	1.38	0.47 ns	−0.07	10.17			
	SEq_V vs. SEq_H	−0.04 ns	−0.38	0.53	0.14 ns	0.00	1.04	0.54 ns	−0.29	10.04			
	SEq_G vs. SEq_H	−0.02 ns	−0.02	0.53	0.06 ns	−0.19	0.85	0.08 ns	−0.22	1.28			
Eucalyptus	SEq_V vs. SEq_D	0.01 ns	0.08	0.09	0.00 ns	−0.09	0.08	0.00 ns	0.06	0.06	0.03 *	−0.02	0.18
	SEq_V vs. SEq_H	0.00 ns	−0.04	0.09	−0.03 ***	−0.09	0.07	0.00 *	−0.05	0.04	−0.25 ***	−0.12	0.87
	SEq_V vs. SEq_D	0.00 ns	−0.02	0.03	−0.01 **	−0.10	0.05	−0.03 ns	−0.31	0.29	−0.03 ns	−0.09	0.49
	SEq_G vs. SEq_H	−0.01 ns	−0.12	0.12	−0.03 ***	0.00	0.12	−0.01 **	−0.11	0.06	−0.28 ***	−0.09	0.86
	SEq_G vs. SEq_D	−0.01 ns	−0.10	0.10	−0.01 ns	−0.02	0.08	−0.03 ns	−0.38	0.30	−0.06 *	−0.07	0.41
	SEq_H vs. SEq_D	0.00 ns	0.02	0.08	0.02 **	−0.01	0.11	−0.02 ns	−0.27	0.26	0.22 **	0.02	1.18
Broad-leaved	SEq_V vs. SEq_D	0.00 ns	−0.01	0.17	−0.03 ns	−0.26	0.37	−0.16 ns	−0.27	3.59			
	SEq_V vs. SEq_H	0.02 ns	0.18	0.39	−0.07 *	−0.55	0.53	−0.30 ns	−0.39	3.56			
	SEq_G vs. SEq_H	0.02 ns	0.19	0.30	−0.04 ns	−0.29	0.51	−0.13 ***	−0.11	0.15			

Note: * Significant at $\alpha = 0.05$; ** Significant at $\alpha = 0.01$; *** Significant at $\alpha = 0.001$; ns, not significant ($\alpha > 0.05$).

3.3. Comparison of the Simultaneous Equations and Independence Model

Compared to the IMs, the performances of the SEqs of all four forest types were slightly worse (Table 3). Although there were a few equations in the SEqs in which the R^2 was larger and the rRMSE smaller than those of the IMs, the R^2 of most SEqs was smaller than that of the IMs, and their rRMSE was larger than that of the IMs. The maximum decrease in R^2 was −3.91%, and the maximum increase in rRMSE was 6.55%. Nevertheless, compared to the R^2 and rRMSE of the IMs, over 90% of the equations in all 43 equations of all 13 SEqs showed that the decreases in R^2 and the increases in rRMSE were not greater than 2%, indicating that the SEqs behaved similarly to the IMs.

Most of the average differences in the estimates between the SEqs and the corresponding IMs were less than ±1.0%, and the maximum average difference did not exceed ±2.0%, indicating that the estimates of the SEqs were close to those of the IMs. Nevertheless, the variations in the differences were large. For example, in the pine forests, the variations in the estimates of H, G, and V between the SEqs with H and G used as the endogenous variables (SEq_V) and the IMs were −4.40%–1.56%, −19.72%–3.44%, and −23.61%–11.94%, respectively; for the SEq_G, the variations in the estimates of the abovementioned attributes were −0.18%–2.62%, −11.98%–26.26%, and −4.09%–29.32%, respectively, as shown in Figure 3.

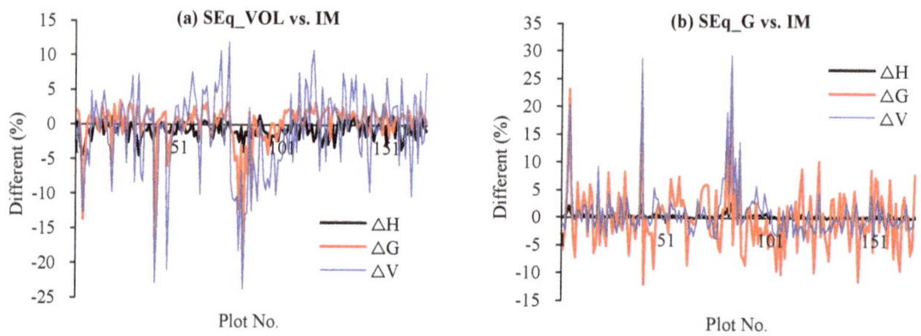

Figure 3. The plot-level differences of the estimates of H, G, and V between the SEqs and the IMs of the pine forests: (**a**) H and G were used as the endogenous variables to estimate V; (**b**) H and V were used as the endogenous variables to estimate G.

The results of the paired *t*-test of the estimates for all forest attributes of all forest types between the SEqs and the IMs showed that there were statistically significant differences ($\alpha = 0.05$) in all attributes, except for the V of the pine forests (Table 5), despite the small average differences.

Although there were some differences in the estimates of forest attributes between the SEqs and IMs, and among the SEqs themselves, the estimations of all forest attributes of all forest types were good for both the IMs and SEqs (Figure 4).

Table 5. Results of paired *t*-test of the mean stand attribute estimates for various forest types between the simultaneous equations and independence model.

Forest Type	Equation/Model for Comparison	Sample Size	ΔH			ΔG			ΔV			ΔD		
			Mean (m)	Mean (%)	Std. (m)	Mean (m^2 ha^{-1})	Mean (%)	Std. (m^2ha^{-1})	Mean (m^3ha^{-1})	Mean (%)	Std. (m^3ha^{-1})	Mean (cm)	Mean (%)	Std. (cm)
Fir	SEq_G vs. IM	139	0.01 nc	−0.27	0.12	−0.43 ***	1.34	0.95	−1.88 ***	1.16	5.42			
	SEq_V vs. IM		0.03 ***	0.43	0.10	−0.40 ***	−1.39	0.68	−1.34 **	−1.07	7.13			
	SEq_H vs. IM		0.02 ns	−0.39	0.39	−0.46 ***	1.63	0.53	−2.23 ***	1.35	5.46			
Pine	SEq_G vs. IM	166	−0.04 ***	0.33	0.04	−0.11 ns	0.04	1.26	−0.82 ns	0.87	6.23			
	SEq_V vs. IM		−0.07 ***	−0.69	0.13	0.15 **	0.15	0.64	0.81 ns	−0.94	9.35			
	SEq_H vs. IM		−0.01 ns	0.31	0.47	−0.14 ns	−0.15	1.05	−0.54 ns	0.65	5.90			
Eucalyptus	SEq_G vs. IM	267	0.04 **	0.44	0.21	−0.11 ***	−0.54	0.30	−1.12 ***	−0.42	4.12	0.04 ***	0.54	0.14
	SEq_V vs. IM		0.05 **	0.52	0.25	−0.12 ***	−0.62	0.28	−1.10 ***	−0.45	4.09	0.04 ***	0.60	0.17
	SEq_H vs. IM		0.05 **	0.56	0.28	−0.09 ***	−0.53	0.26	−0.85 ***	−0.33	4.09	0.05 ***	0.65	0.18
	SEq_D vs. IM		0.05 **	0.27	0.27	−0.11 ***	−0.26	0.29	−1.07 ***	−0.18	4.11	0.07 **	0.46	0.40
Broad-leaved	SEq_G vs. IM	206	−0.03 ***	0.41	0.14	−0.14 *	0.70	0.81	−0.96 ***	1.36	1.51			
	SEq_V vs. IM		−0.03 ***	−0.42	0.05	−0.17 **	−0.97	0.77	−1.13 ***	−1.64	4.00			
	SEq_H vs. IM		−0.05 **	0.60	0.37	−0.10 **	0.41	0.48	−0.83 ***	1.25	1.61			

Note: * Significant at $\alpha = 0.05$; ** Significant at $\alpha = 0.01$; *** Significant at $\alpha = 0.001$; ns, not significant ($\alpha > 0.05$).

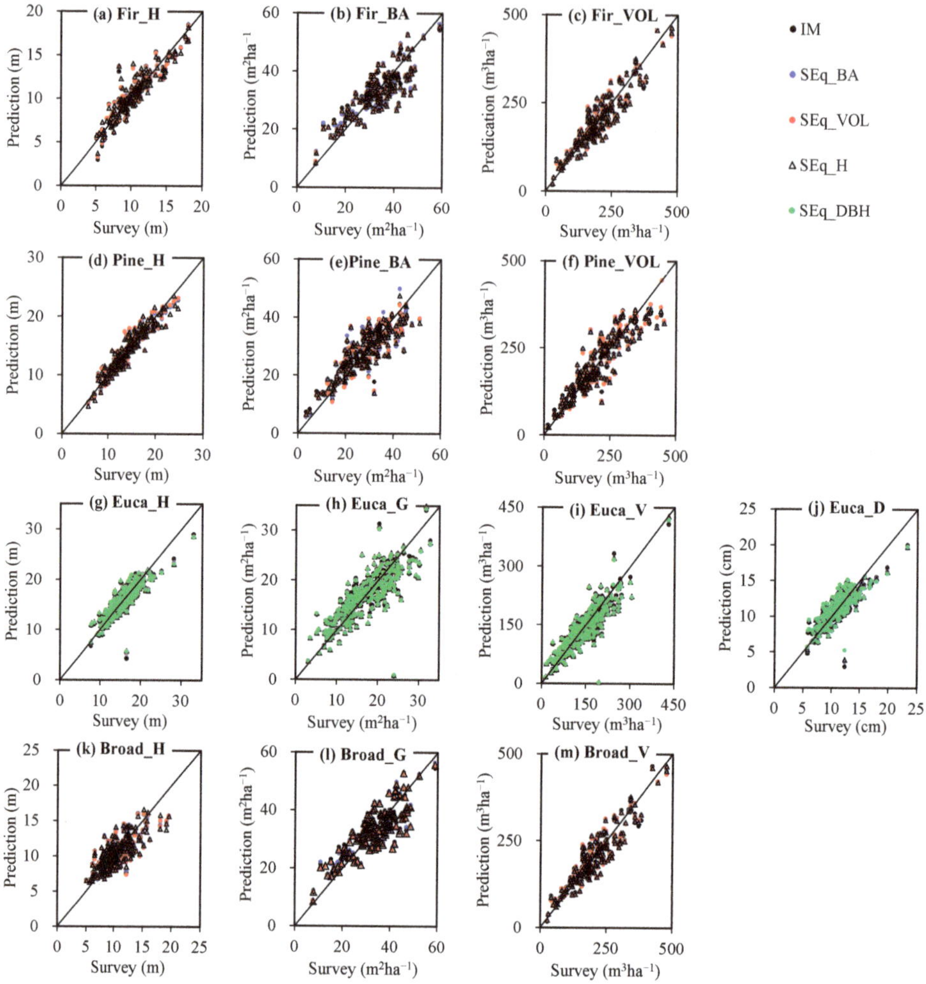

Figure 4. Scatterplots of measured forest attributes vs. LiDAR-predicted forest attributes obtained from simultaneous equations and independence models (the solid line is the 1:1 line) of four forest types: (**a**–**c**) fir forests; (**d**–**f**) pine forests; (**g**–**j**) eucalyptus forests; (**k**–**m**) broad-leaved forests.

4. Discussion

In this study, the error-in-variable simultaneous equation approach was used to construct estimation models of the mean stand height, mean stand diameter, basal area, and stand volume to achieve compatible estimates of forest attributes using airborne LiDAR data. The estimation methods were consistent with forest mensuration and ecology, and the mathematical consistencies were maintained among the estimates of forest attributes. Therefore, the presented approach is suitable for various estimation applications of the pure forest attributes and is useful for updating the forest resource database.

The estimation errors of all equations of the simultaneous equations were slightly larger than those of the independence models (Table 3). For the same attributes, although the average difference of the estimates between the simultaneous equations and the independence models was insignificant, not larger than 2%, there was a significant difference between their means (Table 5). The possible reasons were that during the iteration of SEq, the estimations of the forest attributes (such as $\hat{H}$ and $\hat{G}$ in SEquation (7)) using LiDAR variables as the endogenous variables were affected not only by their own errors but also by the errors of the attributes estimated using these variables as the endogenous variables (e.g., $\hat{V}$ in SEquation (7)). In other words, the estimation errors of all forest attributes were treated synthetically during the calibration procedure of the simultaneous equations. Thus, the equation calibrations were subject to more constraints, which affected the estimation accuracy of all attributes to some extent.

The rRMSEs of the estimation equations for the same forest attribute were close among the SEqs (Table 3). There were few statistically significant differences between the estimates of the same stand attribute among the SEqs of the fir, pine, and broad-leaved forests (Table 4), indicating that the performances of different SEqs were very similar. There were statistically significant differences in the estimates of basal area and diameter at breast height among the simultaneous equations of the eucalyptus forests. The potential reason was that there were four equations in a SEq for the eucalyptus forest, one more than for the other forest types; therefore, the model calibration procedures were subjected to more constraints.

Equations (1)–(4) can be categorized into two classes: Equations (1), (2), and (4) belong to a class in which H and G are used as predictors to calculate V. Equation (3) belongs to another class in which V is calculated by H, D, and G. For the fir, pine, and broad-leaved forests, the SEqs that used H and G as the endogenous variables to estimate V performed slightly better than other SEqs, while the SEq that used DBH, H, and V as the endogenous variables for estimating G was the best in the eucalyptus forests (Table 3). The best SEqs were consistent with the classification of the allometric equations of the V calculations. This is possibly related to the estimation accuracy of all attributes in independent models. The estimation accuracy of V was lower than that of H, D, and G in all independent models for all forest types. Therefore, the best simultaneous equation was the one that estimated the remaining variables with the variables with the highest estimation accuracy as the exogenous variables.

A previous study proposed an allometric equation system to estimate stand attributes. This system estimated the key stand attributes, e.g., the stand volume and Lorey's mean height, using LiDAR variables. Then, these two attributes were used directly to calculate G, QMD, and stand density (N, stems ha^{-1}) using the allometric equations, which achieved good estimation results while maintaining compatibility among forest attributes [45]. In this study, we found that the differences between the V_C directly calculated with $\hat{H}_{IM}$, $\hat{G}_{IM}$, and $\hat{D}_{IM}$ obtained from the IM based on Equations (1)–(4) and the $\hat{V}_{IM}$ obtained from the IM were small. There were no statistically significant differences, except in the eucalyptus forests. The average difference in V_C and the estimate of stand volume ($\hat{V}_{SE}$) obtained from the SEq was also small. However, there were statistically significant differences in their means in most of the equations. Furthermore, the estimation accuracy of other attributes strongly depended on the estimation accuracy of these two key attributes, and the stand volume was not a direct measurement predictor, indicating that the method also had some limitations. In this study, we found that the SEq using LiDAR variables as endogenous variables to estimate H and G, and then utilizing these two estimates as exogenous variables to estimate V, performed the best among several SEqs. Since H and G are directly measurable attributes in the field, these simultaneous equations can serve to boost the estimation accuracy.

In this study, we performed the compatible estimation of airborne LiDAR forest attributes by a multiplicative power model and error-in-variable simultaneous equations and achieved good results. However, the response of the same variable in different tree species is inconsistent. For instance, CC is positively correlated with H of fir, while it is negatively correlated with other tree species (Table 3). Additionally, there are numerous approaches to establishing prediction models of forest attributes using airborne LiDAR data, both parametric and nonparametric. There are various methods for establishing error-in-variable simultaneous equations. Therefore, more experiments are required to fully explore the performance of simultaneous equations in airborne LiDAR forest component attribute estimation.

5. Conclusions

In this paper, simultaneous equations (SEqs) were developed to estimate the mean stand height, mean diameter at breast height, basal area, and stand volume using airborne LiDAR data based on the measurement error model approach. The estimates were also analyzed and compared with the independence models (IMs). The following conclusions were drawn:

(1) Both IMs and SEqs can achieve good estimation results for all forest parameters of all forest types. The SEqs performed slightly worse than the IMs; however, the difference was not obvious.
(2) The SEqs maintain the definite mathematical relationships among various forest attributes, which are consistent with the principle of forest mensuration. The estimation results are useful for forest resource management.
(3) For the Chinese fir, pine, and broad-leaved forests, the SEqs using the mean stand height, and basal area as the endogenous variables to estimate stand volume performed slightly better than the other two SEqs. For the eucalyptus forests, the SEqs with the diameter at breast height, mean stand height, and stand volume as the endogenous variables to estimate basal area (SEq_G) outperformed the other three SEqs.

Author Contributions: C.L.: conceptualization, methodology, calculation, data collection and analysis, writing—original draft and review and editing, project administration, funding acquisition. Z.Y. and X.Z.: Calculation and writing—original draft. M.Z. and Z.L.: Calculation. All authors have read and agreed to the published version of the manuscript.

Funding: This project received financial support from the Forest Department of Guangxi Zhuang Autonomous Region, China (GXLYKJ2016-001).

Data Availability Statement: The original data are available from the corresponding author.

Acknowledgments: This project is a part of the Fifth Forest Management Inventory Project of Guangxi Zhuang Autonomous Region (5th FMI-GX, 2017—2020), China. The airborne LiDAR data acquisition and preprocessing and field data measurement were funded by the finance department of the Guangxi Zhuang Autonomous Region. The authors would like to express their sincere gratitude to Chengling Yang and Yao Liang from the Guangxi Forest Inventory and Planning Institute (FIPI-GX) and to the 120 other field crews who worked on the field measurements. The authors also thank Guangxi 3D Remote Sensing Engineering Technology Co., Ltd., Feiyan Aero Remote Sensing Technology Co., Ltd., Zhongke Remote Sensing Technology Group Co., Ltd., and Guangzhou Jiantong Surveying and Mapping Geographic Information Technology Co., Ltd. These companies were responsible for the acquisition and preprocessing of the airborne LiDAR data utilized herein. We thank the anonymous reviewers and the editor for their insightful suggestions.

Conflicts of Interest: We declare that we have no financial and personal relationships with other people or organizations that could inappropriately influence our work, and that there is no professional or other personal interest of any nature or kind in any product, service and/or company that could be construed as influencing the position presented in, or the review of, this manuscript.

References

1. FAO. *Global Forest Resources Assessment 2020: Main Report*; FAO: Rome, Italy, 2020. [CrossRef]
2. Means, J.E.; Acker, S.A.; Fitt, B.J.; Renslow, M.; Emerson, L.; Hendrix, C.J. Predicting forest stand characteristics with airborne scanning lidar. *Photogramm. Eng. Remote Sens.* **2000**, *66*, 1367–1371.
3. Torre-Tojal, L.; Bastarrika, A.; Boyano, A.; Lopez-Guede, J.M.; Graña, M. Above-ground biomass estimation from LiDAR data using random forest algorithms. *J. Comput. Sci.* **2022**, *58*, 101517. [CrossRef]

4. Coops, N.C.; Tompalski, P.; Goodbody, T.R.; Queinnec, M.; Luther, J.E.; Bolton, D.K.; White, J.C.; Wulder, M.A.; van Lier, O.R.; Hermosilla, T. Modelling lidar-derived estimates of forest attributes over space and time: A review of approaches and future trends. *Remote Sens. Environ.* **2021**, *260*, 112477. [CrossRef]
5. McRoberts, R.E.; Tomppo, E.O.; Næsset, E. Advances and emerging issues in national forest inventories. *Scand. J. For. Res.* **2010**, *25*, 368–381. [CrossRef]
6. Johnson, K.D.; Birdsey, R.; Finley, A.O.; Swantaran, A.; Dubayah, R.; Wayson, C.; Riemann, R. Integrating forest inventory and analysis data into a LiDAR-based carbon monitoring system. *Carbon Balance Manag.* **2014**, *9*, 11. [CrossRef]
7. Straub, C.; Tian, J.; Seitz, R.; Reinartz, P. Assessment of Cartosat-1 and WorldView-2 stereo imagery in combination with a LiDAR-DTM for timber volume estimation in a highly structured forest in Germany. *Forestry* **2013**, *86*, 463–473. [CrossRef]
8. Mora, B.; Wulder, M.A.; White, J.C.; Hobart, G. Modeling stand height, volume, and biomass from very high spatial resolution satellite imagery and samples of airborne LiDAR. *Remote Sens.* **2013**, *5*, 2308–2326. [CrossRef]
9. Fekety, P.A.; Falkowski, M.J.; Hudak, A. Temporal transferability of LiDAR-based imputation of forest inventory attributes. *Can. J. For. Res.* **2015**, *45*, 422–435. [CrossRef]
10. Frank, B.; Mauro, F.; Temesgen, H. Model-based estimation of forest inventory attributes using Lidar: A comparison of the area-based and semi-individual tree crown approaches. *Remote Sens.* **2020**, *12*, 2525. [CrossRef]
11. Asner, G.P.; Clark, J.K.; Mascaro, J.; Galindo García, G.A.; Chadwick, K.D.; Navarrete Encinales, D.A.; Paez-Acosta, G.; Cabrera Montenegro, E.; Kennedy-Bowdoin, T.; Duque, Á.; et al. High-resolution mapping of forest carbon stocks in the Colombian Amazon. *Biogeosciences* **2012**, *9*, 2683–2696. [CrossRef]
12. Morsdorf, F.; Kötz, B.; Meier, E.; Itten, K.I.; Allgöwer, B. Estimation of LAI and fractional cover from small footprint airborne laser scanning data based on gap fraction. *Remote Sens. Environ.* **2006**, *104*, 50–61. [CrossRef]
13. Naesset, E. Accuracy of forest inventory using airborne laser scanning: Evaluating the first nordic full-scale operational project. *Scand. J. Res.* **2004**, *19*, 554–557. [CrossRef]
14. Næsset, E.; Gobakken, T.; Holmgren, J.; Hyyppä, H.; Hyyppä, J.; Maltamo, M.; Nilsson, M.; Olsson, H.; Persson, Å.; Söderman, U. Laser scanning of forest resources: The Nordic experience. *Scand. J. Res.* **2004**, *19*, 482–499. [CrossRef]
15. White, J.C.; Tompalski, P.; Vastaranta, M.; Wulder, M.A.; Saarinen, N.; Stepper, C.; Coops, N.C. *A Model Development and Application Guide for Generating an Enhanced Forest Inventory Using Airborne Laser Scanning Data and an Area-Based Approach*; CWFC Information Report FI-X-018; Natural Resources Canada, Canadian Forest Service, Pacific Forestry Centre: Victoria, BC, Canada, 2017. [CrossRef]
16. Næsset, E.; Bjerknes, K.-O. Estimating tree heights and number of stems in young forest stands using airborne laser scanner data. *Remote Sens. Environ.* **2001**, *78*, 328–340. [CrossRef]
17. Næsset, E. Predicting forest stand characteristics with airborne scanning laser using a practical two-stage procedure and field data. *Remote Sens. Environ.* **2002**, *80*, 88–99. [CrossRef]
18. Næesset, E.; Bollandsas, O.M.; Gobakken, T.; Gregoire, T.G.; Stahl, G. Model-assisted estimation of change in forest biomass over an 11 year period in a sample survey supported by airborne LiDAR: A case study with post-stratification to provide "activity data". *Remote Sens. Environ.* **2013**, *128*, 299–314. [CrossRef]
19. Packalén, P.; Maltamo, M. The k-MSN method for the prediction of species-specific stand attributes using airborne laser scanning and aerial photographs. *Remote Sens. Environ.* **2007**, *109*, 328–341. [CrossRef]
20. Hudak, A.T.; Crookston, N.L.; Evans, J.S.; Hall, D.E.; Falkowski, M.J. Nearest neighbour imputation of species-level, plot-scale forest structure attributes from LiDAR data. *Remote Sens. Environ.* **2008**, *112*, 2232–2245. [CrossRef]
21. Woods, M.; Pitt, D.; Penner, M.; Lim, K.; Nesbitt, D.; Etheridge, D.; Treitz, P. Operational implementation of a LiDAR inventory in Boreal Ontario. *For. Chron.* **2011**, *87*, 512–528. [CrossRef]
22. Penner, M.; Pitt, D.G.; Woods, M.E. Parametric vs. nonparametric LiDAR models for operational forest inventory in Boreal Ontario. *Can. J. Remote Sens.* **2013**, *39*, 426–443. [CrossRef]
23. Penner, M.; Woods, M.; Pitt, D. A comparison of airborne laser scanning and image point cloud derived tree size class distribution models in Boreal Ontario. *Forests* **2015**, *6*, 4034–4054. [CrossRef]
24. Montealegre, A.L.; Lamelas, M.T.; de la Riva, J.A.; García-Martín, A.; Escribano, F. Use of low point density ALS data to estimate stand-level structural variables in Mediterranean Aleppo pine forest. *Forestry* **2016**, *89*, 373–382. [CrossRef]
25. Zhang, Z.; Cao, L.; She, G. Estimating forest structural parameters using canopy metrics derived from airborne LiDAR data in subtropical forests. *Remote Sens.* **2017**, *9*, 940. [CrossRef]
26. Xu, C.; Manley, B.; Morgenroth, J. Evaluation of modelling approaches in predicting forest volume and stand age for small-scale plantation forests in New Zealand with RapidEye and LiDAR. *Int. J. Appl. Earth Obs.* **2018**, *73*, 386–396. [CrossRef]
27. Van Ewijk, K.; Treitz, P.; Woods, M.; Jones, T.; Caspersen, J. Forest site and type variability in ALS-based forest resource inventory attribute predictions over three Ontario forest sites. *Forests* **2019**, *10*, 226. [CrossRef]
28. Bouvier, M.; Durrieu, S.; Fournier, R.A.; Renaud, J.P. Generalizing predictive models of forest inventory attributes using an area-based approach with airborne LiDAR data. *Remote Sens. Environ.* **2015**, *156*, 322–334. [CrossRef]
29. Dube, T.; Siban, M.; Shoko, C.; Mutanga, O. Stand-volume estimation from multi-source data for coppiced and high forest *Eucalyptus* spp. silvicultural systems in Kwa Zulu-Natal, South Africa. *ISPRS J. Photogram.* **2017**, *132*, 162–169. [CrossRef]
30. Giannico, V.; Lafortezza, R.; John, R.; Sanesi, G.; Pesola, L.; Chen, J. Estimating stand volume and above-ground biomass of urban forests using LiDAR. *Remote Sens.* **2016**, *8*, 339. [CrossRef]

31. Yang, T.-R.; Kershaw, J.A., Jr.; Ducey, M.J. The development of allometric systems of equations for compatible area-based LiDAR-assisted estimation. *Forestry* **2021**, *94*, 36–53. [CrossRef]
32. Asner, G.P.; Mascaro, J.; Muller-Landau, H.C.; Vieilledent, G.; Vaudry, R.; Rasamoelina, M.; Hall, J.S.; Van Breugel, M. A universal airborne LiDAR approach for tropical forest carbon mapping. *Oecologia* **2012**, *168*, 1147–1160. [CrossRef]
33. Jenkins, J.C.; Chojnacky, D.C.; Heath, L.S.; Birdsey, R.A. National-scale biomass estimators for United States tree species. *For. Sci.* **2003**, *49*, 12–35.
34. Hill, T.C.; Williams, M.; Bloom, A.A.; Mitchard, E.T.A.; Ryan, C.M. Are inventory based and remotely sensed above-ground biomass estimates consistent? *PLoS ONE* **2013**, *8*, e74170. [CrossRef]
35. Sharma, M.; Oderwald, R.G. Dimensionally compatible volume and taper equations. *Can. J. Res.* **2001**, *31*, 797–803. [CrossRef]
36. Buckman, R.E. *Growth and Yield of Red Pine in Minnesota*; Technical Bulletin 1272; USDA Forest Service: Washington, DC, USA, 1962; 50p, Available online: https://babel-hathitrust-org-s.vpn.gxu.edu.cn:8118/cgi/pt?id=uiug.30112019334355&view=1up&seq=3 (accessed on 10 May 2022).
37. Clutter, J.L. Compatible growth and yield models for loblolly pine. *For. Sci.* **1963**, *9*, 354–371.
38. Fang, Z.; Giley, R.L.; Shiver, B.D. A multivariate simultaneous prediction system for stand growth and yield with fixed and random effects. *For. Sci.* **2001**, *47*, 550–562.
39. Wang, Y.; Huang, S.; Yang, R.C.; Tang, S. Error-in-variable method to estimate parameters for reciprocal base-age invariant site index models. *Can. J. Res.* **2004**, *34*, 1929–1937. [CrossRef]
40. Fu, L.; Lei, Y.; Wang, G.; Bi, H.; Tang, S.; Song, X. Comparison of seemingly unrelated regressions with error-invariable models for developing a system of nonlinear additive biomass equations. *Trees* **2016**, *30*, 839–857. [CrossRef]
41. Zhang, C.; Peng, D.-L.; Huang, G.-S.; Zeng, W.-S. Developing aboveground biomass equations Both compatible with tree volume equations and additive systems for single-trees in Poplar plantations in Jiangsu Province, China. *Forests* **2016**, *7*, 32. [CrossRef]
42. Zeng, W.; Zhang, L.; Chen, X.; Cheng, Z.; Ma, K.; Li, Z. Construction of compatible and additive individual-tree biomass models for *Pinus tabulaeformis* in China. *Can. J. Res.* **2017**, *47*, 467–475. [CrossRef]
43. Sakici, O.E.; Kucuk, O.; Ashraf, M.I. Compatible above-ground biomass equations and carbon stock estimation for small diameter Turkish pine (*Pinus brutia* Ten.). *Environ. Monit. Assess* **2018**, *190*, 285. [CrossRef]
44. Fu, L.; Liu, Q.; Sun, H.; Wang, Q.; Li, Z.; Chen, E.; Pang, Y.; Song, X.; Wang, G. Development of a system of compatible individual tree diameter and aboveground biomass prediction models using error-in-variable regression and airborne LiDAR data. *Remote Sens.* **2018**, *10*, 325. [CrossRef]
45. Yang, Z.; Liu, Q.; Luo, P.; Ye, Q.; Duan, G.; Sharma, R.P.; Zhang, H.; Wang, G.; Fu, L. Prediction of individual tree diameter and height to crown base using nonlinear simultaneous regression and airborne LiDAR data. *Remote Sens.* **2020**, *12*, 2238. [CrossRef]
46. Kangas, A.S. Effect of errors-in-variables on coefficients of a growth model and on prediction of growth. *For. Ecol. Manag.* **1998**, *102*, 203–212. [CrossRef]
47. Tang, S.; Li, Y.; Wang, Y. Simultaneous equations, error-in-variable models, and model integration in systems ecology. *Ecol. Model.* **2001**, *142*, 285–294. [CrossRef]
48. Kinane, S.M.; Montes, C.R.; Albaugh, T.J.; Mishra, D.R. A model to estimate leaf area index in loblolly pine plantations using Landsat 5 and 7 images. *Remote Sens.* **2021**, *13*, 1140. [CrossRef]
49. Liao, Z.; Huang, D. *Forest Inventory Handbook of Guangxi, China*; Forestry Department of Guangxi Zhuang Autonomous Region: Nanning, China, 1986. (In Chinese)
50. Harding, D.J.; Lefsky, M.A.; Parker, G.G.; Blair, J.B. Laser altimeter canopy height profiles methods and validation for closed-canopy, broadleaf forests. *Remote Sens. Environ.* **2001**, *76*, 283–297. [CrossRef]
51. Vauhkonen, J.; Maltamo, M.; McRoberts, R.E.; Næsset, E. Introduction to Forestry Applications of Airborne Laser Scanning. In *Forestry Applications of Airborne Laser Scanning: Concepts and Case Studies*; Managing Forest Ecosystems, 27; Maltamo, M., Næsset, E., Vauhkonen, J., Eds.; Springer: Dordrechtc, The Netherlands, 2014; Volume 464, pp. 1–18.
52. Li, C.; Li, Z. Generalizing predictive models of sub-tropical forest inventory attributes using an area-based approach with airborne LiDAR data. *Sci. Silvae Sin.* **2021**, *57*, 23–35. (In Chinese) [CrossRef]
53. Tang, S.; Wang, Y. A parameter estimation program for the errors-in-variable model. *Ecol. Model.* **2002**, *156*, 225–236. [CrossRef]
54. R Core Team. *R: A Language and Environment for Statistical Computing*; R Foundation for Statistical Computing: Vienna, Austria, 2019; Available online: https://www.R-project.org/ (accessed on 20 August 2020).
55. Zolkos, S.G.; Goetz, S.J.; Dubayah, R. A meta-analysis of terrestrial aboveground biomass estimation using lidar remote sensing. *Remote Sens. Environ.* **2013**, *128*, 289–298. [CrossRef]

MDPI AG
Grosspeteranlage 5
4052 Basel
Switzerland
Tel.: +41 61 683 77 34

Forests Editorial Office
E-mail: forests@mdpi.com
www.mdpi.com/journal/forests

www.ingramcontent.com/pod-product-compliance
Lightning Source LLC
Chambersburg PA
CBHW040802250926
51850CB00007BA/1059
* 9 7 8 3 7 2 5 8 6 7 0 0 4 *